A Programmer's Introduction to Mathematics

Jeremy Kun

Copyright © 2018 Jeremy Kun

All rights reserved. This book or any portion thereof may not be reproduced or used in any manner whatsoever without the express written permission of the publisher except for the use of brief quotations in a book review.

All images used in this book are either the author's original works or in the public domain. In particular, the only non-original images are in the chapter on group theory, specifically the textures from *The Grammar of the Orient*, M.C. Escher's *Circle Limit IV*, and two diagrams in the public domain, sourced from Wikipedia.

First edition, 2018.

pimbook.org

To my wife, Erin.

My unbounded, uncountable thanks goes out to the many people who read drafts at various stages of roughness and gave feedback, including (in alphabetical order by first name), Aaron Shifman, Adam Lelkes, Alex Walchli, Ali Fathalian, Arun Koshy, Ben Fish, Craig Stuntz, Devin Ivy, Erin Kelly, Fred Ross, Ian Sharkey, Jasper Slusallek, Jean-Gabriel Young, João Rico, John Granata, Julian Leonardo Cuevas Rozo, Kevin Finn, Landon Kavlie, Louis Maddox, Matthijs Hollemans, Olivia Simpson, Pablo González de Aledo, Paige Bailey, Patrick Regan, Patrick Stein, Rodrigo Zhou, Stephanie Labasan, Temple Keller, Trent McCormick.

Special thanks to Devin Ivy for a thorough technical review of two key chapters.

Contents

Our Goal		i
1	**Like Programming, Mathematics has a Culture**	**1**
2	**Polynomials**	**5**
	2.1 Polynomials, Java, and Definitions	5
	2.2 A Little More Notation	13
	2.3 Existence & Uniqueness	14
	2.4 Realizing it in Code	22
	2.5 Application: Sharing Secrets	24
	2.6 Cultural Review	27
	2.7 Exercises	28
	2.8 Chapter Notes	31
3	**On Pace and Patience**	**35**
4	**Sets**	**39**
	4.1 Sets, Functions, and Their -Jections	40
	4.2 Clever Bijections and Counting	48
	4.3 Proof by Induction and Contradiction	51
	4.4 Application: Stable Marriages	54
	4.5 Cultural Review	58
	4.6 Exercises	59
	4.7 Chapter Notes	61
5	**Variable Names, Overloading, and Your Brain**	**63**
6	**Graphs**	**69**
	6.1 The Definition of a Graph	69
	6.2 Graph Coloring	71
	6.3 Register Allocation and Hardness	73
	6.4 Planarity and the Euler Characteristic	75
	6.5 Application: the Five Color Theorem	77

6.6	Approximate Coloring	82
6.7	Cultural Review	83
6.8	Exercises	84
6.9	Chapter Notes	85

7 The Many Subcultures of Mathematics — 89

8 Calculus with One Variable — 95
8.1	Lines and Curves	96
8.2	Limits	101
8.3	The Derivative	107
8.4	Taylor Series	111
8.5	Remainders	116
8.6	Application: Finding Roots	118
8.7	Cultural Review	125
8.8	Exercises	125

9 On Types and Tail Calls — 129

10 Linear Algebra — 135
10.1	Linear Maps and Vector Spaces	136
10.2	Linear Maps, Formally This Time	141
10.3	The Basis and Linear Combinations	143
10.4	Dimension	147
10.5	Matrices	149
10.6	Conjugations and Computations	155
10.7	One Vector Space to Rule Them All	157
10.8	Geometry of Vector Spaces	159
10.9	Application: Singular Value Decomposition	164
10.10	Cultural Review	179
10.11	Exercises	179
10.12	Chapter Notes	181

11 Live and Learn Linear Algebra (Again) — 185

12 Eigenvectors and Eigenvalues — 191
12.1	Eigenvalues of Graphs	193
12.2	Limiting the Scope: Symmetric Matrices	195
12.3	Inner Products	198
12.4	Orthonormal Bases	202
12.5	Computing Eigenvalues	205
12.6	The Spectral Theorem	207
12.7	Application: Waves	210
12.8	Cultural Review	225

	12.9	Exercises	226
	12.10	Chapter Notes	229

13 Rigor and Formality — 231

14 Multivariable Calculus and Optimization — 237
- 14.1 Generalizing the Derivative — 237
- 14.2 Linear Approximations — 240
- 14.3 Multivariable Functions and the Chain Rule — 245
- 14.4 Computing the Total Derivative — 246
- 14.5 The Geometry of the Gradient — 250
- 14.6 Optimizing Multivariable Functions — 251
- 14.7 The Chain Rule: a Reprise and a Proof — 260
- 14.8 Gradient Descent: an Optimization Hammer — 263
- 14.9 Gradients of Computation Graphs — 264
- 14.10 Application: Automatic Differentiation and a Simple Neural Network — 267
- 14.11 Cultural Review — 283
- 14.12 Exercises — 283
- 14.13 Chapter Notes — 286

15 The Argument for Big-O Notation — 289

16 Groups — 299
- 16.1 The Geometric Perspective — 301
- 16.2 The Interface Perspective — 305
- 16.3 Homomorphisms: Structure Preserving Functions — 307
- 16.4 Building Blocks of Groups — 310
- 16.5 Geometry as the Study of Groups — 312
- 16.6 The Symmetry Group of the Poincaré Disk — 321
- 16.7 The Hyperbolic Isometry Group as a Group of Matrices — 327
- 16.8 Application: Drawing Hyperbolic Tessellations — 328
- 16.9 Cultural Review — 344
- 16.10 Exercises — 344
- 16.11 Chapter Notes — 349

17 A New Interface — 351

About the Author and Cover — 361

Index — 363

Our Goal

This book has a straightforward goal: to teach you how to engage with mathematics.

Let's unpack this. By "mathematics," I mean the universe of books, papers, talks, and blog posts that contain the meat of mathematics: formal definitions, theorems, proofs, conjectures, and algorithms. By "engage" I mean that for any mathematical topic, you have the cognitive tools to actively progress toward understanding that topic. I will "teach" you by introducing you to—or having you revisit—a broad foundation of topics and techniques that support the rest of mathematics. I say "with" because mathematics requires active participation.

We will define and study many basic objects of mathematics, such as polynomials, graphs, and matrices. More importantly, I'll explain *how to think* about those objects as seasoned mathematicians do. We will examine the hierarchies of mathematical abstraction, along with many of the softer skills and insights that constitute "mathematical intuition." Along the way we'll hear the voices of mathematicians—both famous historical figures and my friends and colleagues—to paint a picture of mathematics as both a messy amalgam of competing ideas and preferences, and a story with delightfully surprising twists and connections. In the end, I will show you how mathematicians think about mathematics.

So why would someone like you[1] want to engage with mathematics? Many software engineers, especially the sort who like to push the limits of what can be done with programs, eventually come to realize a deep truth: mathematics unlocks a *lot* of cool new programs. These are truly novel programs. They would simply be impossible to write (if not inconceivable!) without mathematics. That includes programs in this book about cryptography, data science, and art, but also to many revolutionary technologies in industry, such as signal processing, compression, ranking, optimization, and artificial intelligence. As importantly, a wealth of opportunity makes programming more fun! To quote Randall Munroe in his XKCD comic *Forgot Algebra*, "The only things you HAVE to know are how to make enough of a living to stay alive and how to get your taxes done. All the fun parts of life are optional." If you want your career to grow beyond shuffling data around to meet arbitrary business goals, you should learn the tools that enable you to write programs that captivate and delight you. Mathematics is one of those tools.

Programmers are in a privileged position to engage with mathematics. As a program-

---

[1] Hopefully you're a programmer; otherwise, the title of this book must have surely caused a panic attack.

mer, you eat paradigms for breakfast and reshape them into new ones for lunch. Your comfort with functions, logic, and protocols gives you an intuitive familiarity with basic topics such as boolean algebra, recursion, and abstraction. You can rely on this to make mathematics less foreign, progressing all the faster to more nuanced and stimulating topics. Contrast this to most educational math content aimed at students with no background and focusing on rote exercises and passing tests. As a bonus, programming allows me to provide immediate applications that ground the abstract ideas in code. In each chapter of this book, we'll fashion our mathematical designs into a program you couldn't have written before, to dazzling effect. The code is available on Github,[2] with a directory for each chapter.

All told, this book is *not* a textbook. I won't drill you with exercises, though drills have their place. We won't build up any particular field of mathematics from scratch. Though we'll visit calculus, linear algebra, and many other topics, this book is far too short to cover everything a mathematician ought to know about these topics. Moreover, while much of the book is appropriately rigorous, I will occasionally and judiciously loosen rigor when it facilitates a better understanding and relieves tedium. I will note when this occurs, and we'll discuss the role of rigor in mathematics more broadly.

Indeed, rather than read an encyclopedic reference, you want to become *comfortable* with the process of learning mathematics. In part that means becoming comfortable with discomfort, with the struggle of understanding a new concept, and the techniques that mathematicians use to remain productive and sane. Many people find calculus difficult, or squeaked by a linear algebra course without grokking it. After this book you should have a core nugget of understanding of these subjects, along with the cognitive tools that will enable you dive as deeply as you like.

As a necessary consequence, in this book you'll learn how to read and write proofs. The simplest and broadest truth about mathematics is that it revolves around proofs. Proofs are both the primary vehicle of insight and the fundamental measure of judgment. They are the law, the currency, and the fine art of mathematics. Most of what makes mathematics mysterious and opaque—the rigorous definitions, the notation, the overloading of terminology, the mountains of theory, and the unspoken obligations on the reader—is due to the centrality of proofs. A dominant obstacle to learning math is an unfamiliarity with this culture. In this book I'll show you why proofs are so important, cover the basic methods, and display examples of proofs in each chapter. To be sure, you don't have to understand every proof to finish this book, and you will probably be confounded by a few. Embrace your humility. I hope to convince you that each proof contains layers of insight that are genuinely worthwhile, and that no single person can see the complete picture in a single sitting. As you grow into mathematics, the act of reading even previously understood proofs provides both renewed and increaseed wisdom. So long as you identify the value gained by your struggle, your time is well spent.

I'll also teach you how to read between the mathematical lines of a text, and understand the implicit directions and cultural cues that litter textbooks and papers. As we proceed

---

[2] pimbook.org

through the chapters, we'll gradually become more terse, and you'll have many opportunities to practice parsing, interpreting, and understanding math. All of the topics in this book are explained by hundreds of other sources, and each chapter's exercises include explorations of concepts beyond these pages. In addition, I'll discuss how mathematicians approach problems, and how their process influences the culture of math.

You will not learn everything you want to know in this book, nor will you learn everything this book has to offer in one sitting. Those already familiar with math may find early chapters offensively slow and detailed. Those genuinely new to math may find the later chapters offensively fast. This is by design. I want you to be exposed to as much mathematics as possible, to learn the definitions of central mathematicl ideas, to be introduced to notations, conventions, and attitudes, and to have ample opportunity to explore topics that pique your interest.

A number of topics are conspicuously missing from this book, my negligence of which approaches criminal. Except for a few informal cameos, we ignore complex numbers, probability and statistics, differential equations, and formal logic. In my humble opinion, none of these topics is as fundamental for mathematical computer science as those I've chosen to cover. After becoming comfortable with the topics in this book, for example, probability will be very accessible. The chapter on eigenvalues will include a miniature introduction to differential equations. The chapter on groups will briefly summarize complex numbers. Probability will echo in your brain when we discuss random graphs and machine learning. Moreover, many topics in this book are prerequisites for these other areas. And, of course, as a single human self-publishing this book on nights and weekends, I have only so much time.

The first step on our journey is to confirm that mathematics has a culture worth becoming acquainted with. We'll do this with a comparative tour of the culture of software that we understand so well.

Chapter 1

Like Programming, Mathematics has a Culture

Mathematics knows no races or geographic boundaries; for mathematics, the cultural world is one country.

–David Hilbert

Do you remember when you started to really *learn* programming? I do. I spent two years in high school programming games in Java. Those two years easily contain the worst and most embarrassing code I have ever written. My code absolutely reeked. Hundred-line functions and thousand-line classes, magic numbers, unreachable blocks of code, ridiculous comments, a complete disregard for sensible object orientation, and type-coercion that would make your skin crawl. The code worked, but it was filled with bugs and mishandled edge-cases. I broke every rule in the book, and for all my shortcomings I considered myself a hot-shot (at least, among my classmates!). I didn't know how to design programs, or what made a program "good," other than that it ran and I could impress my friends with a zombie shooting game.

Even after I started studying software in college, it was another year before I knew what a stack frame or a register was, another year before I was halfway competent with a terminal, another year before I appreciated functional programming, and to this day I *still* have an irrational fear of systems programming and networking. I built up a base of knowledge over time, with fits and starts at every step.

In a college class on C++ I was programming a Checkers game, and my task was to generate a list of legal jump-moves from a given board state. I used a depth-first search and a few recursive function calls. Once I had something I was pleased with, I compiled it and ran it on my first non-trivial example. Lo' and behold (even having followed test-driven development!), a segmentation fault smacked me in the face. Dozens of test cases and more than twenty hours of confusion later, I found the error: my recursive call passed a reference when it should have been passing a pointer. This wasn't a bug in syntax or semantics—I understood pointers and references well enough—but a design error. As most programmers can relate, the most aggravating part was that changing four characters (swapping a few ampersands with asterisks) fixed it. Twenty hours of work for four characters! Once I begrudgingly verified it worked, I promptly took the rest of the day off to play Starcraft.

Such drama is the seasoning that makes a strong programmer. One must study the topics incrementally, learn from a menagerie of mistakes, and spend hours in a befuddled stupor before becoming "experienced." This gives rise to all sorts of programmer culture, Unix jokes, urban legends, horror stories, and reverence for the masters of C that make the programming community so lovely. It's like a secret club where you know all the handshakes, but should you forget one, a crafty use of `grep` and `sed` will suffice. The struggle makes you appreciate the power of debugging tools, slick frameworks, historically enshrined hacks, and new language features that stop you from shooting your own foot.

When programmers turn to mathematics, they seem to forget these trials. The same people who invested years grokking the tools of their trade treat new mathematical tools and paradigms with surprising impatience. I can see a few reasons why. One is that they've been taking classes called "mathematics" for far longer than they've been learning to program (and mathematics was always easy!). The forced prior investment of schooling engenders a certain expectation. The problem is that the culture of mathematics and the culture of mathematics education—elementary through lower-level college courses—are completely different.

Even math majors have to reconcile this. I've had many conversations with such students, many of whom are friends, colleagues, and even family, who by their third year decided they didn't really enjoy math. The story often goes like this: a student who was good at math in high school (perhaps because of its rigid structure) reaches the point of a math major at which they must read and write proofs in earnest. It requires an earnest, open-ended exploration that they don't enjoy. Despite being a stark departure from high school math, incoming students are never warned in advance. After coming to terms with their unfortunate situation, they decide that their best option is to hold on until they can return to the comfortable setting of their prior experiences, this time in the teacher's chair.

I don't mean to insult teaching as a profession—I love teaching and understand why one would choose to do it full time. There are many excellent teachers who excel at both the math and the trickier task of engaging aloof teenagers to think critically about it. But this pattern of disenchantment among math teachers is prevalent, and it widens the conceptual gap between secondary and "college level" mathematics. Programmers often have similar feelings, that the math they were once so good at is suddenly impenetrable. It's not a feature of math, but a bug in the education system (and a negative feedback loop!) that gets blamed on math as a subject.

Another reason programmers feel impatient is because they do so many things that relate to mathematics in deep ways. They use graph theory for data structures and search. They study enough calculus to make video games. They hear about the Curry-Howard correspondence between proofs and programs. They hear that Haskell is based on a complicated math thing called category theory. They even use mathematical results in an interesting way. I worked at a "blockchain" company that implemented a Bitcoin wallet, which is based on elliptic curve cryptography. The wallet worked, but the implementer didn't understand why. They simply adapted pseudocode found on the internet. At the risk of a dubious analogy, it's akin to a "script kiddie" who uses hacking tools as black boxes, but has little idea how they work. Mathematicians are on the other end of the spectrum, caring almost exclusively about why things work the way they do.

While there's nothing inherently wrong with using mathematics as a black box, especially the sort of applied mathematics that comes with provable guarantees, many programmers *want* to understand why they work. This isn't surprising, given how much time engineers spend studying source code and the internals of brittle, technical systems. Systems that programmers rely on, such as dependency management, load balancers, search engines, alerting systems, and machine learning, all have rich mathematical foundations. We're naturally curious about how they work and how to adapt them to our needs.

Yet another hindrance to mathematics is that it has no centralized documentation. Instead it has a collection of books, papers, journals, and conferences, each with discrepancies of presentation, citing each other in a haphazard manner. A theorem presented at a computer science conference can be phrased in completely unfamiliar terms in a dynamical systems journal—even though they boil down to the same facts! In subfields like network science that straddle disciplines, one often sees "translation tables" for jargon.

Dealing with this is not easy. Students of mathematics solve these problems with knowledgeable teachers. Working mathematicians just "do it." They work out the translation details themselves with coffee and contemplation. Advanced books also lean toward terseness, despite being titled as "elementary" or an "introduction." They opt not to redefine what they think the reader must already know. The purest fields of mathematics take a sort of pretentious pride in how abstract and compact their work is (to the point where many students spend weeks or months understanding a single chapter!).

What programmers would consider "sloppy" notation is one symptom of the problem, but there there are other expectations on the reader that, for better or worse, decelerate the pace of reading. Unfortunately I have no solution here. Part of the power and expressiveness of mathematics is the ability for its practitioners to overload, redefine, and omit in a suggestive manner. Mathematicians also have thousands of years of "legacy" math that require backward compatibility. Enforcing a single specification for all of mathematics—a suggestion I frequently hear from software engineers—would be horrendously counterproductive.

Indeed, ideas we take for granted today, such as algebraic notation, drawing functions in the Euclidean plane, and summation notation, were at one point actively developed technologies. Each of these notations had a revolutionary effect, not just on science, but also, to quote Bret Victor, on our capacity to "think new thoughts." One can even draw a line from the proliferation of algebraic notation and the computational questions it raised to the invention of the computer.[1] Borrowing software terminology, algebraic notation is

---

[1] Leibniz, one of the inventors of calculus, dreamed of a machine that could automatically solve mathematical problems. Ada Lovelace (up to some irrelevant debate) designed the first program for computing Bernoulli numbers, which arise in algebraic formulas for computing sums of powers of integers. In the early 1900's Hilbert posed his Tenth Problem on algorithms for computing solutions to Diophantine equations, and later his Entscheidungsproblem, which was solved concurrently by Church and Turing and directly led to Turing's

among the most influential and scalable technologies humanity has ever invented. And as we'll see in Chapter 10 and Chapter 16, we can find algebraic structure hiding in exciting places. Algebraic notation helps us understand this structure not only because we can compute, but also because we can visually see the symmetries in the formulas. This makes it easier for us to identify, analyze, and encapsulate structure when it occurs.

Finally, the best mathematicians study concepts that connect decades of material, while simultaneously inventing new concepts which have no existing words to describe them. Without flexible expression, such work would be impossible. It reduces cognitive load, a theme that will follow us throughout the book. Unfortunately, it only does so for the readers who have *already* absorbed the basic concepts of discussion. By contrast, good software practice encourages code that is simple enough for anyone to understand. As such, the uninitiated programmer often has a much larger cognitive load when reading math than when reading a program.

Taken together, mathematical notation is closer to spoken language than to code. It can reduce one's mental burden via rigorous rules applied to an external representation, coupled with context and convention. All of this, the notation, the differences among subfields, the tradeoff between expressiveness and cognitive load, has grown out of hundreds of years of mathematical progress.

Equipped with this understanding, that mathematics has culturally relevant reasons for its strange practices, let's begin our journey through the mists of math with renewed openness.

Read on, and welcome to the club.

code-breaking computer.

Chapter 2
Polynomials

We are not trying to meet some abstract production quota of definitions, theorems and proofs. The measure of our success is whether what we do enables people to understand and think more clearly and effectively about mathematics.

–William Thurston

We begin with polynomials. In studying polynomials, we'll reveal some of the implicit assumptions behind mathematical definitions, work carefully through two nontrivial proofs, and learn about how to "share secrets" using something called *polynomial interpolation*.

To whet your appetite, this secret sharing scheme allows one to encode a secret message in 10 parts so that any 6 can be used to reconstruct the secret, but with fewer than 6 pieces it's impossible to determine even a single bit of the original message. The numbers 10 and 6 are just examples, and the scheme we'll present works for any pair of integers. This almost magical application turns out to be possible using nothing more than polynomials.

2.1 Polynomials, Java, and Definitions

We need to start with the definition of a polynomial. The problem, if you're the sort of person who struggled with math, is that reading the definition as a formula will make your eyes glaze over. In this chapter we're going to overcome this.

The reason I'm so confident is that I'm certain you've overcome the same obstacle in the context of programming. For example, my first programming language was Java. And my first program, which I didn't write but rather copied verbatim, was likely similar to this monstrosity.

```
/******************************************
 *  Compilation:  javac HelloWorld.java
 *  Execution:    java HelloWorld
 *
 *  Prints "Hello, World".
 ******************************************/
public class HelloWorld {
    public static void main(String[] args) {
        // Prints "Hello, World" to stdout on the terminal.
        System.out.println("Hello, World");
    }
}
```

It was roughly six months before I understood what all the different pieces of this program did, despite the fact that I had written 'public static void main' so many times I had committed it to memory. One nice thing about programming is that you don't have to understand a code snippet before you can start using it. But at *some point,* I stopped to ask, "what do those words actually mean?" That's the step when my eyes stop glazing over. That's the same procedure we need to invoke for a mathematical definition, preferably faster than six months.

Now I'm going to throw you in the thick of the definition of a polynomial. But stay with me! I want you to start by taking out a piece of paper and literally copying down the definition (the entire next paragraph), character for character, as one would type out a program from scratch. This is not an idle exercise. Taking notes by hand uses a part of your brain that both helps you remember what you wrote, and helps you *read* it closely. Each individual word and symbol of a mathematical definition affects the concept being defined, so it's important to parse everything slowly.

Definition 2.1. A single variable *polynomial with real coefficients* is a function f that takes a real number as input, produces a real number as output, and has the form

$$f(x) = a_0 + a_1 x + a_2 x^2 + \cdots + a_n x^n,$$

where the a_i are real numbers. The a_i are called *coefficients* of f. The *degree* of the polynomial is the integer n.

Let's analyze the content of this definition in three ways. First, *syntactically,* which also highlights some general features of written definitions. Second, *semantically,* where we'll discuss what a polynomial should represent as a concept in your mind. Third, we'll inspect this definition *culturally,* which includes the unspoken expectations of the reader upon encountering a definition in the wild.

Syntax

A *definition* is an English sentence or paragraph in which italicized words represent the concepts being defined. In this case, Definition 2.1 defines three things: a *polynomial with real coefficients* (the function f), *coefficients* (the numbers a_i), and a polynomial's *degree* (the integer n).

A proper mathematical treatment might also define what a "real number" is, but we simply don't have the time or space.[1] For now, think of a real number as a floating point number without the emotional baggage that comes from trying to fit all decimals into a finite number of bits.

An array of numbers a, which in most programming languages would be indexed using square brackets like a[i], is almost always indexed in math using subscripts a_i. For two-dimensional arrays, we place the indices comma separated in the subscript, i.e. $a_{i,j}$ is equivalent to a[i][j]. Hence, the coefficients are just an array of real numbers.

To say f "has the form" means that f is restricted to some choice of the unbound variables in its formula. In this case those are, in order:

1. A choice of names for all the variables involved. The definition has chosen f for the function, x for the input variable name (usually called the "variable," but we won't overload that term for now), a for the array of coefficients, and n for the degree. One can choose other names as desired.

2. A value for the degree.

3. A value for the array of coefficients $a_0, a_1, a_2, \ldots, a_n$.

Specifying all of these results in a concrete polynomial.

Semantics

Let's start with a simple example polynomial, where I pick g for the function name, t for the input name, b for the coefficients, and define $n = 3$, and $b_0, b_1, b_2, b_3 = 2, 0, 4, -1$. By definition, g has the form

$$g(t) = 2 + 0t + 4t^2 + (-1)t^3.$$

Letting zero be zero, we take some liberties and usually write g more briefly as $g(t) = 2 + 4t^2 - t^3$. As you might expect, g is a function you can evaluate, and evaluating it at an input $t = 2$ means substituting 2 for t and doing the requisite arithmetic to get

$$g(2) = 2 + 4(2^2) - 2^3 = 10.$$

According to the definition, a polynomial is a function that is written in a certain form. The concept of a polynomial is a bit more general. It is any function of a single numeric input that can be expressed using only addition and multiplication and constants. This conceptual understanding allows for more general representations. For example, the following "is" a polynomial even if we haven't expressed it strictly to the letter of Definition 2.1.

---

[1] If you're truly interested in how real numbers are defined from scratch, Chapter 29 of Spivak's text *Calculus* is devoted to a gold-standard treatment. You might be ready for it after working through a few chapters of this book, but be warned: it was reserved for the end of a long book on calculus! Spivak even starts Chapter 29 with, "The mass of drudgery which this chapter necessarily contains..."

$$f(x) = x^5 - x - 1$$

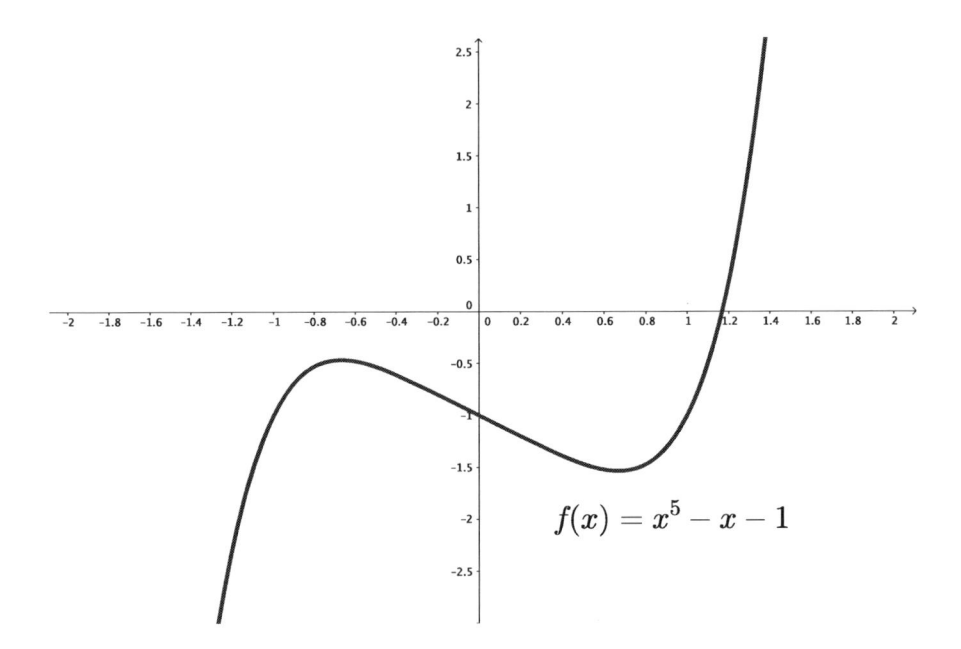

Figure 2.1: A polynomial as a curve in the plane.

$$f(x) = (x-1)(x+6)^2$$

You recover the precise form of Definition 2.1 by algebraically simplifying and grouping terms. Indeed, the form described in Definition 2.1 is not ideal for every occasion! For example, if you want to evaluate a polynomial quickly on a computer, you might represent the polynomial so that evaluating it doesn't redundantly compute the powers $t^1, t^2, t^3, \ldots, t^n$. One such scheme is called Horner's method.

In any case, the abstract concept of a polynomial $g(t)$ doesn't depend on the choices you use to write it down, so long as one can get from your representation to a standard form. Though I said earlier the variable names are part of the syntactic data of a polynomial, they're really only the data of a *particular representation* of a polynomial. I don't need to remind you, dear programmer, that variable names are a matter of syntax, not semantics.

There are other ways to think about polynomials, and we'll return to polynomials in future chapters with new and deeper ideas about them. Here are some previews of that.

The first is that a polynomial, as with any function, can be represented as a set of pairs called *points*. That is, if you take each input t and pair it with its output $f(t)$, you get a set of tuples $(t, f(t))$, which can be analyzed from the perspective of set theory. We will return to this perspective in Chapter 4.

Second, a polynomial's graph can be plotted as a curve in space, so that the horizontal direction represents the input and the vertical represents the output. Figure 2.1 shows a plot of one part of the curve given by the polynomial $f(x) = x^5 - x - 1$.

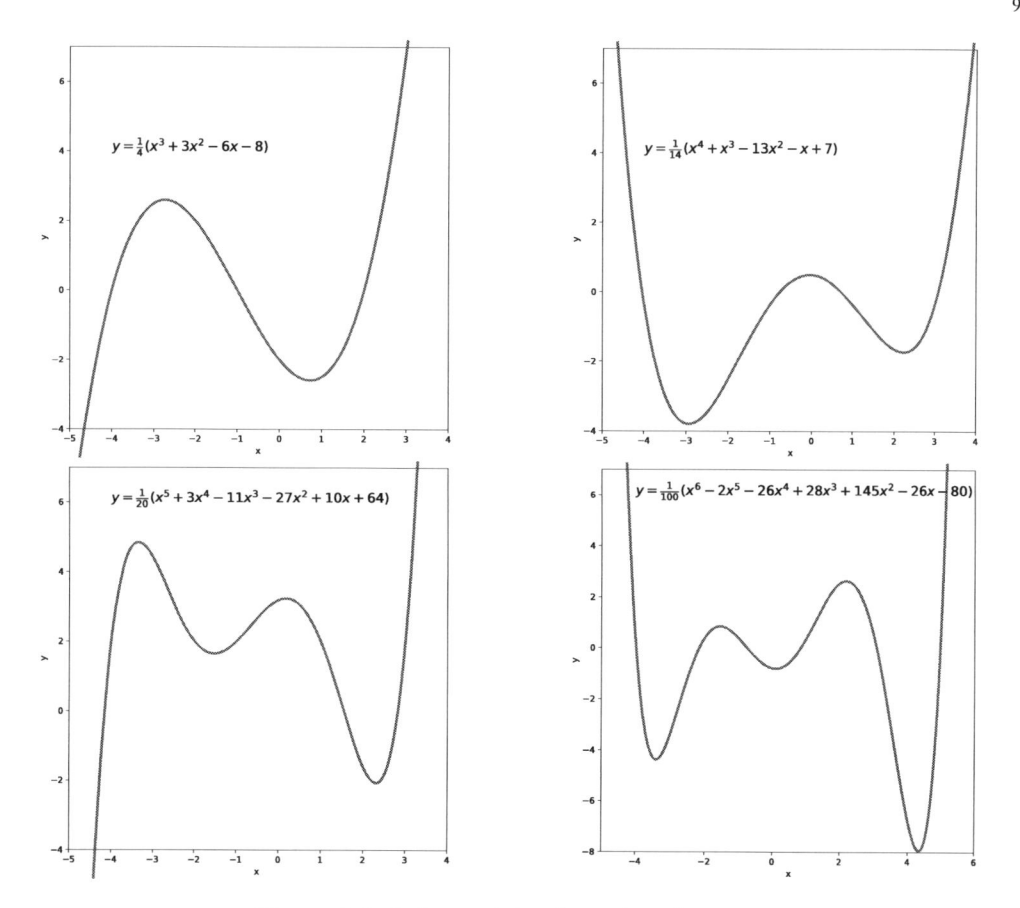

Figure 2.2: Polynomials of varying degrees.

Using the curves they "carve out" in space, polynomials can be regarded as geometric objects with geometric properties like "curvature" and "smoothness." In Chapter 8 we'll return to this more formally, but until then one can guess how they might faithfully describe a plot like the one in Figure 2.1. The connection between polynomials as geometric objects and their algebraic properties is a deep one that has occupied mathematicians for centuries. For example, the degree gives some information about the shape of the curve. For example, Figure 2.2 shows plots of generic polynomials of degrees 3 through 6. As the degree goes up, so does the number of times the polynomial "changes direction" between increasing and decreasing. Turning this into a mathematically rigorous theorem requires more nuance, but a pattern is clear.

Finally, polynomials can be thought of as "building blocks" for complicated structures. That is, polynomials are families of increasingly expressive objects, which get more complex as the degree increases. This idea is the foundation of the application for this chapter (sharing secrets), and it will guide us to Taylor polynomials as a hammer for every nail in Chapters 8 and 14.

Polynomials occur with stunning ubiquity across mathematics. It makes one wonder exactly why they are so central, but to reiterate, polynomials encapsulate the full expressivity of addition and multiplication. As programmers, we know that even such simple operations as binary AND, OR, and NOT, when combined arbitrarily, yield the full gamut of algorithms. Polynomials fill the same role for arithmetic. Indeed, polynomials with multiple variables can represent AND, OR, and NOT, if you restrict the values of the variables to be zero and one (interpreted as *false* and *true*, respectively).

$$\begin{aligned} \text{AND}(x, y) &= xy \\ \text{NOT}(x) &= 1 - x \\ \text{OR}(x, y) &= 1 - (1-x)(1-y) \end{aligned}$$

Any logical condition, again assuming the inputs are binary, can be represented using a combination of these three polynomials. Polynomials are expressive enough to capture all of boolean logic. This suggests that even single-variable polynomials *should* have strikingly complex behavior. The rest of the chapter will display bits of that dazzling performance.

Culture

The most important cultural expectation, one every mathematician knows, is that the second you see a definition in a text **you must immediately write down examples.** Generous authors provide examples of genuinely new concepts, but an author is never obligated to do so. The unspoken rule is that the reader may not continue unless the reader understands what the definition is saying. That is, you aren't expected to *master* the concept, most certainly not at the same speed you read it. But you should have some idea going forward of what the defined words refer to.

The best way to think of this is like testing in software. You start with the simplest possible tests, usually setting as many values as you can to zero or one, then work your way up to more complicated examples. Later, when you get stuck on some theorem or proof—an occupational hazard faced by gods and mortals alike—you return to those examples and test how the claims in the proof apply to them. This is how one builds so-called "mathematical intuition." In the long term, one uses that intuition to speed up the process of absorbing new ideas.

So let's write down some definitions of polynomials according to Definition 2.1, starting from literally the simplest possible thing. To make you pay attention, I'll slip in some examples that are not polynomials and your job is to run them against the definition. Take your time, and you can check your answers in the Chapter Notes.

$$f(x) = 0$$
$$g(x) = 12$$
$$h(x) = 1 + x + x^2 + x^3$$
$$i(x) = x^{1/2}$$
$$j(x) = \frac{1}{2} + x^2 - 2x^4 + 8x^8$$
$$k(x) = 4.5 + \frac{1}{x} - \frac{5}{x^2}$$
$$l(x) = \pi - \frac{1}{e}x^5 + e\pi^3 x^{10}$$
$$m(x) = x + x^2 - x^\pi + x^e$$

Like software testing, examples weed out pesky edge cases and clarify what is permitted by the definition. For example, the exponents of a polynomial must be nonnegative integers, though I only stated it implicitly in the definition.

When reading a definition, one often encounters the phrase "by convention." This can be in regard to a strange edge case or a matter of taste. A common example is the factorial $n! = 1 \cdot 2 \cdots n$, where $0! = 1$ by convention. This makes formulas cleaner and provides a natural default value of an "empty product," an idea programmers understand when choosing a base case for a loop that computes the product of a (possibly empty) list of numbers.

For polynomials, convention strikes when we inspect the example $f(x) = 0$ given above. What is the degree of f? On one hand, it makes sense to say that the zero polynomial has degree $n = 0$ and $a_0 = 0$. On the other hand, it also makes sense (in a strict, syntactical sense) to say that f has degree $n = 1$ with $a_0 = 0$ and $a_1 = 0$, or $n = 2$ with three zeros. But we don't want a polynomial to have multiple distinct possibilities for degree. Indeed, this would allow $f(x)$ to have *every* positive degree (by adding extra zeros), depriving the word "degree" of a consistent interpretation.

To avoid this, we amend Definition 2.1 so that the last coefficient a_n is required to be nonzero. But then the function $f(x) = 0$ is not allowed to be a polynomial! So, by convention, we define a special exception, the function $f(x) = 0$, as the *zero polynomial*. By convention, the zero polynomial is defined to have degree -1. One recurring theme is that every time a definition includes the phrase "by convention," it becomes a special edge-case in the resulting program.

Dealing with this edge case made us think hard about the right definition for a polynomial, but it was mostly a superficial change. Other times, as we will confront head on in Chapter 8 when we define limits, dealing with an edge case reveals the soul of a concept. It's curious how mathematical books tend to start with the final product, instead of the journey to the right definition. Perhaps teaching the latter is much harder and more time consuming, with fewer tangible benefits. But in advanced mathematics, deep understanding comes in fits and starts. Often, no such distilled explanation is known.

In any case, examples are the primary method to clarify the features of a definition. Having examples in your pocket as you continue to read is important, and *coming up with the examples yourself* is what helps you internalize a concept.

It is a bit strange that mathematicians choose to write definitions with variable names by example, rather than using the sort of notation one might use to define a programming language syntax. Using a loose version of Backus-Naur form (BNF), which is a mostly self-explanatory language for describing syntax, I might define a polynomial as:

```
coefficient = number
variable = 'x'
term = coefficient * variable ^ int
polynomial = term
           | term + polynomial
```

The problem is that this definition doesn't tell you what polynomials are all about. It doesn't communicate anything to the reader about the semantics of the definition, but rather how a computer should parse it. While Definition 2.1 isn't perfect—I still had to explain the semantics—it signals that a polynomial is a function of a single input. BNF only provides a sequence of named tokens. This theme, that most mathematics is designed for human-to-human communication, will follow us throughout the book. Mathematical discourse is about getting a vivid idea from your mind into someone else's mind.

That's why an author usually starts with a conceptual definition like Definition 2.1 many pages before discussing a programmatic representation of a polynomial. It's why mathematicians will seamlessly convert between representations—such as the functional, set-theoretic, and geometric representations I described earlier—as if mathematics were the JavaScript type system on methamphetamines. In Java you have to separate an interface from the class which implements it, and in C++ templates are distinct from their usage. In math, much of conceptual understanding happens at the level of interfaces and templates, while particular representations are used for computation.

I want to make this extremely clear because in mathematics it's implicit. My math teachers in college and grad school *never explicitly* discussed why one would use one definition over another, because somehow along the arduous journey through a math education, the folks who remained understood it.

Polynomials may seem frivolous to illustrate the difference between an object-as-abstract-concept and the representational choices that go into understanding a definition, but the same pattern lurks behind more complicated definitions. First the author will start with the best conceptual definition—the one that seems to them, with the hindsight of years of study, to be the most useful way to communicate the idea behind the concept. For us that's Definition 2.1. Often these definitions seem totally useless from a programming perspective.

Then ten pages later (or a hundred!) the author introduces another definition, often a data definition, which turns out to be *equivalent* to the first. Any properties defined in the first definition automatically hold in the second and vice versa. But the *data definition* is the one that allows for nice programs. You might think the author was crazy not to start with the data definition, but it's the conceptual definition that sticks in your mind, generalizes, and guides you through proofs.

This interplay between intuitive and data definitions will take center stage in Chapter 10, our first exposure to linear algebra. We'll see that so-called *linear maps* are equivalent to matrices in a formal sense. While linear maps are easy to conceptualize, the corresponding operations on matrices are complicated and best suited for a computer. But a mathematician would argue you can't see the elegance or truly grok linear algebra if you only ever see a matrix without conceptualizing it as a linear map. In linear algebra, the line between interface and implementation is crisp. Even better, few areas of math are as widely applicable.

It's also worth noting that the multiplicity of definitions arose throughout history. Polynomials have been studied for many centuries, but parser-friendly forms of polynomials weren't needed until the computer age. Likewise, algebra was studied before the graphical representations of Descartes allowed us to draw polynomials as curves. Other perspectives on polynomials were developed to enable useful approximations and calculations on the positions of planets, the path of projectiles, and many other tasks. We'll get a taste of this in Chapter 8. Each new perspective and definition was driven by an additional need. As a consequence, what's thought of as the "best" definition of a concept can change. Throughout history math has been shaped and reshaped to refine, rigorize, and distill the core insights, often to ease calculations in fashion at the time.

In any case, the point is that we will fluidly convert between the many ways of thinking about polynomials: as expressions defined abstractly by picking a list of numbers, or as functions with a special structure. Effective mathematics is flexible in this way.

2.2   A Little More Notation

When defining a function, one often uses the compact arrow notation $f : A \to B$ to describe the allowed inputs and outputs. All possible inputs are collectively called the *domain*, and all possible outputs are called the *range*. There is one caveat I'll explain via programming. Say you have a function that doubles the input, such as

```
int f(int x) {
   return 2*x;
}
```

The possible inputs include all integers, and the *type* of the output is also "integer." But it's obvious that 3 is not a possible output of this particular function.

In math we disambiguate this with two words. *Range* is the set of actual outputs of a function, and the "type" of outputs is called the *codomain*. So the notation $f : A \to B$ specifies the domain A and codomain B, whereas the range depends on the semantics of f. When one introduces a function, as programmers do with type signatures and function headers, we state the notation $f : A \to B$ first, and the actual function definition second.

Because mathematicians were not originally constrained by ASCII, they developed other symbols for types. The symbol for the set of real numbers is $\mathbb{R}$. The font is called "blackboard-bold," and it's the standard font for denoting number systems. Applying the arrow notation, a polynomial is $f : \mathbb{R} \to \mathbb{R}$. A common phrase is to say a polynomial is "over the reals" to mean it has real coefficients. As opposed to, say, a polynomial over the integers that has integer coefficients.

Most famous number types have special symbols. The symbol for integers is $\mathbb{Z}$, and the positive integers are denoted by $\mathbb{N}$, often called the *natural numbers*.[2] There is an amusing dispute of no real consequence between logicians and other mathematicians on whether zero is a natural number, with logicians demanding it is.

Finally, I'll use the $\in$ symbol, read "in," to assert or assume membership in some set. For example $q \in \mathbb{N}$ is the claim that q is a natural number. It is literally short hand for the phrase, "q is in the natural numbers," or "q is a natural number." It can be used in a condition (preceded by "if"), an assertion (preceded by "suppose"), or a question.

2.3 Existence & Uniqueness

Having seen some definitions, we're ready to develop the main tool we need for secret sharing: the existence and uniqueness theorem for polynomials passing through a given set of points.

First, a word about existence and uniqueness. Existence proofs are classic in mathematics, and they come in all shapes and sizes. Basically, mathematicians like to take interesting properties they see on small objects, write down the property in general, and then ask things like, "Are there arbitrarily large objects with this property?" or, "Are there infinitely many objects with this property?" It's like in physics: when you come up with some equations that govern the internal workings of a star you might ask: would these equations support arbitrarily massive stars?

One simple example is quite famous: whether there are infinitely many pairs of prime numbers of the form $p, p + 2$. For example, 11 and 13 work, but 23 is not part of such a pair.[3] Perhaps surprisingly, it is an open question whether there are infinitely many such pairs. The assertion that there are is called the Twin Prime Conjecture.

In some cases you get lucky, and the property you defined is specific enough to single out a *unique* mathematical object. This is what will happen to us with polynomials. Other times, the property (or list of properties) you defined are too restrictive, and there are no mathematical objects that can satisfy it. For example, Kleinberg's Impossibility Theorem for Clustering lays out three natural properties for a clustering algorithm (an algorithm that finds dense groups of points in a geometric dataset) and proves that no algorithm can satisfy all three simultaneously. See the Chapter Notes for more on this. Though such theorems are often heralded as genius, more often than not mathematicians avoid impossibility by turning small examples into broad conjectures.

That's how we'll approach existence and uniqueness for polynomials. Here is the theo-

---

[2] The Z stands for Zahlen, the German word for "numbers."
[3] See how I immediately wrote down examples?

rem we'll prove, stated in its most precise form. Don't worry, we'll go carefully through every bit of it, but try to read it now.

Theorem 2.2. *For any integer $n \geq 0$ and any list of $n+1$ points $(x_0, y_0), (x_1, y_1), \ldots, (x_n, y_n)$ in $\mathbb{R}^2$ with $x_0 < x_1 < \cdots < x_n$, there exists a unique degree n polynomial $p(x)$ such that $p(x_i) = y_i$ for all i.*

The one piece of new notation is the exponent on $\mathbb{R}^2$. This just means "pairs" of real numbers, each of which is in $\mathbb{R}$. Likewise, $\mathbb{Z}^3$ would be triples of integers, and $\mathbb{N}^{10}$ tuples of size ten, each entry of which is a natural number.

A briefer, more informal way to state the theorem: there is a unique degree n polynomial passing through a choice of $n+1$ points.[4] Now just like with definitions, the first thing we need to do when we see a new theorem is write down the simplest possible examples. In addition to simplifying the theorem, it will give us examples to work with while going through the proof. Write down some examples now. As mathematician Alfred Whitehead said, "We think in generalities, but we live in details."

Back already? I'll show you examples I'd write down, and you can compare your process to mine. The simplest example is $n = 0$, so that $n + 1 = 1$ and we're working with a single point. Let's pick one at random, say $(7, 4)$. The theorem asserts that there is a unique degree zero polynomial passing through this point. What's a degree zero polynomial? Looking back at Definition 2.1, it's a function like $a_0 + a_1 x + a_2 x^2 + \cdots + a_d x^d$ (I'm using d for the degree here because n is already taken), where we've chosen to set $d = 0$. Setting $d = 0$ means that f has the form $f(x) = a_0$. So what's such a function with $f(7) = 4$? There is no choice but $f(x) = 4$. It should be clear that it's the only degree zero polynomial that does this. Indeed, the datum that defines a degree-zero polynomial is a single number, and the constraint of passing through the point $(7, 4)$ forces that one piece of data to a specific value.

Let's move on to a slightly larger example which I'll allow you to work out for yourself before going through the details. When $n = 1$ and we have $n + 1 = 2$ points, say $(2, 3), (7, 4)$, the theorem claims a unique degree 1 polynomial f with $f(2) = 3$ and $f(7) = 4$. Find it by writing down the definition for a polynomial in this special case and solving the two resulting equations.[5]

Alright. A degree 1 polynomial has the form

$$f(x) = a_0 + a_1 x.$$

Writing down the two equations $f(2) = 3, f(7) = 4$, we must simultaneously solve:

---

[4] To say a function $f(x)$ "passes" through a point (a, b) means that $f(a) = b$. When we say this we're thinking of f as a geometric curve. It's 'passing' through the point because we imagine a dot on the curve moving along it. That perspective allows for colorful language in place of notation.

[5] If you're more than comfortable solving basic systems of equations, you may want to skip ahead to Section 2.3. This introductory chapter is intended to be much more gradual than the average math book.

$$a_0 + a_1 \cdot 2 = 3$$
$$a_0 + a_1 \cdot 7 = 4$$

If we solve for a_0 in the first equation, we get $a_0 = 3 - 2a_1$. Substituting that into the second equation we get $(3 - 2a_1) + a_1 \cdot 7 = 4$, which solves for $a_1 = 1/5$. Plugging this back into the first equation gives $a_0 = 3 - 2/5$. This has forced the polynomial to be exactly

$$f(x) = \left(3 - \frac{2}{5}\right) + \frac{1}{5}x = \frac{13}{5} + \frac{1}{5}x.$$

Geometrically, a degree 1 polynomial is a line. So despite all our work above, we're just stating a fact we already know, that there is a unique line between any two points. Well, it's not *quite* the same fact. What is different about this scenario? The statement of the theorem said, "$x_0 < x_1 < \cdots < x_n$". In our example, this means we require $x_0 < x_1$. So this is where we run a sanity check. What happens if $x_0 = x_1$? Think about it, and if you can't tell then you should try to prove it wrong: try to find a degree 1 polynomial passing through the points $(2, 3), (2, 5)$.

The problem could be that there is *no* degree 1 polynomial passing through those points, violating existence. Or, the problem might be that there are *many* degree 1 polynomials passing through these two points, violating uniqueness. It's your job to determine what the problem is. And despite it being pedantic, you should work straight from the definition of a polynomial! Don't use any mnemonics or heuristics you may remember; we're practicing reading from precise definitions.

In case you're stuck, let's follow our pattern from before. If we call $a_0 + a_1 x$ our polynomial, saying it passes through these two points is equivalent to saying that there is a simultaneous solution to the following two equations $f(2) = 3$ and $f(2) = 5$.

$$a_0 + a_1 \cdot 2 = 3$$
$$a_0 + a_1 \cdot 2 = 5$$

What happens when you try to solve these equations like we did before? Try it.

What about for three points or more? Well, that's the point at which it might start to get difficult to compute. You can try by setting up equations like those I wrote above, and with some elbow grease you'll solve it. Such things are best done in private so you can make plentiful mistakes without being judged for it.

Now that we've worked out two examples of the theorem in action, let's move on to the proof. The proof will have two parts, existence and uniqueness. That is, first we'll show that a polynomial satisfying the requirements exists, and then we'll show that if two polynomials both satisfied the requirements, they'd have to be the same. In other words, there can only be one polynomial with that property.

Existence of Polynomials Through Points

We will show existence by direct construction. That is, we'll "be clever" and find a general way to write down a polynomial that works. Being clever sounds scary, but the process is actually quite natural, and it follows the same pattern as we did for reading and understanding definitions: you start with the simplest possible example (but this time the example will be generic) and then you work up to more complicated examples. By the time we get to $n = 2$ we will notice a pattern, that pattern will suggest a formula for the general solution, and we will prove it's correct. In fact, once we understand how to build the general formula, the proof that it works will be trivial.

Let's start with a single point (x_1, y_1) and $n = 0$. I'm not specifying the values of x_1 or y_1 because I don't want the construction to depend on my arbitrary specific choices. I must ensure that $f(x_1) = y_1$, and that f has degree zero. Simply enough, we set the first coefficient of f to y_1, the rest zero.

$$f(x) = y_1$$

On to two points. Call them $(x_1, y_1), (x_2, y_2)$ (note the variable is just plain x, and my example inputs are $x_1, x_2, \dots$). Now here's an interesting idea: I can write the polynomial in this strange way:

$$f(x) = y_1 \frac{x - x_2}{x_1 - x_2} + y_2 \frac{x - x_1}{x_2 - x_1}$$

Let's verify that this works. If I evaluate f at x_1, the second term gets $x_1 - x_1 = 0$ in the numerator and so the second term is zero. The first term, however, becomes $y_1 \frac{x_1 - x_2}{x_1 - x_2} = y_1 \cdot 1$, which is what we wanted: we gave x_1 as input and the output was y_1. Also note that we have explicitly disallowed $x_1 = x_2$ by the conditions in the theorem, so the fractions will never be $0/0$.

Likewise, if you evaluate $f(x_2)$ the first term is zero and the second term evaluates to y_2. So we have both $f(x_1) = y_1$ and $f(x_2) = y_2$, and the expression is a degree 1 polynomial. How do I know it's degree one when I wrote f in that strange way? For one, I could rewrite f like this:

$$f(x) = \frac{y_1}{x_1 - x_2}(x - x_2) + \frac{y_2}{x_2 - x_1}(x - x_1),$$

and simplify with typical algebra to get the form required by the definition:

$$f(x) = \frac{x_1 y_2 - x_2 y_1}{x_1 - x_2} + \left(\frac{y_1 - y_2}{x_1 - x_2}\right) x$$

What a headache! Instead of doing all that algebra I, could observe that no powers of x appear in the formula for f that are larger than 1, and we never multiply two x's together. Since these are the only ways to get degree bigger than 1, we can skip the algebra and be confident that the degree is 1.

The key to the above idea, and the reason we wrote it down in that strange way, is so that each constraint (i.e. $f(x_1) = y_1$) could be isolated in its own term, while all the other terms evaluate to zero. For three points $(x_1, y_1), (x_2, y_2), (x_3, y_3)$ we just have to beef up the terms to maintain the same property: when you plug in x_1, all terms except the first evaluate to zero and the fraction in the first term evaluates to 1. When you plug in x_2, the second term is the only one that stays nonzero, and likewise for the third. Here is the generalization that does the trick.

$$f(x) = y_1 \frac{(x-x_2)(x-x_3)}{(x_1-x_2)(x_1-x_3)} + y_2 \frac{(x-x_1)(x-x_3)}{(x_2-x_1)(x_2-x_3)} + y_3 \frac{(x-x_1)(x-x_2)}{(x_3-x_1)(x_3-x_2)}$$

Now you come in. Evaluate f at x_1 and verify that the second and third terms are zero, and that the first term simplifies to y_1. The symmetry in the formula should convince you that the same holds true for x_2, x_3 without having to go through all the steps two more times.

Again, it's clear that the polynomial we defined is degree 2, because each term consists of a product of two degree-1 terms like $(x - x_i)$ and taking their product gives at most x^2. This has saved me the effort of rearranging that nonsense to get something in the form of Definition 2.1.

The general formula for $(x_1, y_1), \ldots, (x_n, y_n)$ should follow the same pattern. Add up a bunch of terms, and for the i-th term you multiply y_i by a fraction you construct according to the rule: the numerator is the product of $x - x_j$ for every j except i, and the denominator is a product of all the $(x_i - x_j)$ for the same js as the numerator. It works for the same reason that our formula works for three terms above. In fact, the process is clear enough that you could write a program to build these polynomials quite easily, and we'll walk through such a program together at the end of the chapter.

Here is the notation version of the process we just described in words. It's a mess, but we'll break it down.

$$f(x) = \sum_{i=0}^{n} y_i \cdot \left(\prod_{j \neq i} \frac{x - x_j}{x_i - x_j} \right)$$

What a mouthful! I'll assume the $\sum, \prod$ symbols are new to you. They are read semantically as "sum" and "product," or typographically as "sigma" and "pi". They essentially represent loops of arithmetic. That is, if I have a statement like $\sum_{i=0}^{n}(\text{expr})$, it is equivalent to the following code snippet.

```
int i;
sometype theSum = defaultValue;

for (i = 0; i <= n; i++) {
   theSum += expr(i);
}

return theSum;
```

I wrote it this way because `defaultValue` is whatever the conventional 'zero object' is in that setting. For adding numbers the zero object is zero, for concatenating lists it's the empty list, and for adding polynomials it's the zero polynomial. It can get much more exotic with more advanced mathematics, which we'll see in Chapter 16 when we study groups. The point is that the $\sum$ notation does not imply a specific type of the thing being "summed." It's just a shorthand for the symbol +, and when you know what things are being added, there's always a contextually relevant zero.

Moreover, explaining $\sum$ using code allows me to define $\prod$ by analogy: you just replace += with *= and reinterpret the "default value" as what makes sense for multiplication. Functional programmers will know this pattern well, because both are a "fold" (or "reduce") function with a particular choice of binary operation and initial value.

The notation $\prod_{j \neq i}$ adds three caveats. First, recall that in this context i is fixed by the outer loop, so j is the looping variable (unfortunately, the reader is required to keep track of scope when it comes to nested sums and products). Second, the bounds on j are not stated; we have to infer them from the context. There are two hints: we're comparing j to i, so it should probably have the same range as i unless otherwise stated, and we can see where in the expression we're using j. We're using it as an index on the x's. Since the x indices go from 0 to n, we'd expect j to have that range. It might seem totally nonrigorous to a programmer, but if mathematicians consider it "easy" to infer the intent of a notation, then it is considered rigorous enough.[6]

Though it sometimes makes me cringe to say it, give the author the benefit of the doubt. When things are ambiguous, pick the option that doesn't break the math. In this respect, you have to act as both the tester, the compiler, and the bug fixer when you're reading math. The best default assumption is that the author is far smarter than we are, and if we the reader don't understand something, it's likely a user error and not a bug. In the occasional event that the author is wrong, it's more often than not a simple mistake or typo, to which an experienced reader would say, "The author obviously meant 'foo' because otherwise none of this makes sense," and continue unscathed.

Finally, the $j \neq i$ part is an implied filter on the range of j. Inside the for loop you add an extra if statement to skip that iteration if $j = i$. Read out loud, $\prod_{j \neq i}$ would be "the product over j not equal to i." If we wanted to write out the product-nested-in-a-sum as a nested loop, it would look like this:

---

[6] Another reason is that mathematicians get tired of writing these "obvious" details over and over again.

```
int i, j;
sometype theSum = defaultSumValue;

for (i = 0; i <= n; i++) {
  othertype product = defaultProductValue;

  for (j = 0; j <= n; j++) {
    if (j != i) {
      product *= foo(i, j);
    }
  }

  theSum += bar(i) * product;
}

return theSum;
```

$$f(x) = \sum_{i=0}^{n} \text{bar}(i) \left(\prod_{j \neq i} \text{foo}(i, j) \right)$$

Compare the math and code, and make sure you can connect the structural pieces. Often the inner parentheses are omitted, with the default assumption that everything to the right of a $\sum$ or $\prod$ is in the body of that loop.

If the formula on the right still seems impenetrable, take solace in your own experience: the reason you find the left side so easy to read is that you've spent years building up the cognitive pathways in your brain for reading code. You can identify what's filler and what's important; you automatically filter out the noise in the syntax. Over time, you'll achieve this for mathematical formulas, too. You'll know how to zoom in to one expression, understand what it's saying, and zoom out to relate it to the formula as a whole. Everyone struggles with this, myself included.

One additional difficulty of reading mathematics is that the author will almost never go through these details for the reader. It's a rather subtle point to be making so early in our journey, but it's probably the first thing you notice when you read math books. Instead of doing the details, a typical proof of the existence of these polynomials looks like this.

Proof. Let $(x_1, y_1), \ldots, (x_n, y_n)$ be a list of points with no two x_i the same. To show existence, construct $f(x)$ as

$$f(x) = \sum_{i=0}^{n} y_i \prod_{j \neq i} \frac{x - x_j}{x_i - x_j}$$

Clearly the constructed polynomial $f(x)$ has degree n because each term has degree n. For each i, plugging in x_i kills[7] all but the i-th term in the sum, and the i-th term clearly evaluates to y_i, as desired.

... Uniqueness part (we'll complete this proof in the next section) ...

□

---

[7] "Kills" is an informal term for "this thing evaluates to zero."

The square □ is called a *tombstone* and marks the end of a proof.

The proof writer gives a relatively brief overview and you are expected to fill in the details to your satisfaction. It sucks, but if you do what's expected of you—that is, write down examples of the construction before reading on—then you build up those neural pathways, and eventually you realize that the explanation is as simple and clear as it can be. Until then, your job is to evaluate the statements made in the proof on your examples. Practice allows you to judge how much work you need to put into understanding a construction or definition before continuing. And, more importantly, you'll understand it more thoroughly for all your testing.

Uniqueness of Polynomials Through Points

Now for the uniqueness part. This is a straightforward proof, but it relies on a special fact about polynomials. We'll state the fact as a theorem that we won't prove. Some terminology: a *root* of a polynomial $f : \mathbb{R} \to \mathbb{R}$ is a value z for which $f(z) = 0$.

Theorem 2.3. *The zero polynomial is the only polynomial over $\mathbb{R}$ of degree at most n which has more than n distinct roots.*

On to the proof. It works by supposing we actually have *two* polynomials f and g, both of degree n, passing through the desired set of points $(x_1, x_2), \ldots, (x_{n+1}, y_{n+1})$. We don't assume we know anything else about the polynomials ahead of time. They could be different, or they could be the same. If you wrote down two different looking polynomials with the two properties, they might just *look* different (maybe one is in factored form). So the proof operates by making no other assumptions, and showing that actually f and g have to be the same.

So suppose f, g are two such polynomials. Let's look at the polynomial $(f - g)(x)$, which we define as $(f - g)(x) = f(x) - g(x)$. Note that $f - g$ is a polynomial because, if the coefficients of f are a_i and the coefficients of g are b_i, the coefficients of $f - g$ are $c_i = a_i - b_i$ (extending with zeros as necessary so the degrees match). It is crucial to this proof that $f - g$ is a polynomial.

What do we know about $f - g$? It's degree is certainly *at most* n, because you can't magically produce a coefficient of x^7 if you subtract two polynomials whose highest-degree terms are x^5. Moreover, we know that $(f - g)(x_i) = 0$ for all i. Recall that x is the generic input variable, while x_i are the input values of the specific list of points $(x_1, y_1), \ldots, (x_{n+1}, y_{n+1})$ that f and g are assumed to agree on. Indeed, for every i, $f(x_i) = g(x_i) = y_i$, so subtracting them gives zero.

Now we apply Theorem 2.3. If we call d the degree of $f - g$, we know that $d \leq n$, and hence that $f - g$ can have no more than n roots unless it's the zero polynomial. But there are $n + 1$ points x_i where $f - g$ is zero! Theorem 2.3 implies that $f - g$ must be the zero polynomial, meaning f and g have the same coefficients.

Just for completeness, I'll write the above argument more briefly and put the whole proof of the theorem together as it would show up in a standard textbook. That is, extremely tersely.

Theorem 2.4. *For any integer* $n \geq 0$ *and any list of* $n + 1$ *points* $(x_0, y_0), (x_1, y_1), \ldots, (x_n, y_n)$ *in* $\mathbb{R}^2$ *with* $x_0 < x_1 < \cdots < x_n$, *there exists a unique degree* n *polynomial* $p(x)$ *such that* $p(x_i) = y_i$ *for all* i.

Proof. Let $(x_1, y_1), \ldots, (x_n, y_n)$ be a list of points with no two x_i the same. To show existence, construct $f(x)$ as

$$f(x) = \sum_{i=0}^{n} y_i \left(\prod_{j \neq i} \frac{x - x_j}{x_i - x_j} \right)$$

Clearly the constructed polynomial $f(x)$ is degree $\leq n$ because each term has degree n. For each i, plugging in x_i kills all but the i-th term in the sum, and the i-th term clearly evaluates to y_i, as desired.

To show uniqueness, let $g(x)$ be another such polynomial. Then $f - g$ is a polynomial with degree at most n which has all of the $n+1$ values x_i as roots. This implies that $f - g$ is the zero polynomial, or equivalently that $f = g$.

□

We spent quite a few pages expanding the details of a ten-line proof. This is par for the course. When you encounter a mysterious or overly brief theorem or proof it becomes your job to expand and clarify it as needed. Much like with reading programs written by others, as your mathematical background and experience grows you'll need less work to fill in the details.

Now that we've shown the existence and uniqueness of a degree n polynomial passing through a given list of $n + 1$ points, we're allowed to give "it" a name. It's called the *interpolating polynomial* of the given points. The verb *interpolate* means to take a list of points and find the unique minimum-degree polynomial passing through them.

2.4 Realizing it in Code

For the sake of concreteness, let's write a Python program that interpolates points. I'm going to assume the existence of a polynomial class that accepts as input a list of coefficients (in the same order as Definition 2.1, starting from the degree zero term) and has methods for adding, multiplying, and evaluating at a given value. All of this code, including my own version of the polynomial class, is available at this book's Github repository.[8] Note the polynomial class is not intended to be perfect. I'm certainly leaving the code open to floating point rounding errors and other such things. The point of the code is not to be industry-strength, but to help you understand the constructions we've seen in the chapter. On to the code.

Here are some examples of constructing polynomials.

[8] See pimbook.org.

```
# special syntax for the zero polynomial
zero = Polynomial([])

f = Polynomial([1, 2, 3])  # 1 + 2 x + 3 x^2
g = Polynomial([-8, 17, 0, 5])  # -8 + 17 x + 5 x^3
```

Now we write the main interpolate function. It uses the yet-to-be-defined function `singleTerm` that computes a single term of the interpolating polynomial for a given degree. Note we use Python list comprehensions, for which `[EXPRESSION for x in myList]` is a shorthand expression for the following.

```
output_list = []

for x in my_list:
   output_list.append(EXPRESSION)

# the list comprehension expression evaluates to this list
output_list
```

Now the interpolate function:

```
def interpolate(points):
    """ Return the unique degree n polynomial passing through the given n+1 points.
    """
    if len(points) == 0:
        raise ValueError('Must provide at least one point.')

    x_values = [p[0] for p in points]
    if len(set(x_values)) < len(x_values):
        raise ValueError('Not all x values are distinct.')

    terms = [single_term(points, i) for i in range(0, len(points))]
    return sum(terms, ZERO)
```

The first two blocks check for the edge cases: an empty input or repeating x-values. Finally, the last block creates a list of terms, each one being a term of the sum from the proof of Theorem 2.2. The return statement sums all the terms, with the second argument being the starting value for the sum, in this case the zero polynomial. Now for the `singleTerm` function.

```
def single_term(points, i):
    """ Return one term of an interpolated polynomial.

    Arguments:
      - points: a list of (float, float)
      - i: an integer indexing a specific point
    """
    theTerm = Polynomial([1.])
    xi, yi = points[i]

    for j, p in enumerate(points):
        if j == i:
            continue
        xj = p[0]
        theTerm = theTerm * Polynomial(
            [-xj / (xi - xj), 1.0 / (xi - xj)]
        )

    return theTerm * Polynomial([yi])
```

We had to break up the degree-1 polynomial $(x - x_j)/(x_i - x_j)$ into its coefficients, which are $a_0 = -x_j/(x_i - x_j)$ and $a_1 = 1/(x_i - x_j)$. The rest computes the product over the relevant terms. Some examples:

```
>>> points1 = [(1, 1)]
>>> points2 = [(1, 1), (2, 0)]
>>> points3 = [(1, 1), (2, 4), (7, 9)]
>>> interpolate(points1)
1.0
>>> interpolate(points2)
2.0 + -1.0 x^1
>>> f = interpolate(points3)
>>> f
-2.666666666666666 + 3.9999999999999996 x^1 + -0.3333333333333334 x^2
>>> [f(xi) for (xi, yi) in points3]
[1.0, 3.999999999999999, 8.999999999999993]
```

Ignoring the rounding errors, we can see the interpolation is correct.

2.5 Application: Sharing Secrets

Next we'll use polynomial interpolation to "share secrets" in a secure way. Here's the scenario. Say I have five daughters, and I want to share a secret with them, represented as a binary string and interpreted as an integer. Perhaps the secret is the key code for a safe which contains my will. The problem is that my daughters are greedy. If I just give them the secret one might do something nefarious, like forge a modified will that leaves her all my riches at the expense of the others.

Moreover, I'm afraid to even give them *part* of the key code. They might be able to brute force the rest and gain access. Any daughter of mine will be handy with a computer! Even worse, three of the daughters might get together with their pieces of the key code and then they'd really have a good chance of guessing the rest and excluding the other two daughters.[9] So what I really want is a scheme that has the following properties.

1. Each daughter gets a "share," i.e., some string unique to them.

2. If any four of the daughters gets together, they cannot use their shares to reconstruct the secret.

3. If all five of the daughters get together, they can reconstruct the secret.

In fact, I'd be happier if I could prove, not only that any four out of the five daughters couldn't pool their shares to determine the secret, but that they'd provably have *no information at all* about the secret. They can't even determine a single bit of information about the secret, and they'd have an easier time breaking open the safe with a jackhammer.

The magical fact is that there is such a scheme. Not only is it possible, but it's possible no matter how many daughters I have (say, n), and no matter what minimum size group I want to allow to reconstruct the secret (say, k). So I might have 20 daughters,[10] and I may want any 14 of them to be able to reconstruct the secret, but prevent any group of 13 or fewer from doing so.

Polynomial interpolation gives us all of these guarantees. Here is the scheme. First represent your secret s as an integer. Now construct a random polynomial $f(x)$ so that $f(0) = s$. We'll say in a moment what degree d to use for $f(x)$. If we know d, generating f is easy. Call $a_0, \ldots, a_{d+1}$ the coefficients of f. Set $a_0 = s$ and randomly pick the other coefficients. If you have n people, the shares you distribute are values of $f(x)$ at $f(1), f(2), \ldots, f(n)$. In particular, to person i you give the point $(i, f(i))$.

What do we know about subsets of points? Well, if any k people get together, they can construct the unique degree $k - 1$ polynomial $g(x)$ passing through all those points. The question is, will the resulting $g(x)$ be the same as $f(x)$? If so, they can compute $g(0) = f(0)$ to get the secret!

This is where we pick d, to control how many shares are needed. If we want k to be the minimum number of shares needed to reconstruct the secret, we make our polynomial degree $d = k - 1$. Then if k people get together and reconstruct $g(x)$, they can appeal to Theorem 2.2 to be sure that $g(x) = f(x)$. For example, a degree 3 polynomial would prevent any trio of people from reconstructing $f(x)$, but allow 4 people to reconstruct the secret. A degree 17 polynomial would stop any group of size ≤ 17 from obtaining $f(x)$.

Let's be more explicit and write down an example. Say we have $n = 5$ daughters, and we want any $k = 3$ of them to be able to reconstruct the secret. Then we pick a polynomial $f(x)$ of degree $d = k - 1 = 2$. If the secret is 109, we generate f as

$$f(x) = 109 + \text{random} \cdot x + \text{random} \cdot x^2$$

---

[9] My family clearly has issues.
[10] I've been busy.

Note that if you're going to actually use this to distribute secrets that matter, you need to be a bit more careful about the range of these random numbers. For the sake of this example let's say they're random 10-bit integers, but in reality you'd want to do everything with modular arithmetic. See the Chapter Notes for further discussion.

Next, we distribute one point to each daughter as their share.

$$(1, f(1)), (2, f(2)), (3, f(3)), (4, f(4)), (5, f(5))$$

To give concrete numbers to the examples, if

$$f(x) = 109 - 55x + 271x^2,$$

then the secret is $f(0) = 109$ and the shares are

$$(1, 325), (2, 1083), (3, 2383), (4, 4225), (5, 6609).$$

The polynomial interpolation theorem tells us that with any three points we can completely reconstruct $f(x)$, and then plug in zero to get the secret.

For example, using our polynomial interpolation algorithm, if we feed in the first, third, and fifth shares we reconstruct the polynomial exactly:

```
>>> points = [(1, 325), (3, 2383), (5, 6609)]
>>> interpolate(points)
109.0 + -55.0 x^1 + 271.0 x^2
>>> f = interpolate(points); int(f(0))
109
```

At this point you should be asking yourself: how do I know there's not some other way to get $f(x)$ (or even just $f(0)$) if you have fewer than k points? You should clearly understand the claim being made. It's not just that one can reconstruct $f(0)$ when given enough points on f, but also that *no algorithm* can reconstruct $f(0)$ with fewer than k points.

Indeed it's true, and I'll make two little claims to show why. Say f is degree d and you have d points (just one fewer than the theorem requires to reconstruct). The first claim is that there are infinitely many different degree d polynomials passing through those same d points. Indeed, if you pick any new x value, say $x = 0$, and any y value, and you add (x, y) to your list of points, then you get an interpolated polynomial for that list whose "decoded secret" is different. Moreover, for each choice of y you get a *different* interpolating polynomial (this is due to Theorem 2.3).

The second claim is a consequence of the first. If you only have d points, then not only can $f(0)$ be different, but it can be *anything you want it to be!* For *any* value y that you think might be the secret, there is a choice of a new point that you could add to the list to make y the "correct" decoded value $f(0)$.

Let's think about this last claim. Say your secret is an English sentence s = "Hello, world!" and you encode it with a degree 10 polynomial $f(x)$ so that $f(0)$ is a binary representation of s, and you have the shares $f(1), \ldots, f(10)$. Let y is the binary representation of the string "Die, rebel scum!" Then I can take those same 10 points, $f(1), f(2), \ldots, f(10)$, and I can make a polynomial passing through them *and* for which $y = f(0)$. In other words, your knowledge of the 10 points give you no information to distinguish between whether the secret is "Hello world!" or "Die, rebel scum!" Same goes for the difference between "John is the sole heir" and "Joan is the sole heir," a case in which a single-character difference could change the entire meaning of the message.

To drive this point home, let's go back to our small example secret 109 and encoded polynomial

$$f(x) = 109 - 55x + 271x^2$$

I give you just two points, $(2, 1083), (5, 6609)$, and a desired "fake" decrypted message, 533. The claim is that I can come up with a polynomial that has $f(2) = 1083$ and $f(5) = 6609$, and also $f(0) = 533$. Indeed, we already wrote the code to do this! Figure 2.3 demonstrates this with four different "decoded secrets."

```
>>> points = [(2, 1083), (5, 6609)]
>>> interpolate(points + [(0, 533)])
533.0 + -351.7999999999999 x^1 + 313.4 x^2
>>> f = interpolate(points + [(0, 533)]); int(f(0))
533.0
```

You should notice that the coefficients of the fake secret polynomial are no longer integers, but this problem is fixed when you do everything with modular arithmetic instead of floating point numbers (again, see the Chapter Notes).

This scheme raises some interesting security questions. For example, if the secret is, say, the *text* of a document instead of the key-code to a safe, and if one of the daughters sees the shares of two others before revealing her own, she could compute a share that produces whatever "decoded message" she wants, such as a will giving her the entire inheritance!

This property of being able to decode any possible plaintext given an encrypted text is called *perfect secrecy*, and it's an early topic on a long journey through mathematical cryptography.

2.6 Cultural Review

1. A mathematical concept usually has multiple definitions. We prefer to work with the conceptual definition that is easiest to maintain in our minds, and we ften don't say when we switch between two representations.

2. Whenever you see a definition, you must immediately write down examples. They are your test cases and form a foundation for intuition.

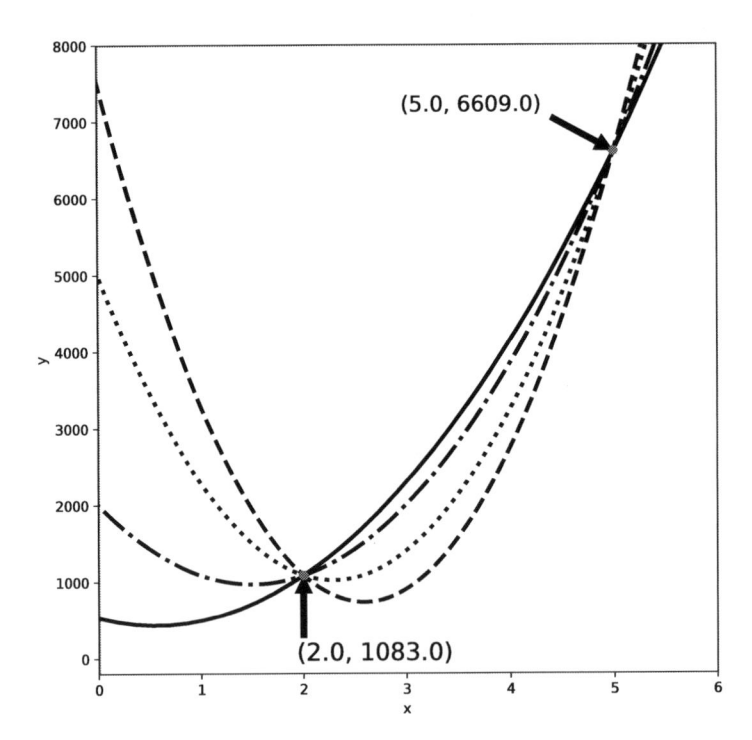

Figure 2.3: A plot of four different curves that agree on the two points $(2, 1083), (5, 6609)$, but have a variety of different "decoded secret" values.

3. In mathematics, we place a special emphasis on the communication of ideas from human to human.

2.7 Exercises

2.1 Prove the following:

1. If f is a degree-2 polynomial and g is a degree-1 polynomial, then their product $f \cdot g$ is a degree 3 polynomial.

2. Generalize the above: if f is a degree-n polynomial and g is a degree-m polynomial, then their product $f \cdot g$ has degree $n + m$.

3. Does the above fact work when f or g are the zero polynomial, using our convention that the zero polynomial has degree -1? If not, can you think of a better convention?

2.2 Write down examples for the following definitions:

• Two integers a, b are said to be *relatively prime* if their only common divisor is 1. Let n be a positive integer, and define by $\varphi(n)$ the number of positive integers less than n that are relatively prime to n.

- A polynomial is called *monic* if its leading coefficient a_n is 1.
- A *factor* of a polynomial f is a polynomial g of smaller degree so that $f(x) = g(x)h(x)$, for some polynomial h. It is said that f can be "factored" into g and h. Note that g and h must both have real coefficients and be of smaller degree than f.
- Two polynomials are called *relatively prime* if they have no (polynomial) factors in common. A polynomial is called *irreducible* if it cannot be factored into smaller polynomials. The *greatest common divisor* of two polynomials f, g is the monic polynomial of largest degree that is a factor of both f and g.

2.3 Verify the following theorem using the examples from the previous exercise. If a, n are relatively prime integers, then $a^{\varphi(n)}$ has remainder 1 when dividing by n. This result is known as Euler's theorem (pronounced "OY-lurr"), and it is the keystone of the RSA cryptosystem.

2.4 A number x is called *algebraic* if it is the root of a polynomial whose coefficients are rational number (fractions of integers). Otherwise it is called *transcendental*. Numbers like $\sqrt{2}$ are algebraic, while numbers like π and e are famously not algebraic. The golden ratio is the number $\phi = \frac{1+\sqrt{5}}{2}$. Is it algebraic? What about $\sqrt{2} + \sqrt{3}$?

2.5 Prove the product and sum of algebraic numbers is algebraic. Despite the fact that π and e are *not* algebraic, it is not known whether $\pi + e$ or πe are algebraic. Prove that they cannot *both* be algebraic.

2.6 Let $f(x) = a_0 + a_1 x + \cdots + a_n x^n$ be a degree n polynomial, and suppose it has k real roots $r_1, \ldots, r_n$.[11] Prove Vieta's formulas, which are

$$\sum_{i=1}^{n} r_i = -\frac{a_{n-1}}{a_n}$$

$$\prod_{i=1}^{n} r_i = (-1)^n \frac{a_0}{a_n}.$$

Hint: if r is a root, then $f(x)$ can be written as $f(x) = (x - r)g(x)$ for some smaller degree $g(x)$. This formula (and its extensions) shows how the coefficients of a polynomial encode information about the roots.

2.7 Look up a proof of Theorem 2.3. There are many different proofs. Either read one and understand it using the techniques we described in this chapter (writing down examples and tests), or, if you cannot, then write down the words in the proofs that you don't understand and look for them later in this book.

2.8 Bezier curves are single-variable polynomials that draw a curve controlled by a given set of "contol points." The polynomial separately controls the x and y coordinates of the

---
[11] This also works for possibly complex roots.

Bezier curve, allowing for complex shapes. Look up the definition of quadratic and cubic Bezier curves, and understand how it works. Write a program that computes a generic Bezier curve, and animates how the curve is traced out by the input. Bezier curves are most commonly seen in vector graphics and design applications as the "pen tool."

2.9 It is a natural question to ask whether the roots of a polynomial f are sensitive to changes in the coefficients of f. Wilkinson's polynomial, defined below, shows that it is:

$$w(x) = \prod_{i=1}^{20}(x-i)$$

The coefficient of x^{19} in $w(x)$ is -210, and if it's decreased by 2^{-23} the position of many of the roots change by more than 0.5. Read more details online, and find an explanation of why this polynomial is so sensitive to changes in its coefficients.[12]

2.10 Write a web app that implements the distribution and reconstruction of the secret sharing protocol using the polynomial interpolation algorithm presented in this chapter, using modular arithmetic modulo and a 32-bit modulus p.

2.11 The extended Euclidean algorithm computes the greatest common divisor of two numbers, but it also works for polynomials. Write a program that implements the Euclidean algorithm to compute the greatest common divisor of two monic polynomials. Note that this requires an algorithm to compute polynomial long division as a subroutine.

2.12 Perhaps the biggest disservice in this chapter is ignoring the so-called Fundamental Theorem of Algebra, that every single-variable monic polynomial of degree k can be factored into linear terms $p(x) = (x - a_1)(x - a_2) \cdots (x - a_k)$. The reason is that the values a_i are not necessarily real numbers. They might be complex. Moreover, all of the proofs of the Fundamental Theorem are quite hard. In fact, one litmus test for the "intellectual potency" of a new mathematical theory is whether it provides a new proof of the Fundamental Theorem of Algebra! There is an entire book dedicated to these often-repeated proofs.[13] Sadly, we will completely avoid complex numbers in this book, with the exception of a few exercises in Chapter 16 for the intrepid reader. Luckily, there is a "baby" fundamental theorem, which says that every single-variable polynomial can be factored into a product of linear and degree-2 terms

$$p(x) = (x-a_1)(x-a_2)\cdots(x-a_m)(x^2 + b_{m+1}x + a_{m+1})\cdots(x^2 + b_k + a_k),$$

where none of the quadratic terms can be factored into smaller degree-1 terms. One of the most famous mathematicians of all time, Carl Friedrich Gauss, provided the first

---

[12] In "The Perfidious Polynomial," Wilkinson wrote, "I regard [the discovery of this polynomial] as the most traumatic experience in my career as a numerical analyst."
[13] Fine & Rosenberger's "The Fundamental Theorem of Algebra."

proof that this decomposition is possible as his doctoral thesis in 1799. As part of this exercise, look up some different proofs of the Fundamental Theorem, but instead of trying to understand them, take note of the different areas of math that are used in the proofs.

2.8 Chapter Notes

Which are Polynomials?

The polynomials were $f(x), g(x), h(x), j(x)$, and $l(x)$. The reason i is not a polynomial is because $\sqrt{x} = x^{1/2}$ does not have an integer power. Similarly, $k(x)$ is not a polynomial because its terms have negative integer powers. Finally, $m(x)$ is not because its powers, π, e, are not integers. Of course, if you were to define π and e to be particular constants that happened to be integers, then the result would be a polynomial. But without any indication, we assume they're the famous constants.

Twin Primes

The Twin Prime Conjecture, the assertion that there are infinitely many pairs of prime numbers of the form $p, p + 2$, is one of the most famous open problems in mathematics. Its origin is unknown, though the earliest record of it in print is in the mid 1800's in a text of de Polignac. In an exciting turn of events, in 2013 an unknown mathematician named Yitang Zhang[14] published a breakthrough paper making progress on Twin Primes.

His theorem is not about Twin Primes, but a relaxation of the problem. This is a typical strategy in mathematics: if you can't solve a problem, make the problem easier until you can solve it. Insights and techniques that successfully apply to the easier problem often work, or can be made to work, on the harder problem.

Zhang successfully solved the following relaxation of Twin Primes, which had been attempted many times before Zhang.

Theorem. *There is a constant M, such that infinitely many primes p exist such that the next prime q after p satisfies $q - p \leq M$.*

if M is replaced with 2, then you get Twin Primes. The thinking is that perhaps it's easier to prove that there are infinitely many primes pairs with distance 6 of each other, or 100. In fact, Zhang's paper established it for M approximately 70 million. But it was the first bound of its kind, and it won Zhang a MacArthur "genius award" in addition to his choice of professorships.

As of this writing, subsequent progress, carried out by some of the world's most famous mathematicians in an online collaboration called the Polymath Project, brought M down to 264. Assuming a conjecture in number theory called the Elliott-Halberstam conjecture, they reduced this constant to 6.

---

[14] Though he had a Ph.D, Zhang had worked in a motel, as a delivery driver, and at a Subway sandwich shop when he was unable to find an academic job.

Impossibility of Clustering

A *clustering algorithm* is a program f that takes as input:

- A list of points S,
- A distance function d that describes the distance between two points $d(x, y)$ where x, y are in S,

and produces as output a *clustering* of S, i.e., a choice of how to split S into non-overlapping subsets. The individual subsets are called "clusters."

The function d is also required to have some properties that make it reasonably interpretable as a "distance" function. In particular, all distances are nonnegative, $d(x, y) = d(y, x)$, and the distance between a point and itself is zero.

The Kleinberg Impossibility Theorem for Clustering says that no clustering algorithm f can satisfy all of the following three properties, which he calls *scale-invariance, richness,* and *consistency*.[15]

- **Scale-invariance**: The output of f is unchanged if you stretch or shrink all distances in d by the same multiplicative factor.

- **Richness**: Every partition of S is a possible output of f, (for some choice of d).

- **Consistency**: The output of f on input (S, d) is unchanged if you modify d by shrinking the distances between points in the same cluster and enlarging the distances between points in different clusters.

One can interpret this theorem as an explanation (in part) for why clustering is a hard problem for computer science. While there are hundreds of clustering algorithms to choose from, none quite "just works" the way we humans intuitively want one to. This may be, as Kleinberg suggests, because our naive brains expect these three properties to hold, despite the fact that they are mutually exclusive.

It also suggests that the "right" clustering function depends more on the application you use it for, which raises the question: how can one pick a clustering function with principle?

It turns out, if you allow the required *number* of output clusters to be an input to the clustering algorithm, you can avoid impossibility and instead achieve uniqueness. For more, see the 2009 paper "A Uniqueness Theorem for Clustering" of Zadeh and Ben-David. The authors proceeded to study how to choose a clustering algorithm "in principle" by studying what properties uniquely determine various clustering algorithms; meaning if you want to do clustering in practice, you have to think hard about exactly what properties your application needs from a clustering. Suffice it to note that this process is a superb example navigating the border separating impossibility, existence, and uniqueness in mathematics.

---

[15] Of incidental interest to readers of this book, Jon Kleinberg also developed an eigenvector-based search ranking algorithm that was a precursor to Google's PageRank algorithm.

More on Secret Sharing

The secret sharing scheme presented in this chapter was originally devised by Adi Shamir (the same Shamir of RSA) in a two-page 1979 paper called "How to share a secret." In this paper, Shamir follows the themes elucidated in this book and chooses not to remind the reader how the interpolating polynomial is constructed.

He does, however, mention that in order to make this scheme secure, the coefficients of the polynomial must be computed using modular arithmetic. Here's what is meant by that, and note that we'll return to understand this in Chapter 16 from a much more general perspective.

Given an integer n and a *modulus* p (in our case a prime integer), we represent n "modulo" p by replacing it with its remainder when dividing by p. Most programming languages use the % operator for this, so that $a = n\%p$ means a is the remainder of n/p. Note that if $n < p$, then $n\%p = n$ is its own remainder. The standard notation in mathematics is to use the word "mod" and the $\equiv$ symbol (read "is equivalent to"), as in

$$a \equiv n \mod p.$$

The syntactical operator precedence is a bit weird here: "mod" is not a binary operation, but rather describes the entire equation, as if to say, "everything here is considered modulo p."

We chose a prime p for the modulus because doing so allows you to "divide." Indeed, for a given n and prime p, there is a unique k such that $(n \cdot k) \equiv 1 \mod p$. Again, an interesting example of existence and uniqueness. Note that it takes some work to find k, and the extended Euclidean algorithm is the standard method. When evaluating a polynomial function like $f(x)$ at a given x, the output is taken modulo p and is guaranteed to be between 0 and p.

Modular arithmetic is important because (1) it's faster than arithmetic on arbitrarily large integers, and (2) when evaluate $f(x)$ at an unknown integer x not modulo p, the size of the output and knowledge of the degree of f can give you some information about the input x. In the case of secret sharing, seeing the sizes of the shares reveals information about the coefficients of the underlying polynomial, and hence information about $f(0)$, the secret. This is unpalatable if we want perfect secrecy.

Moreover, when you use modular arithmetic you can prove that picking a uniformly random $(d+1)$-th point in the secret sharing scheme will produce a uniformly random decoded "secret" $f(0)$. That is, uniformly random between 0 and p. Without bounding the allowed size of the integers, it doesn't make sense to have a "uniform" distribution. As a consequence, it is harder to define and interpret the security of such a scheme.

Finally, from discussions I've had with people using this scheme in industry, polynomial interpolation is not fast enough for modern applications. For example, one might want to do secret sharing between three parties at streaming-video rates. Rather, one should use so-called "linear" secret sharing schemes, which are based on systems of linear equations. Such schemes are best analyzed from the perspective of linear algebra, the topic of Chapter 10.

Chapter 3

On Pace and Patience

You enter the first room of the mansion and it's completely dark. You stumble around bumping into the furniture but gradually you learn where each piece of furniture is. Finally, after six months or so, you find the light switch, you turn it on, and suddenly it's all illuminated. You can see exactly where you were. Then you move into the next room and spend another six months in the dark. So each of these breakthroughs, while sometimes they're momentary, sometimes over a period of a day or two, they are the culmination of, and couldn't exist without, the many months of stumbling around in the dark that precede them.

–Andrew Wiles on what it's like to do mathematics research.

We learned a lot in the last chapter. One aspect that stands out is just how *slow* the process of learning unfamiliar math can be. I told you that every time you see a definition or theorem, you had to stop and write stuff down to understand it better. But this isn't all that different from programming. Experienced coders know when to fire up a REPL or debugger, or write test programs to isolate how a new feature works.

The main difference for us is that mathematics has no debugger or REPL. There is no reference implementation. Mathematicians often get around this hurdle by conversation, and I encourage you to find a friend to work through this book with. As William Thurston writes in his influential essay, "On Proof and Progress in Mathematics," mathematical knowledge is embedded in the minds and the social fabric of the community of people thinking about a topic. Books and papers support this, but the higher up you go, the farther the primary sources stray from textbooks.

If you are reading this book alone, you have to play the roles of the program writer, the tester, and the compiler. The writer for when you're conjuring new ideas and asking questions; the tester for when you're reading theorems and definitions; and the compiler to check your intuition and hunches for bugs. This often slows reading mathematics down to a crawl, for novices and experts alike. Mathematicians always read with a pencil and notepad handy.

When you first read a theorem, you expect to be confused. Let me say it again: the **rule** is that you are confused, the **exception** is that everything is clear. Mathematical culture requires being comfortable being almost continuously in a state of little to no understanding. It's a humble life, but once you nail down what exactly is unclear, you can make progress toward understanding. The easiest way to do this is by writing down lots of examples, but it's not always possible to do that. We've already seen an example, a theorem about the *impossibility* of having a nonzero polynomial with more roots than its degree.

In the quote at the beginning of this chapter, Andrew Wiles discusses what it's like to do mathematical research, but the same analogy holds for learning mathematics. Speaking with experienced mathematicians and reading their books makes you feel like an idiot. Whatever they're saying is the most basic idea in the world, and you barely stumble along. My favorite dramatic embodiment of this feeling is an episode of a YouTube series called Kid Snippets in which children are asked to pretend to be in a math class, while adult actors act it out using dubbed voices.[1] The older child tries to explain to the younger child how to subtract, and the little kid just doesn't get it. Aside from being absolutely hilarious, the video has a deep and probably unintentional truth, that the more mathematics you try to learn the more you feel like the poor student! The video especially resonates when, toward the end, the teacher asks, "Do you get it now?" and the student pauses and slowly says, "Yes." That yes is the fledgling mathematician saying, "I obviously don't understand, but I've accepted it and will try to understand it later."

I've been in the student's shoes a thousand times. Indeed, if I'm *not* in those shoes at least once a day then it wasn't a productive day! I say at least a dozen stupid things daily and think countlessly many more stupid thoughts in search of insight. It's a rare moment when I think, "I'm going to solve this problem I don't already know how to solve," and there is no subsequent crisis. Even in reading what should be basic mathematical material (there's a huge list of things that I am embarrassed to be ignorant about) I find myself mentally crying out, "How the hell does that statement follow!?"

I had a conversation with an immensely talented colleague, a *far* more talented mathematician than I, in which she said (I paraphrase), "If I spend an entire day and all I do is understand this one feature of this one object that I didn't understand before, then that's a *great* day." We all have to build up insight over time, and it's a slow and arduous process. In Andrew Wiles's analogy, my friend is still in the dark room, but she's feeling some object precisely enough to understand that it's a vase. She still has no idea where the light switch is, and the vase might give her no indication as to where to look next. But if piece by piece she can construct a clear enough picture of the room in her mind, then she will find the switch. What keeps her going is that she knows enough little insights will lead her to a breakthrough worth having.

Though she is working on far more complicated and abstract mathematics than you are likely to, we must all adopt her attitude if we want to learn mathematics. If it sounds like all of this will take *way* too much of your time (all day to learn a single little thing!), remember two things. First, my colleague works on much more abstract and difficult mathematics than the average programmer interested in mathematics would encounter. She's looking for the meta-insights that are many levels above the insights found in this

---

[1] You can watch it at http://youtu.be/KdxEAt91D7k

book. As we'll see in Chapter 11, insights are like a ladder, and every rung is useful. Second, the more you practice reading and absorbing mathematics, the better you get at it. When my colleague says she spent an entire day understanding something, she efficiently applied tools she had built up over time. She knows how to cycle through applicable proof techniques, and how to switch between different representations to see if a different perspective helps. She has a bank of examples to bolster her. Her time budget just balances out to a day because of the difficulty of her work.

But most importantly, she's being inquisitive! Her journey is led as much by her task as by her curiosity. As mathematician Paul Halmos said in his book, "I Want to be a Mathematician,"

> *Don't just read it; fight it! Ask your own questions, look for your own examples, discover your own proofs.*

Mathematician Terence Tao expands on this in his essay, "Ask yourself dumb questions—and answer them!"

> *When you learn mathematics, whether in books or in lectures, you generally only see the end product—very polished, clever and elegant presentations of a mathematical topic. However, the process of discovering new mathematics is much messier, full of the pursuit of directions which were naive, fruitless or uninteresting.*
>
> *While it is tempting to just ignore all these "failed" lines of inquiry, actually they turn out to be essential to one's deeper understanding of a topic, and (via the process of elimination) finally zeroing in on the correct way to proceed.*
>
> *So one should be unafraid to ask "stupid" questions, challenging conventional wisdom on a subject; the answers to these questions will occasionally lead to a surprising conclusion, but more often will simply tell you why the conventional wisdom is there in the first place, which is well worth knowing.*

So you'll get confused. We all do. A good remedy is finding the right pace to make steady progress. And when in doubt, start slow.

Chapter 4
Sets

God created infinity, and man, unable to understand infinity, created finite sets.

– Gian-Carlo Rota

In this chapter we'll lay foundation for the rest of the book. Most of the chapter is devoted to the mathematical language of sets and functions between sets. Sets and functions serve not only as the basis of most mathematics related to computer science, but also as a common language shared between all mathematicians. Sets are the modeling language of math. The first, and usually simplest, way to convert a real world problem into math involves writing down the core aspects of that problem in terms of sets and functions. Unfortunately set theory has a lot of new terminology, The parts that are new to you are best understood by writing down lots of examples.

After converting an idea into the language of sets, you may use the many existing tools and techniques for working with sets. As such, the work one invests into understanding these techniques pays off across all of math. It's largely the same for software: learning how to decompose a complex problem into simple, testable, maintainable functions pays off no matter the programming language or problem you're trying to solve. The same goes for the process of modeling business rules in software in a way that is flexible as the business changes. Sets are a fundamental skill.

At the end of the chapter we'll see the full modeling process for an application called *stable marriages*, which is part of an interdisciplinary field of mathematics and economics called *market design*. In economics, there are occasionally markets in which money can't be used as a medium of exchange. In these instances, one has to find some other mechanism to allow the market to function efficiently. The example we'll see is the medical residency matching market, but similar ideas apply to markets like organ donation and housing allocation. As we'll see, the process of modeling these systems so they can be analyzed with mathematics requires nothing more than fluency with sets and functions. The result is a Nobel-prize winning algorithm used by thousands of medical students every year, and the algorithm.

4.1 Sets, Functions, and Their -Jections

A set is a collection of unique objects. You've certainly seen sets before in software. In Python they are simply called "sets." In Java they go by `HashSet`, and in C++ by `unordered_set`. Functionally they are all equivalent: a collection of objects without repetition. While set implementations often have a menagerie of details—such as immutability of items, collision avoidance techniques, complexity of storing/lookup—mathematical sets "just work." In other words, we don't care how items enter and leave sets, and mutability is not a concern because we aren't hashing anything to look it up.

The first thing we need to know about sets is how to describe them. Most of our ways will be implicit, and the simplest way is with words. For example, I can describe the set of integers divisible by seven, or the set of primes, or the set of all syntactically correct Java programs.[1] Often the goal of analyzing a mathematical object is to come up with a more useful description of a set than the implicit one, but implicit definitions are a great starting point for studying a set one doesn't understand.

A more familiar way to describe sets is with set-builder notation. Fans of functional programming styles are cheering as they read, because a formal version of set-builder notation exists in many programming languages as *comprehension* syntax. For example, if we wanted to define the set of all nonnegative numbers divisible by seven, we could do that as

$$S = \{x : x \in \mathbb{N}, x \text{ is divisible by } 7\}$$

The notation reads like the sentence in words, where the colon stands for "such that." I.e., "The set of values x such that x is in $\mathbb{N}$ and x is divisible by 7." Sometimes a vertical pipe | is used in place of the colon. The symbols separate the constructive expression from the membership conditions (it's not an output-input pipe as in shell scripting). As with sets of numbers, the $\in$ symbol denotes membership in a set, and the objects in a set are often called *elements*.

In a language with infinite list comprehensions, say Haskell, the above would be implemented as follows:

```
[x | x <- [1..], mod x 7 == 0]
```

Of course, lists made with list comprehensions need not have unique elements, while mathematical sets must, but the notation is similar enough. In set-builder notation you can add whatever conditions you like after the colon, even if you don't know how to compute them! The left hand side of the colon may also be an expression, as in

---

[1] There are some strange "meta" things you are not allowed to describe as sets, such as the set of all sets. It turns out this is not a set, and it caused a lot of grief to early 20th century mathematicians who really care about the logical foundations of mathematics. This book omits these topics, since all of our sets will be comfortably finite or concrete like $\mathbb{R}$.

$$\{(x, 2x+1) : 0 \leq x < 10\}$$

Now we turn to some definitions you may already be familiar with. If not, remember it's your job to write down examples. In either case, mathematical texts typically define something once *and only once*. I will occasionally repeat definitions that are used across chapters, but generally authors will not. You're expected to have understood a definition to an appropriate degree of comfort before continuing.

Definition 4.1. The *cardinality* or *size* of a set A, denoted $|A|$, is the number of elements in A when that number is finite, and otherwise we say A has infinite cardinality.[2] The set of cardinality zero is called the *empty set*.

Definition 4.2. A set B is a *subset* of another set A if every element $b \in B$ is also an element of A. This relationship is denoted $B \subset A$. Two sets are said to be equal if they contain the same elements. Equivalently, two sets A, B are equal if both $A \subset B$ and $B \subset A$. Set equality is denoted by $A = B$.[3]

Most of the time proving one set is a subset of another is trivial, but not always. The standard technique is to fix b to be an arbitrary element of B, and use whatever characteristic defines B to show that $b \in A$ as well. Here's a brief example: the set of integers divisible by 57 is a subset of the set of integers divisible by 3, because any number b divisible by 57 has the form $b = 57 \cdot k = 3 \cdot (19 \cdot k)$, which means it's also divisible by 3. No alarms and no surprises.

If I have a binary boolean-valued operator like $\in$, then putting a slash through it like $\notin$ denotes the negation of that claim or query. Other slashed operators include $\neq, \not\subset, \not\sim$.

Definition 4.3. Given two sets A and B, the *complement* of B in A is the set $\{a \in A : a \notin B\}$. The complement is denoted either by $A \setminus B$ or $A - B$, and sometimes B^C when $B \subset A$ and A is clear from context.

You can already see I'm starting to be creatively flexible with set-builder notation. Here $a \in A$ might be interpreted as a boolean-valued expression, suggesting the set has only boolean-valued members. However, reading it as a sentence makes sense of it instead as an assertion: "The set of a in A such that a is not in B." Writing it more verbosely, $\{a : a \in A \text{ and } a \notin B\}$ is extra work without significant gain for the reader. If you prefer the verbose version, it's likely because you've spent so long phrasing your thoughts to be machine readable. Appeal to your inner voice here, not your inner type-checker.

---

[2] See the exercises for more about infinite sets.
[3] Mathematicians are divided on whether $A \subset B$ allows A to be equal to B. Some authors insist, drawing from the $\leq$ and $<$ notation for numbers, that only $A \subseteq B$ allows for $A = B$, and they call $\subset$ the "strict subset" operator. I have never heard a convincing argument that the matter warrants debate, and so I opt for the briefest: $\subset$ allows equality. This book never requires a strict subset operator, but if it did I would use $\subsetneq$.

Definition 4.4. Given sets A, B, their *union*, denoted $A \cup B$ is the set $\{x : x \in A \text{ or } x \in B\}$. The *intersection*, denoted $A \cap B$, is the set $\{x : x \in A \text{ and } x \in B\}$.[4]

If you want some practice working with basic set definitions, prove that for any two sets A, B, the following containments hold: $A \cap B \subset A$ and $A \subset A \cup B$.

Definition 4.5. The *product* of two sets A, B denoted $A \times B$, is the set of all ordered pairs of elements in a and elements in b. In set-builder notation it is:

$$A \times B = \{(a, b) : a \in A \text{ and } b \in B\}$$

The parentheses denote a tuple, i.e., an ordered list allowing repetition.

The product is the usual way we turn the real line $\mathbb{R}$ into the real plane $\mathbb{R}^2$. That is, $\mathbb{R}^2$ is *defined* to be $\mathbb{R} \times \mathbb{R}$ and $\mathbb{R}^3 = \mathbb{R} \times \mathbb{R} \times \mathbb{R}$. Unpacking this, there is a little confusion over where the parentheses go. That is, should it be $(\mathbb{R} \times \mathbb{R}) \times \mathbb{R}$ or $\mathbb{R} \times (\mathbb{R} \times \mathbb{R})$? These give rise to two different sets. The first is

$$(\mathbb{R} \times \mathbb{R}) \times \mathbb{R} = \{((a, b), c) : a \in \mathbb{R}, b \in \mathbb{R}, c \in \mathbb{R}\}$$

and the second is

$$\mathbb{R} \times (\mathbb{R} \times \mathbb{R}) = \{(a, (b, c)) : a \in \mathbb{R}, b \in \mathbb{R}, c \in \mathbb{R}\}$$

We want these sets to be considered the same. Indeed, the difference between the two is the kind of distinction that programmers are very familiar with, because compilers will refuse to proceed unless the parentheses align. But mathematicians, for reasons we'll see shortly,[5] brush aside the difference and just say they're the "same" set, and they're both equivalent to

$$(\mathbb{R} \times \mathbb{R}) \times \mathbb{R} = \mathbb{R} \times (\mathbb{R} \times \mathbb{R}) = \{(a, b, c) : a \in \mathbb{R}, b \in \mathbb{R}, c \in \mathbb{R}\}$$

We will return later in this chapter, and again in Chapters 9 and 16 when complexity will beg for a rigorous and useful abstraction called the *quotient*, to understand why it's okay to call these two sets "the same." For now, simply define an n-fold product to collapse pairs into tuples of length n:

$$\mathbb{R}^n = \underbrace{\mathbb{R} \times \cdots \times \mathbb{R}}_{n \text{ times}} = \{(a_1, \ldots, a_n) : a_i \in \mathbb{R} \text{ for every } i\}$$

This notation can be used for any set. Next we define functions as special subsets of a product.

---
[4] Hence the name of my blog, Math $\cap$ Programming.
[5] There's a bijection!

Definition 4.6. Let A, B be sets, and let F be a subset of $A \times B$. We say that F is a *function* if it satisfies the following property: for each $a \in A$, there is a unique pair $(a, b) \in F$ (an input must have exactly one output). The set A is called the *domain* of F and B is called the *codomain* of F. To denote this, we use the arrow notation $f : A \to B$.

We've left it to you to give examples for most definitions so far, but this one needs some help because it disagrees with how we usually think of functions. Indeed, we think of functions computationally as *mappings* from inputs to outputs. So much so that the words *function* and *map* are interchangeable synonyms in math! But this definition of a function is just some set. I'm going to convince you that the distinction is merely a matter of notation and language.

Let's illustrate it with a simple example. Say your function F is the set of pairs of positive integers and their squares.

$$F = \{(1, 1), (2, 4), (3, 9), (4, 16), \dots\} = \{(x, x^2) : x \in \mathbb{N}\}.$$

It's a subset of $\mathbb{N} \times \mathbb{N}$. Now we can add a bit of notation: instead of saying that $(3, 9) \in F$ we use the mapping notation $F(3) = 9$. With this, we could describe F the way we wanted to all along, as $F(x) = x^2$. The conditions in the definition ensure that every input x has *some* output $F(x)$, and that each input x has *only* one output $F(x)$. Providing a concrete algorithm to compute the output from the input makes these conditions trivial, as is the case with squared integers, but an algorithm is not needed to define a function.

So why go through all the trouble of defining functions in terms of sets? Part of this is historical. The concept of sets as a modeling tool has probably existed for as long as mathematics, but it was primarily used in its language form ("I declare, considereth only those heavenly numbers whose factorisation into prymes containeth nary a repeated factor!"). Indeed, the notation $y = f(x)$ was invented in the 1700's by Leonhard Euler. It was not until the late 19th century that mathematicians formally studied sets, and proposed them as a logical foundation for all of mathematics. To do so requires restating all existing concepts in terms of sets. Definition 4.6 does this for functions. Similar definitions exist defining, e.g., integers and ordered tuples in terms of sets. However, our initial definition of a set was compltely imprecise. There is a more precise definition, but it is the sort that only a logician would love. In brief, its base concepts are just the empty set and set containment, and a restricted choice of ways to build sets from smaller sets. Using this one can define numbers, functions—even all of calculus—from "first principles." To instill this idea in future mathematicians, many introductory proof textbooks define everything in terms of sets, and do formal proofs to a degree of precision most mathematicians avoid in their day to day work.

In theory, mathematicians quite like the idea that everything can be reduced to sets. Actually doing it in practice will drive you mad. It's like writing all your programs in pure binary. We need abstractions to keep ourselves productive, and we take comfort in the idea that, if we wanted, we could peel back the layers to reveal the raw instructions. Defining the entirety of mathematics in sets is like "bare metal" programming, but without any of the speed benefits of the finished program. Someone did it once, and we have a record of their work that has been checked many times (and we could check it if we wanted to), and social consensus is good enough.

For us, the special notation for functions highlights our conceptual emphasis. We think of functions differently than regular sets, with a semantic input-output dependence that set notation doesn't natively convey. For most of the book, we'll ditch the mindset that all ideas *must* be modeled as a set. Rather, set-builder notation is a tool to help clarify an idea, or iron out a particularly tricky crease in a definition.

Now we turn to a few useful definitions about subsets of inputs and outputs of a function. A seasoned programmer is less likely to be familiar with the remainder of the definitions in this chapter, but we will rely on them throughout the book.

Definition 4.7. Given a function $f : A \to B$, we define the *image* of f (or the image of A under f) as the set

$$f(A) = \{f(a) : a \in A\}$$

This is denoted $f(A)$ to signify that we're putting everything in A through f, though it is also denoted $\text{im}(f)$ or just $\text{im } f$. If $C \subset A$ is a subset, we can similarly define the image of C, denoted $f(C)$, as $f(C) = \{f(c) : c \in C\}$. The image of A is equivalent to the range of a function with domain A, but we use a different word so we can speak of the image of a particular subset as well.

As a shorthand for "there exists," mathematicians often use the symbol $\exists$. So an equivalent definition of $\text{im } f$ is

$$\text{im } f = \{b \in B : \exists a \in A \text{ with } f(a) = b\}.$$

We won't rely heavily on the $\exists$ notation, but it is quite common. Now we move on to the preimage, the set of inputs mapping to a specified set of outputs.

Definition 4.8. Let A, B be sets and $f : A \to B$ a function. Let $b \in B$. The *preimage* of b under f, denoted $f^{-1}(b)$ is the set $\{a \in A : f(a) = b\}$. Likewise, if $C \subset B$ is a subset, then $f^{-1}(C)$ is defined to be $\{a \in A : f(a) \in C\}$.

For $F(x) = x^2$ as a mapping $\mathbb{R} \to \mathbb{R}$, the preimage of 4 is $F^{-1}(4) = \{-2, 2\}$.

The next three definitions are quite special.

Definition 4.9. A function $f : A \to B$ is called a *injection* (adjectivally, is *injective*) if whenever $a, a' \in A$ are different elements of A, then $f(a), f(a')$ are different elements of B.

An injection "injects" a copy of A inside B by way of f, so that no two elements of A get mapped to the same thing in B. For example, $F(x) = x^2$ is an injection from $\mathbb{N} \to \mathbb{N}$, but if we defined it for all integers $\mathbb{Z} \to \mathbb{Z}$ it would not be injective because, by way of counterexample, $(-4)^2 = 4^2 = 16$. Figure 4.1 is the picture you should have in your head whenever you think of an injection. To put injectivity another way, $f : A \to B$ is an injection exactly when the preimage of every element $b \in B$ has size 0 or 1.

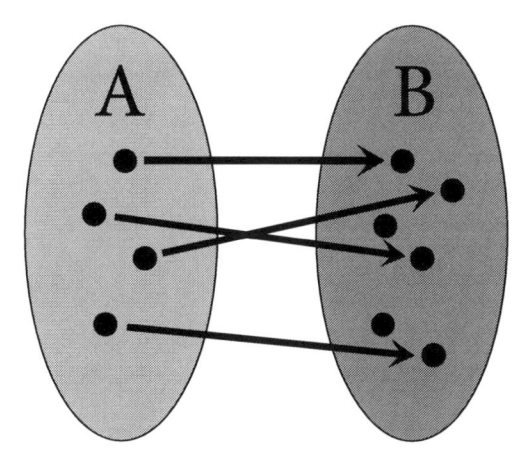

Figure 4.1: An example of an injection, where different inputs are mapped to different outputs. The dots are elements of the set, and the arrows show the mapping. This example is also a non-surjection.

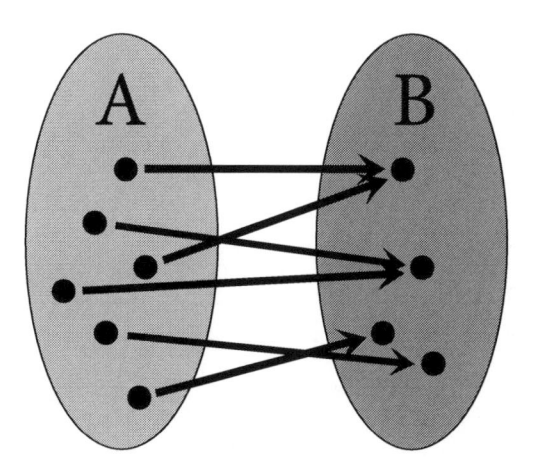

Figure 4.2: An example of an surjection, where every element of the codomain is hit by some element of the domain mapped through f. The dots are elements of the set, and the arrows show the mapping. This example is also a non-injection.

Definition 4.10. A function $f : A \to B$ is called an *surjection* (adjectivally, is *surjective*) if for every $b \in B$, there is some $a \in A$, with $f(a) = b$. In other words, f is surjective if $\operatorname{im} f = B$.

Surjections "hit everything" in B by things mapped from A. So our squaring function on integers $F(x) = x^2$ is not a surjection, because 2 has no integer square root. However, if we redefined it for positive *real* numbers it would be: every positive real number has a positive square root. Figure 4.2 shows a picture. To phrase it in terms of preimages, a surjection $f : A \to B$ has the property that every $b \in B$ has a nonempty preimage $f^{-1}(b)$.

Another bit of notation, just like $\exists$ meaning "there exists," the symbol $\forall$ is a shorthand for "for all." I remember it by the backwards E standing for Exists, while the upside-down A stands for All. So the surjective property can be written hyper-compactly as

$$\forall b \in B, \exists a \in A \text{ such that } f(a) = b.$$

The symbols $\forall, \exists$ are called *quantifiers* and an expression in which every variable is bound by a quantifier is called "fully quantified."

I will shy away from such dense notation in this book, though it will come in handy when we study Calculus in Chapter 8. While this example is not particularly difficult to parse, unrestrained use of $\forall, \exists$ can quickly spin out of control. Just as programmers shouldn't try to cram a lot of complex logic into a single line of code, bad mathematical writers cram a lot of these quantifiers onto a single line of math. That being said, familiarity with the symbols is broadly assumed, especially in the field of logic.

Finally, $f : A \to B$ is called a *bijection* if it is both a surjection and an injection. Adjectivally f is called *bijective*. A bijection is also called a one-to-one correspondence.[6] Bijections are nice because they can be used to say that two sets have the same cardinality (size), and it makes sense for infinite sets. So if there is a bijection $A \to B$ then $|A| = |B|$. Likewise if there is an injection $A \to B$ then $|A| \leq |B|$ and the opposite works for surjections. See the exercises for more on this. Figure 4.3 shows the typical picture for a bijection.

Being a bijection $f : A \to B$ means every $b \in B$ has a preimage of size exactly 1. In particular, this means the idea of an "inverse" to f makes sense: it just maps b to the unique element of the preimage. One denotes this map $f^{-1} : B \to A$. More abstractly an inverse function for f is a function $g : B \to A$ satisfying both $g(f(a)) = a$ for every $a \in a$ and $f(g(b)) = b$ for every $b \in B$.

Only bijections have inverses, and computing the inverse function given only a description of a function can be notoriously difficult. Indeed, most of cryptography rests on the assumption that some functions are computationally infeasible to invert. On the other hand, in linear algebra it is feasible, though often expensive, to compute the inverse of a matrix. As such, it can often be worthwhile to study the notion of an inverse

---

[6] Injections are sometimes called *one-to-one* and surjections are sometimes called *onto*. I won't use those terms in this book, but they are common.

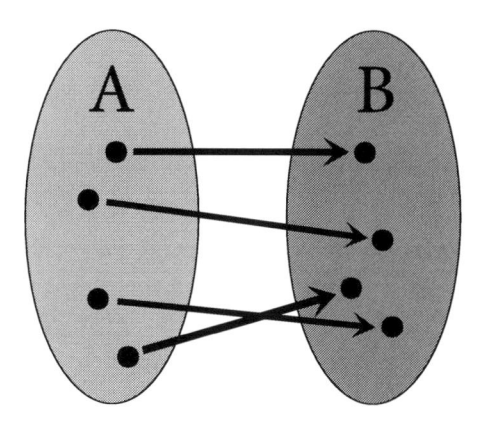

Figure 4.3: An example of an bijection, which is both an injection and a surjection.

in generality. This can grease the wheels of a complicated proof in an advanced setting, but more importantly it separates the aspects of a topic concerning sets with additional structure that are intrinsic to that structure from those that are "merely" questions about the underlying sets.

Indeed, here are two propositions we'll use much later in our study of linear algebra concerning the existence and structure of inverses. If you feel emotionally drained by all the definitions in this chapter so far, feel free to skip these and come back when we refer to them in Chapter 12.

Proposition 4.11. *Inverses are unique.*

Proof. Let $f : A \to B$ be a bijection and suppose that both $g_1 : B \to A$ and $g_2 : B \to A$ are inverses. We will show $g_1(b) = g_2(b)$ for every $b \in B$. Fix any $b \in B$. Let a be an element of A such that $f(a) = b$. Then $g_1(b) = g_1(f(a)) = a$ and the same reasoning proves $g_2(b) = a$. So g_1 and g_2 are the same function.

$\square$

Proposition 4.12. *Let A, B be sets and $f : A \to B$ a bijection. Suppose $g : B \to A$ is a function satisfying $g(f(a)) = a$ for every $a \in A$ (just one of the two requirements to be an inverse). Then $f(g(b)) = b$ for every $b \in B$, and g is the inverse of f.*

Proof. It's crucial here that f is surjective (otherwise the theorem is not true!). Given $b \in B$, we need to show that $f(g(b)) = b$. Start by choosing an $a \in A$ for which $f(a) = b$. Then $g(b) = g(f(a)) = a$. Apply f to both sides to get $f(g(b)) = f(a) = b$, as desired.

$\square$

Before we move on I have to explain an earlier comment. I said we call $(\mathbb{R} \times \mathbb{R}) \times \mathbb{R} = \mathbb{R} \times (\mathbb{R} \times \mathbb{R})$ by "brushing aside" the differences between the two. There is a rigorous way to do this, but I'll only explain half of the rigor right now. The essential reason is because there is a bijection $(\mathbb{R} \times \mathbb{R}) \times \mathbb{R} \to \mathbb{R} \times (\mathbb{R} \times \mathbb{R})$ that maps $((a, b), c)$ to $(a, (b, c))$. Often when mathematicians want to "call" two things the same, they'll come up with such a bijection, and say the two things on either side of such a bijection should be considered the same. It's kind of like a typecast that's built into the programming language, but it's always reversible here. The formal idea is called a "quotient," which we'll make formal in Chapter 9.

4.2 Clever Bijections and Counting

Now that we have the basic language of sets to model our problems, on to some problems. Say you want to count the size of a set. Since sets can be defined implicitly, it may not be obvious how. A useful tool used all over math is the trick of coming up with a clever bijection. This can transform a seemingly difficult counting problem into an elegantly trivial one.[7]

I'm going to give you many examples of that through the rest of this chapter. The first requires no formulas. Say you are running a tournament of tennis players. The tournament is *single-elimination*, meaning when two players finish a match the winner stays in the tournament and the loser is out. As the tournament host, you need to know various things like how long it will take to finish the tournament, how many games to run simultaneously, etc. One basic data required to reason about these harder questions is: how many games will be played in total? That is, given a set of games (pairs of players) generated by this complicated process, we want to count its size.

Say you start with a thousand players. Let's entertain how one might naively compute this. In the first round of the tournament, each player is paired up with another and 500 games are played. In the second round there are 500 remaining players, and they again pair off to play 250 games. In the third, 125 games. But then in the fourth round you get some weird edge cases, because there are an odd number of players and one must sit out. Fine, you keep going, diligently tracking the players who sit out, and eventually you get to a number. You should try this yourself, and verify that the answer is 999 games. Isn't that a weird coincidence? We got 1 less than the total number of players. Does this pattern hold for other tournament sizes?

The answer is yes. To prove it, here's where we show the trick of finding a clever bijection. It will make you feel like our computation was a complete waste of time, but if you did the exercise you'll appreciate the elegance of this method that much more.

The primary observation is that every player loses exactly one game. So if we want to count the number of games, we can instead count the number of losers. But there is only one player who is not a loser: the winner. Hence 999 games.

---

[7] Here's a neat fact I learned from John D. Cook: in the Middle Ages, people studied a "quadrivium" of mathematical arts: arithmetic, geometry, music, and astronomy. This followed the "trivium" of grammar, rhetoric and logic. So when I say a result is "trivial," I'm not trying to insult anyone, but rather informing that no new ideas are needed above basic logic. The best and most pleasing mathematics takes a hard-seeming problem, and rephrases it in a clever way so that the proof is trivial.

Let's rephrase that elegant argument in the language of sets, if X is the set of games and Y is the set of players, then we can define a function $f : X \to Y$ by calling $f(x)$ the loser of game x. Then the image $f(X)$ is the subset $L \subset Y$ of losers. In fact, f is an injection (two different games have different losers), and f defines a bijection between X and L. This means that X and L have the same size, and the fact that there is only one winner of the entire tournament means that $|L| = |Y| - 1$. So if there are n players then there will always be $n - 1$ games.

To make sure you understand this argument, extend it to the case of a double-elimination tournament. In double-elimination, you are ousted from the tournament once you've lost two games, and a player who loses one game might still ultimately win the tournament. In this case you won't have an injection, but a so-called "double-cover" of the set of players. What I mean by double-cover is that every $y \in Y$ has a preimage $f^{-1}(y) = \{x \in X : f(x) = y\}$ of size exactly 2. Another minor complication here is that the winner may have lost zero games or one game. So you can't count the number of games exactly, but you can compute tight bounds instead.

This general strategy for counting has applications any time you need to count or estimate the size of a set. Imagine you want to estimate the number of homeless people in a city, a problem the Census Bureau faces regularly. You have to find a clever way to implicitly count them by observing the residual effects of their actions. This is precisely looking for functions between sets that are close to bijections, or double- or triple-covers of the set you want to count.

Here is another magnificent example of finding a clever bijection. Given a set X let's define the quantity $\binom{X}{2}$, read "X choose two," to be the set of all unordered pairs of distinct elements of X. I.e.,

$$\binom{X}{2} = \{\{x, y\} : x, y \in X \text{ and } x \neq y\}.$$

If X is a finite set of size $n = |X|$, we denote the size of $\binom{X}{2}$ by $\binom{n}{2}$. Obviously this value doesn't depend on the particular elements in X, just its size, so it's okay to use n in place of X. Colloquially, $\binom{n}{2}$ is the number of ways to choose two objects from a set of n objects.[8] So the question is, can we come up with an arithmetic formula for $\binom{n}{2}$ in terms of n? You may have already seen the answer: it's $\frac{n(n-1)}{2}$. But we'll show by way of a bijection that it's equal to the quantity

$$1 + 2 + \cdots + n - 1.$$

In fact, the bijection is easiest to understand by the picture in Figure 4.4. Here's how we read this picture. We're setting $n = 7$ and calling the lightly shaded balls Y, and calling the n squares in the last row X. The picture shows how to define a bijection $g : Y \to \binom{X}{2}$: given any ball $y \in Y$, you draw two diagonals as in the picture and you get $g(y)$ as the pair of squares at the end of both diagonals. The picture should convince

---

[8] In general, $\binom{n}{k}$ is the number of different ways to choose k objects from an n object set.

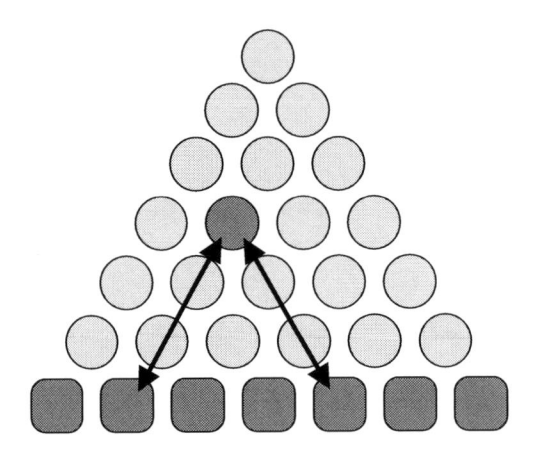

Figure 4.4: A picture proof that $|\binom{X}{2}| = 1 + 2 + \cdots + n - 1$. Each pair of squares in the bottom row corresponds to a unique ball in the pyramidal arrangement above it.

you that two different choices of balls give you different diagonals, so that means g is an injection. Likewise, if you give me a pair of squares $x_1, x_2 \in X$, the inner diagonals meet at a ball y that maps under g back to (x_1, x_2). So g is a surjection, and together with being an injection this makes g a bijection.

Now we count: how many balls and squares are there? The last row has $n - 1 = 6$ squares, and each row has one fewer ball than the row underneath it, so $|Y| = 1 + 2 + \cdots + n - 1$. Moreover, X has n balls in it, so $|\binom{X}{2}| = \binom{n}{2}$. The bijection tells us that these two values must be equal.

You might be wondering: how can we use a picture as the central part of our proof? Didn't we only prove that this bijection works for $n = 7$? Technically you're right: no mathematician would consider a picture as a rigorous proof in of itself. However, when the goal is to communicate the central nugget of wisdom in a proof, a small example with all the essential features of a general proof is often good enough. Consider one alternative. You could actually make the balls points inside $\mathbb{R}^2$ somewhere. You'd need a generic way to construct coordinates for them, and a generic way to describe the diagonals. That's a huge pain in the ass for something so simple! Every mathematician would agree it *could* be done but it would be a colossal waste of time to actually do it.

This is a common feature of more advanced mathematics. Mathematicians are constantly reading papers, and there is rarely enough time to verify all the details of every argument. If you're not an official reviewer of the paper before it's been published, it is usually enough to be convinced that something *should* be true, especially if the details are messy but clear, while focusing on the high level picture. An example with all the essential features of a general solution is an effective substitute. And this doubles for readers of mathematics too: finding a simple example with the essential features of a general solution, and testing claims on the example, is one of the best ways to read a proof!

4.3 Proof by Induction and Contradiction

Next we're going to see two rigorous methods of proof that are part of the basic syntax of proofs. The first is induction, but you're likely familiar with it by a different name: recursion.

We understand recursion: a function is defined in such a way that it invokes itself for some smaller set of parameters, with a "base case" to process the smallest allowed parameters. The classic example is the Fibonacci sequence fib(n), defined recursively as

$$\text{fib}(n) = \text{fib}(n-1) + \text{fib}(n-2),$$

with fib(0) = fib(1) = 1. Most programmers have implemented some version of this function early on in their career, since it is a common instrument to teach recursion.[9]

Likewise, induction is a proof technique that allows you to prove a statement by invoking the same statement for smaller parameters, with a similar base case. One difficulty is identifying when and where induction is likely to be used. It's usually when someone is trying to prove a statement which holds for all natural numbers (or all positive integers above some number). So a statement might look like, "For all integers $n \geq 6$, the statement $P(n)$ is true." A proof by induction operates in two steps:

1. First show the *base case*, in this case that $P(6)$ is true.

2. Second, do the *inductive step*, where one uses the assumption that $P(n)$ is true to prove that $P(n+1)$ is true. Equivalently, one can use $P(n-1)$ to prove $P(n)$.

Just like with recursion, you get a chain of proofs: $P(6)$ implies $P(7)$ implies ... implies $P(n)$ for any n you like. One bit of terminology: one often invokes the *inductive hypothesis*, which is the assumption that $P(n)$ is true. It's helpful when $P(n)$ is cumbersome to restate.

Let's use induction for a second proof that $\binom{n}{2} = 1 + 2 + \cdots + n - 1$.

Proof. Call the statement to be proved $P(n)$. We prove this by induction for $n \geq 2$. For the base case[10] $n = 2$, we need to prove

$$P(2) : \binom{2}{2} = 1$$

We argue $\binom{2}{2}$ is trivially 1. There is only one way to choose two items from a set of two items. Now assume the inductive hypothesis $P(n)$ holds:

---

[9] It also displays some of the variety of programming approaches. Fibonacci sequences can be computed in-place with an array, using recursion (and hopefully memoization), or with a closed-form formula. Each has advantages and disadvantages that show how we think about tradeoffs in software.

[10] When $n = 0$ or 1 we are asking how many ways there are to choose two things from a set of fewer than two things. According to our definition this is zero (which you saw if you wrote your test cases starting from the simplest ones), and one usually calls an empty summation to be zero. But the first n that's not "vacuously" true is $n = 2$ so we start there.

$$P(n) : \binom{n}{2} = 1 + 2 + \cdots + n - 1.$$

We must now prove that $P(n+1)$ follows, i.e.:

$$P(n+1) : \binom{n+1}{2} = 1 + 2 + \cdots + n.$$

Take the set $X = \{1, 2, \ldots, n+1\}$ of size $n+1$, and consider the set $\binom{X}{2}$ of ways to pick two elements from X. Note that we are using numbers as elements of X instead of "arbitrary objects." We might have instead called them "ball 1, ball 2, ball 3" and discuss how many ways to select two balls from a bin. For simplicity of notation we'll just use the numbers themselves. Now X is a set of size $\binom{n+1}{2}$ and we want to express it in terms of our known formula for $\binom{n}{2}$. Pick any element of X, say $n+1$, and define Y to be the set that remains after removing that element from X.

$$Y = X - \{n+1\} = \{1, 2, \ldots, n\}.$$

Now I will split up the elements of $\binom{X}{2}$ into two parts: the parts where both chosen elements are in Y, and the part where one of the two chosen elements is $n+1$. Since there are no other options and no overlap between the two options, I can add the sizes of both parts to count the size of $\binom{X}{2}$.

The first part, where both chosen elements are in Y, has size $\left|\binom{Y}{2}\right| = \binom{n}{2}$, which by the inductive hypothesis is $1 + 2 + \cdots + n - 1$. The second part, where one of the chosen elements is guaranteed to be $n+1$, has size n by the following reasoning: if you *had* to choose $n+1$ as one of your two elements, then there are only n remaining choices for your second element, and so there are only n ways to do this.[11]

Adding up the sizes of the two parts gives exactly

$$1 + 2 + \cdots + (n-1) + n,$$

which is what we set out to prove.

$\square$

Is the proof still a bit murky? Go back and set $n = 4$, $X = \{1, 2, 3, 4, 5\}$, and then write down the elements of $\binom{X}{2}$. Follow the steps through the inductive step of the proof on this example, and your understanding of the general case will feel like an epiphany.

Interestingly, proof by induction has a a bad reputation in mathematics. The reason is that proofs by induction often convey little insight to the reader. Our inductive proof that $1 + 2 + \cdots + (n-1) = \binom{n}{2}$ is decidedly less intuitive than the picture proof. As the mathematician Gian-Carlo Rota once said, "If we have no idea why a statement is

---

[11] If you read this part carefully, you'll notice I'm defining a bijection. One can require the pairs (a, b) in $\binom{X}{2}$ are written so that $a < b$, and then the mapping is $f(a, b) = a$, the *projection* onto the first coordinate. Then f is a bijection between Y and the subset $\{(a, b) \in \binom{X}{2} : b = n+1\}$.

true, we can still prove it by induction." Be that as it may, induction is a central tool for proving theorems.

The second proof technique is called "proof by contradiction." There's a simple puzzle I often use to illustrate the technique.

You're at a party. You're chatting with your friend, and out of curiosity you ask how many friends he has at the party. He counts them up, there are five, and you realize that you also have five friends at the party. What a coincidence! Putting on your mathematician hat, you poll everyone at the party and you're shocked to find that a few other people also have five friends at the party. The puzzle is: is this true of *every* party? Maybe not five exactly, but will there *always* be at least two people with the same number of friends who are at the party?

Before I give the solution by contradiction, let's iron out what I mean by "friendship." I insist that friendship is symmetric: you can't be friends with someone who is not friends with you. And moreover you can't be friends with yourself.[12]

You'll appreciate the answer to this problem best if you spend some time trying to solve it first.

Back already? Okay, so the answer is yes, there will always be a pair of people with the same number of friends. The technique we use to prove it is called *proof by contradiction*. It works by assuming the opposite of what you want to prove it true, and using that assumption to deduce nonsense.

Proof. Suppose for the sake of contradiction that there is some party where everybody has a different number of friends at the party. Say the party has n people, then everyone must have between zero and $n - 1$ friends. Since there are n people and n different numbers between zero and $n - 1$, we can map each person to the number of friends they have, and this map will be a bijection. Now here comes the contradiction: that means someone must have zero friends at the party, and someone must have $n - 1$ friends, i.e., someone must be friends with everyone. But the person who is friends with everyone must be friends with the person that has no friends! The only way to resolve this contradiction is if the original assumption is actually false. That is, there must be two people with the same number of friends.

□

This is how every proof by contradiction goes, but they're usually a bit more concise. They always start with, "Suppose to the contrary" to signal the method. And there is no warning when the contradiction will come. A proof writer usually just states the contradiction and follows it with "which is a contradiction," ending the proof.[13]

---

[12] Looking forward to Chapter 6 on graph theory, we're saying that the social connections at our "party" form a simple, undirected graph.

[13] A professor of mine had a funny refrain to end his proofs by contradiction. If, say, x was assumed to be prime, he'd arrive at a contradiction and say, "and this is very embarrassing for x because it was claiming to be prime."

The point of a proof by contradiction is to get an object with a property that you can work with. If you're trying to prove that *no* object with some special property exists, a proof by contradiction gives you an instance of such an object, and you can use its special property to go forward in the proof. In this case the object was a special friendship count among partygoers, and in the next section we'll apply the same logic to "marriages."

4.4   Application: Stable Marriages

Now we're ready to apply the tools in this chapter to develop a Nobel Prize-winning algorithm for the *stable marriage problem*. The problem is set up as follows. Say you have n men and n women. Your end goal is to choose who should marry whom. Same-sex marriages are excluded, not for political or religious reasons but because it's a more difficult problem. So if we call M the men and W the women, our output will be a bijection $M \to W$ describing the marriages (or equivalently $W \to M$). I will freely switch between "bijection" and "marriage" in this section.

Of course, we don't just want *any* bijection. This is where the "stable" part comes in. We want to choose the marriage so that everyone is happy in some sense. Let's make this precise. Say that each man has a ranking of the women, mathematically a bijection $W \to \{1, 2, \ldots, n\}$, with 1 being the most preferred and n being the least. In other words, if we call the bijection p then $p(w) < p(x)$ means that this particular man prefers woman w over woman x. Likewise, each woman has a ranking of the men $M \to \{1, 2, \ldots, n\}$. Now we obviously can't ensure that every woman gets her top choice and vice versa; the men could all prefer the same woman. So we need a subtler notion of happiness: that no (man, woman) pair mutually prefer each other over their assigned partners.

Marriages are a colorful, if somewhat silly, setting for this problem. Realistic applications of this algorithm involves a different sort of marriage, the assignment of a student to an apprenticeship. A widely known example is medical residency,[14] in which medical students work in a hospital before becoming a doctor. This is the perfect example of a market in which money should not play a part. As a society we want all our hospitals filled with talented apprentices. We don't want the students with the richest parents or best connections to get the most prestigious positions in the best cities, while poorer areas suffer. We want to spread the talent around. So we need a market with a protocol that respects student and hospital preferences in a way that no (student, hospital) pair is incentivized to make their own arrangements. This version of the problem is a natural extension of the marriage version. So we'll explore marriages in depth here, and dive into medical residency matching in the exercises.

Define a *ranking function* as a bijection between $\{1, 2, \ldots, n\}$ and either M or W. Before I state what "not cheating" means mathematically for the marriage problem, I encourage you to write down a small example of sets M, W of size $n = 4$, rankings $\text{rank}_w(m)$ for each $w \in W$ and $\text{rank}_m(w)$ for each $m \in M$, and a candidate marriage $f : M \to W$. I'll call the marriage from the women's perspective $f^{-1} : W \to M$.

---

[14] In the US, it's the National Medical Residency Matching Program.

What I mean by "no mutually desired cheating" is the following.

Definition 4.13. A bijection $f : M \to W$ is called *stable* there is no pair $m \in M$ and $w \in W$ such that the following two conditions hold:

1. $f(m) \neq w$, i.e., the two are not matched by f.

2. The pair m and w mutually prefer each other over their assigned mathces. I.e., both $\text{rank}_m(w) < \text{rank}_m(f(m))$ and $\text{rank}_w(m) < \text{rank}_w(f^{-1}(w))$.

In other words, the bijection is called stable if there is no pair of people with mutual incentive to cheat on their assigned spouses. This is not to say cheating *can't* happen, but if it does one of the two involved will be "lowering their standards."

The algorithmic question is, given lists of preferences as input, can we find a stable marriage? Can we even guarantee a stable marriage will exist for any set of preferences? The answer to both questions is yes, and it uses an algorithm called *deferred acceptance*.

Here is an informal description of the algorithm. It goes in rounds. In each round, each man "proposes" to the highest-preferred woman that has not yet rejected him. On the other side, each woman holds a reference to a man at all times. If a woman gets new proposals in a round, she immediately rejects every proposer except her most preferred, but does not accept that proposal. She "defers" the acceptance of the proposal until the very end.

The rejected men are sad, but in the next round they recover and propose to their next most preferred woman, and again the women reject all but one. The men keep proposing until every man is tentatively held by some woman, or until all women have rejected them. That is not a happy place to imagine. But actually, the theorem that we'll prove says that this process always ends with each woman holding onto a man, and no men are left out; the set of women's held picks forms a stable bijection.

Before we prove that the algorithm works, let's state it more formally in Python code. A complete working program is available on this book's Github repository.[15] In the interest of generality, I've defined classes `Suitor` and `Suited` to differentiate: `Suitor`s propose to `Suited`s.

```
class Suitor:
    def __init__(self, id, preference_list):
        self.preference_list = preference_list
        self.index_to_propose_to = 0
        self.id = id

    def preference(self):
        return self.preference_list[self.index_to_propose_to]

    def post_rejection(self):
        self.index_to_propose_to += 1
```

---
[15] See pimbook.org

The `Suitor` class is simple. Instances are uniquely identified by an `id`, which I'm defining to be the index in a global list of `Suitor`s. A `Suitor` has a `preference_list`, which is a list of integers so that the i-th entry in the list is the index of a global list of `Suited`s containing the i-th most preferred Suited. In other words, the id of the most preferred Suited is the first entry in `preference_list`. The `index_to_propose_to` variable simultaneously counts the number of rejections and which index in the `preference_list` to use for the next proposal.

A bit more complicated is the `Suited` class:

```
class Suited:
    def __init__(self, id, preference_list):
        self.preference_list = preference_list
        self.held = None
        self.current_suitors = set()
        self.id = id

    def reject(self):
        """Return the subset of Suitors in self.current_suitors to reject,
        leaving only the held Suitor in self.current_suitors.
        """
        if len(self.current_suitors) == 0:
            return set()

        self.held = min(self.current_suitors,
                    key=lambda suitor: self.preference_list.index(suitor.id))
        rejected = self.current_suitors - set([self.held])
        self.current_suitors = set([self.held])

        return rejected

    def add_suitor(self, suitor):
        self.current_suitors.add(suitor)
```

Here `current_suitors` are the new proposals in a given round, and `held` is the Suited's held pick. In the method `reject`, a Suited looks at all her current suitors, chooses the minimum according to the index in her `preference_list`, and returns all but the best as rejected `Suitor`s.

Finally, we have the main routine for the deferred acceptance algorithm.

```
def stable_marriage(suitors, suiteds):
    """ Construct a stable marriage between Suitors and Suiteds. """
    unassigned = set(suitors)

    while len(unassigned) > 0:
        for suitor in unassigned:
            next_to_propose_to = suiteds[suitor.preference()]
            next_to_propose_to.add_suitor(suitor)
        unassigned = set()

        for suited in suiteds:
            unassigned |= suited.reject()

        for suitor in unassigned:
            suitor.post_rejection() # have some ice cream

    return dict([(suited.held, suited) for suited in suiteds])
```

The dictionary at the end is the type we use to represent a bijection. Now let's prove this algorithm always produces a stable marriage.

Theorem 4.14. *The deferred acceptance algorithm always terminates, and the bijection produced at the end is stable.*

Proof. We argue that the algorithm will terminates by *monotonicity*. Here's what I mean by that: say you have a sequence of integers $a_1, a_2, \ldots$ which is *monotonic increasing*, meaning that $a_1 < a_2 < \cdots$. Say moreover that you know none of the a_i are larger than 50—a_i is *bounded* from above—but each $a_{i+1} \geq a_i + C$ for some constant $C > 0$. Then it's trivial to see that either the sequence stops before it hits 50, or eventually it hits 50.

To show an algorithm terminates, you can cleverly choose an integer a_t for each step t, and show that a_t is monotonic increasing (or decreasing) and bounded. Then show that if the algorithm hits the bound then it's forced to finish, and otherwise it finishes on its own.

For the deferred acceptance algorithm we have a nice sequence. For round t set a_t to be the sum of all the Suitor's index_to_propose_to variables. Recall that this variable also represents the number of rejections of each Suitor. Since there are exactly n preferences in the list and exactly n Suitors, we get the bound $a_t \leq n^2$ (each Suitor could be at the very end of their list; come up with an example to show this can happen!).

Moreover, in each round one of two things happens. Either no Suitor is rejected by a Suited and by definition the algorithm finishes, or someone is rejected and their index_to_propose_to variable increases by 1, so $a_{t+1} \geq a_t + 1$. Now in the case where all the Suitors are at the end of their lists, that means that every Suited was proposed to by every Suitor. In other words, each of the Suiteds gets their top pick: they only reject when they see a better option, and they got to consider all proposals! Clearly the algorithm will stop in this case.

Now that we've shown the algorithm will stop, we need to show the bijection f produced as output is stable. The definition of stability says there is no Suitor m and

Suited w with mutual incentive to have an affair, so for contradiction's sake we'll suppose that the f output by the algorithm does have such a pair, i.e., for some m, w, $\text{pref}_m(w) < \text{pref}_m(f(m))$ and $\text{pref}_w(m) < \text{pref}_w(f^{-1}(w))$.

What had to happen to w during the algorithm? Well, m ended up with $f(m)$ instead of w, and if $\text{pref}_m(f(m)) > \text{pref}_m(w)$, then m must have proposed to w at some earlier round. Likewise, the `held` pick of w only increases in quality when w rejects a Suitor, but w ended up with some Suitor $f^{-1}(w)$ while $\text{pref}_w(m) < \text{pref}_w(f^{-1}(w))$. So at some point in between being proposed to by m and choosing to hold on to $f^{-1}(w)$, w had to go the wrong way in her preference list, contradicting the definition of the algorithm. □

We close with an example run:

```
>>> suitors = [
    Suitor(0, [3, 5, 4, 2, 1, 0]),
    Suitor(1, [2, 3, 1, 0, 4, 5]),
    Suitor(2, [5, 2, 1, 0, 3, 4]),
    Suitor(3, [0, 1, 2, 3, 4, 5]),
    Suitor(4, [4, 5, 1, 2, 0, 3]),
    Suitor(5, [0, 1, 2, 3, 4, 5]),
]
>>> suiteds = [
    Suited(0, [3, 5, 4, 2, 1, 0]),
    Suited(1, [2, 3, 1, 0, 4, 5]),
    Suited(2, [5, 2, 1, 0, 3, 4]),
    Suited(3, [0, 1, 2, 3, 4, 5]),
    Suited(4, [4, 5, 1, 2, 0, 3]),
    Suited(5, [0, 1, 2, 3, 4, 5]),
]
>>> stable_marriage(suitors, suiteds)
{
    Suitor(0): Suited(3),
    Suitor(1): Suited(2),
    Suitor(2): Suited(5),
    Suitor(3): Suited(0),
    Suitor(4): Suited(4),
    Suitor(5): Suited(1),
}
```

4.5 Cultural Review

1. Sets and functions between sets are a modeling language for mathematics.

2. Bijections show up everywhere, and they're a central tool of understanding the same object from two different perspectives.

3. Mathematicians usually accept silent type conversions between sets when it makes sense to do so, i.e., when there is a very clear and natural bijection between the two sets.

4. Induction is just another name for recursion, but applied to proofs.

5. A picture or example that captures the spirit of a fully general proof is often good enough.

4.6 Exercises

4.1 Write down examples for the following definitions. A set A (finite or infinite) is called *countable* if there is a surjection $\mathbb{N} \to A$. The *power set* of a set A, denoted 2^A, is the set of all subsets of A. For two sets A, B, we denote by B^A the set of all functions from A to B. This makes sense with the previous notation 2^A if we think of "2" as the set of two elements $2 = \{0, 1\}$, and think of a function $f : A \to \{0, 1\}$ as describing a subset $C \subset A$ by sending elements of C to 1 and elements of $A - C$ to 0. In other words, the subset defined by f is $C = f^{-1}(1)$.

4.2 Prove De Morgan's law for sets, which for $A, B \subset X$ states that $(A \cap B)^C = A^C \cup B^C$, and $(A \cup B)^C = A^C \cap B^C$. Draw the connection between this and the corresponding laws for negations of boolean formulas (e.g., `not (a and b) == (not a) or (not b)`).

4.3 Look up a formula online for the quantity $\binom{n}{k}$, the number of ways to choose k elements from a set of size n, in terms of factorials $m! = 1 \cdot 2 \cdot 3 \cdots\cdots m$. Find a proof that explains why this formula is true.

4.4 Look up a statement of the *pigeonhole principle*, and research how it is used in proofs.

4.5 Prove that $\mathbb{N} \times \mathbb{N}$ is countable.

4.6 Suppose that for each $n \in \mathbb{N}$ we picked a countable set A_n. Prove that the union of all the A_n is countable. Hint: use the previous problem and write the elements of all the A_n in a grid.

4.7 Is there a bijection between $2^\mathbb{N}$ and the interval $[0, 1]$ of real numbers x with $0 \le x \le 1$? Is there a bijection between $(0, 1] = \{x \in \mathbb{R} : 0 < x \le 1\}$ and $[1, \infty) = \{x \in \mathbb{R} : x \ge 1\}$?

4.8 I would be remiss to omit Georg Cantor from a chapter on set theory. Cantor's Theorem states that the set of real numbers $\mathbb{R}$ is not countable. The proof uses a famous technique called "diagonalization." There are many expositions of this proof on the internet ranging in difficulty. Find one that you can understand and read it. The magic of this theorem is that it means there is more than one kind of infinity, and some infinities are bigger than others.

4.9 The *principle of inclusion-exclusion* is a technique used to aid in counting the size of a set. Look of a description of this principle (it is a family of theorems) and find ways it is used to help count.

4.10 There is a large body of mathematics work related to configurations of sets with highly symmetric properties. Let n, k, t be integers. A *Steiner system* is a family F of size-k subsets of an n-element set, say $\{1, \ldots, n\}$, such that every size-t subset is in exactly one member of F. For example, for $(n, k, t) = (7, 3, 2)$, the corresponding Steiner system is a choice of triples in $\{1, 2, 3, 4, 5, 6, 7\}$, such that every pair of numbers is in exactly one of the chosen triples. Find such a $(7, 3, 2)$-system.

4.11 Using the previous exercise, a Steiner system may not exist for every choice of $n > k > t$. Prove that if an (n, k, t)-system exists, then so must an $(n-1, k-1, t-1)$-system. Determine under what conditions a Steiner $(n, 3, 2)$-system exists.

4.12 Continuing the previous exercise, the non-existence of Steiner systems for some choices of n suggests a modified problem of finding a minimal size family F of size-k subsets such that every t-size subset is in *at least* one set in F. For (n, k, t) arbitrary, find a lower bound on the size of F. Try to come up with an algorithm that gets close to this lower bound for small values of k, t.

4.13 A generalization of Steiner systems are called *block designs*. A block design F is again a family of size-k subsets of $X = \{1, \ldots, n\}$ covering all size-t subsets, but also with parameters controlling: the number of sets in F that contain each $x \in X$, and the number of sets covering each size-t subset (i.e., it can be more than one). Block designs are used in the theory of experimental design in statistics when, for example, one wants to test multiple drugs on patients, but the outcome could be confounded by which subset of drugs each patient takes, as well as which order they are taken in, among other factors. Research how block designs are used to mitigate these problems.

4.14 A *Sperner family* is a family F of subsets of $\{1, \ldots, n\}$ for which no member of F is a subset of any other member of F. Sperner's theorem gives an upper bound on the maximum size of a Sperner family. Find a proof of this theorem. There are multiple proofs, though one of them has at its core an inequality called the Lubell–Yamamoto–Meshalkin inequality, which is proved using a double-counting argument (and Exercise 4.3).

4.15 The formal mathematical foundations for set theory are called the Zermelo-Fraenkel axioms (also called ZF-set theory, or ZFC). Research these axioms and determine how numbers and pairs are represented in this "bare metal" mathematics. Look up Russell's paradox, and understand why ZF-set theory avoids it.

4.16 A *fuzzy set* $S \subset X$ is a function $m_S : X \to [0, 1]$ that measures the (possibly partial) membership of an $x \in X$ in the set S. One can think of $m_S(x)$ as representing the "confidence," or "probability" that an x is in S. Research fuzzy sets, and determine how one measures the cardinality of a fuzzy set.

4.17 Write a program that extends the deferred acceptance algorithm to the setting of "marriages with capacity." That is, imagine now that instead of men and women we have medical students and hospitals. Each hospital may admit multiple students as residents, but each student attends a single hospital. Find the most natural definition for what a stable marriage is in this context, and modify the algorithm in this chapter to find stable marriages in this setting. Then implement it in code. See the chapter notes for historical notes on this algorithm.

4.18 Come up with a version of stable marriages that includes the possibility of same-sex marriage. This variant is sometimes called the stable roommate problem. In this setting, there is simply a pool of people that must be paired off, and everybody ranks everyone else. Perform the full modeling process: write down the definitions, design an algorithm, prove it works, and implement it in code.

4.19 Is the stable marriage algorithm biased? Come up with a concrete measure of how "good" a bijection is for the men or the women collectively, and determine if the stable marriage algorithm is biased toward men or women for that measure.

4.7 Chapter Notes

Residency Matching

Medical residency matching was the setting for one of the major accomplishments of Alvin Roth, currently an economics professor at Stanford. He applied this and related algorithms to kidney exchange markets and schooling markets. Along with Lloyd Shapley, one of the original designers of the deferred acceptance algorithm, their work designing and implementing these systems in practice won the 2012 Nobel Prize in economics. Measured by a different standard, their work on kidney markets has saved thousands of lives, put students in better schools, and reduced stress among young doctors.

Roth gives a fascinating talk[16] about the evolution of the medical residency market before he stepped in, detailing how students and hospitals engaged in a maniacal day-long sprint of telephone calls, and all the ways unethical actors would try to game the protocol in their favor.

Marriage

Please don't treat marriage as an allocation problem in real life. I hope it's clear that the process of doing mathematics—and the modeling involved in converting real world problems to sets—involves deliberately distilling a problem down to a tractable core. This often involves ignoring features that are quite crucial to the real world. A quote often attributed to Albert Einstein speaks truth here, that "a problem should be made as simple as possible, but no simpler." Indeed, the unstated hope is that by analyzing the simplified, distilled problem, one can gain insights that are applicable to the more complex, realistic problem. Don't remove the core of the problem when phrasing it in mathematics, but remove as much as you need to make progress. Then gradually restore complexity until

[16] https://youtu.be/wvG5b2gmk70

you have solved the original problem, or fail to make more progress. Marriage is used as a communication device for this particular simplification. It's not the problem being solved.

The idea that one can reduce complex human relationships to a simple allocation problem is laughable, and borderline offensive. In the stable marriage problem the actors are static, unchanging symbols that happen to have preferences. In reality, the most important aspect of human relationships is that people can grow and improve through communication, introspection, and hard work.

Chapter 5
Variable Names, Overloading, and Your Brain

Math is the art of giving the same name to different things.

– Henri Poincaré

Programmers often complain about how mathematicians use single-letter variable names, how they overload and abuse notation, and how the words they use to describe things are essentially nonsense words made up for the sole purpose of having a new word. This causes bizarre sentences like "Map each co-monad to the Hom-set of quandle endomorphisms of X." I just made that up, by the way, though each word means something individually. One question programmers rarely ask is why mathematicians do this. Is it to feign complexity? Historical precedence? A hint of malice?

Of course there are bad writers out there, along with people who like to sound smart. There is certainly a somewhat unhealthy pattern of mathematicians who think a dose of emotional and intellectual pain is the best way to learn. But that's true of every field. I want to take a quick moment to explain the mathematician's perspective. As you've probably guessed by now, a central issue is culture. I won't try to convince you that this is the only explanation, but rather show you a different reasonable angle on the debate.

In producing mathematics, the mathematician has two goals: discover insight about a mathematical thing, and then communicate that truth to others in an intuitive and elegant way. While the second goal implies that mathematicians do care about style, what makes a proof or mathematical theory elegant is first and foremost the degree to which it facilitates understanding.

On the other hand, good software is measured (after it's deemed to work) by maintainability, extensibility, modularization, testability, robustness, and a whole host of other metrics which are primarily *business* metrics. You care about modularization because you want to be able to delegate work to many different programmers without stepping on each others toes. You want extensibility because customers never know what features they actually want until you finish designing the features they later decide are no good. You want to ensure that your software is idiot-proof because your company just hired three idiots! These metrics are good targets because they save time and money.

Mathematicians don't experience these scaling problems to the same degree of tedium because mathematics isn't a business. Mathematics isn't idiot-proof because the success of a mathematical theory doesn't depend on whether the next idiot that comes along understands it.[1] In fact, mathematical sophistication in the business world is extraordinary. And while having tests (providing worked-out examples) is a sign of a good mathematical writer, there's no manager staking their job or a salary bonus on the robustness of a bit of notation. If someone gets confused reading your paper, it doesn't siphon out the window the same way it does at Twitter during an outage. There's just not the same sense of urgency in mathematics.

I should make a side note that saying "mathematics isn't a business" is overly naive. Mathematicians need to make money just like everyone else, and this manifests itself in some strange practices in academic journals, conferences, and the multitude of committees that decide who is worth hiring and giving tenure. Mathematicians, like folks in industry, bend over backwards to game (or accommodate) the system. But all of that is academia. What I'm talking about is established mathematics which has been around for decades, or even centuries, which has been purified of political excrement. This applies to basically every topic in this book.

That's not to say that mathematics isn't designed to scale. To the contrary, the invention of algebraic notation was one of humanity's first massively scalable technologies. On the other end of the spectrum, category theory—which you can think of as a newer foundation for math roughly based on a new notation that goes beyond what sets and functions can offer—provides the foundation for much of modern pure mathematics. It's considered by many as a major advancement.

Rather than being designed to scale to millions of average users, mathematics aims to scale far up the ladder of abstraction. Algebra—literally, the marks on paper—boosted humanity from barely being able to do arithmetic through to today's machine learning algorithms and cryptographic protocols. Sets, which were only invented in the late 1800's, hoisted mathematical abstraction even further. Category theory is a relative rocket fuel boosting one through the stratosphere of abstraction (for better or worse).

The result of this, as the argument goes, is that mathematicians have optimized their discourse for more relevant metrics. Indeed, it's optimized for maximizing efficiency and minimizing cognitive load *after deep study*.

Let me map out a few areas where this shows up:

- Variable names

- Operator overloading

- Sloppy notation

Variable names. Variable names are designed to transmit a lot of information: types, behavior, origin, and more. Every mathematician knows that n is a natural number, and

---

[1] I mean this in a practical sense, not a social sense. If your math is so hard to understand that nobody but you learns it, it will be lost to history. But from a practical standpoint, calculus doesn't stop being a good foundation for a video game engine just because the programmer doesn't understand the math.

that f is a function. Or at least, they know that when they see these letters out of context, they should at least behave like a natural number and a function. Seeing $n(f)$ out of context would momentarily startle me, though I can imagine situations making it appropriate.[2] Similarly, if f is a function and you can use f to construct another function in a "canonical" (forced, unique) way, then a mathematician might typically adorn f with a star like f^*. Two related objects often inhabit the same letter with a tick, like x and x'. Even if you forget what they represent, you know they're related.

Every field of mathematics has its own little conventions that help save time. This is especially true since mathematics is often done in real time (talking with colleagues in front of a blackboard, or speaking to a crowd). The time it takes to write f^* while saying *out loud* "the canonical induced homomorphism,[3]" is much faster than writing down `InducedHomomorphismF` in ten places. And then when you need an h^* to compose f^*h^*, half of the characters help you distinguish it from h^*f^*. Whereas determining the order of

```
InducedHomomorphismF.compose(InducedHomomorphismH)
```

is harder with more characters, and Gauss forbid you have to write down an identity about the composition of three of these things! A single statement would fill up an entire blackboard, and you'd never get to the point of your discussion.

More deeply, there is often nothing more a name can do to elucidate the nature of a mathematical object. Does saying f^* really tell you less about what an object is than something like `InducedF`? It's related to f, its definition is somehow "induced," and what? The further up the ladder of abstraction you go, the more contrived these naming conventions would get. Rather than say, for example, `FirstCohomologyGroupOfInvertibleSubsheavesOfX`, you say $H^1(X, O^*)$ because you would rather claw your eyes out than read the first thing, which could easily be *just one part* of a larger expression, with maybe ten more similar copies of the notation. For example, here is an actual snippet from a chapter of a graduate algebra textbook cheeily titled, "Algebra: Chapter 0."

$$\nu_L : \mathsf{L}_0 \mathcal{F}(M) = H^0(\mathsf{C}(\mathcal{F})(P^\bullet)) \to \mathcal{F}(M)$$

It is a bit ridiculous that L and L refer to different mathematical things, despite being the same letter. Here L is an object and L (short for "left") describes a kind of function. But this is a trade-off: use long words that make it difficult to put everything you want to say in front of your face at the same time—thus making it harder to reason—or use fonts and foreign alphabets to differentiate concepts. Sans-serif is for one purpose, the curly-scripty font is for another.

---

[2] For example, n could represent some integer-valued property of a function, like the so-called *winding number*
[3] For example. You don't need to know what a homomorphism is.

The claim that a variable name in mathematics can do what programmers claim they must is naive. In fact, because the expression $H^1(X, O^*)$ is so important in this field called algebraic geometry, it was further shortened to $\text{Pic}(X)$ named after Picard who studied them. But it might take decades to get to the point where you realize this object is worth giving a name, and in the mean time you just can't use 80 character names and expect to get things done.

One reason mathematicians can get away with single-character variable names is that they spend so much time studying them. When a mathematician comes up with a new definition, it's usually the result of weeks of labor, if not months or years! Moreover, these objects aren't just variables in some program whose output or process is the real prize. The variables represent the cool things! It's as if you returned to rewrite and recheck and retest the same twenty-line program every day for a month. You'd have such an intimate understanding of every line that you could recite them all while drunk or asleep. You could recognize the program even if it were minified. Now imagine that the intimate understanding of every line of that program was the basis of every program you wrote for the next year, and you see how ingrained this stuff is in the mind of a mathematician.

Mathematicians don't just write a proof and file it away under "great tool; didn't read." They constantly revisit the source. It's effective to gild meaning and subtext into the bones of single letters, because after years you don't have to think about it any more. It eliminates the need to keep track of types. Clearly f is a function, z is probably a complex variable, and everyone knows that $\aleph_0$ is the countably infinite cardinal. If you use b and β in the same place, I will know that they are probably related, or at least play analogous roles in two different contexts, and that will jump-start my understanding in a way that descriptive variable names do not.

Operator overloading. Much of what I said above for variable names holds for operator overloading too. One key feature that stands out for operator overloading is that it highlights the intended nature of an operation.

We'll get to this more in Chapter 9, but mathematicians use just a handful of boolean logic operations for almost everything. The standard inequalities, equalities, and weird ones that look like $\cong$ or $\simeq$ that are supposed to represent equality "up to some differences that we don't care about." In Java terms, mathematicians regularly roll their own .equals() methods, with proofs that their notions behave. Specifically, they prove it satisfies the properties required of an *equivalence relation*, which is the mathematical version of saying "equals agrees with hashing and toString."

And so typically mathematicians will drop whatever the original operator symbol was and replace it with the equal sign. We'll see this in detail in Chapters 9 and 16, but the same idea goes behind the reuse of standard arithmetic operations like addition and multiplication: it's so that we can know even out of context what behavior to expect from the operation. For example, it is considered bad form to use the $+$ operator for an operation that doesn't satisfy $a + b = b + a$ for every choice of a and b, because this is true of addition.

With this in mind it's the mathematician's turn to criticize programmers. For example, reading programming style guides has always amused me. It makes sense for a company to impose a style guide on their employees (especially when your IDE is powerful enough to auto-format your programs) because you want your codebase to be uniform. In the same way, a mathematician would never change notational convention in the same paper, unless the point of the paper is to introduce a new notation. But to have a programming language designer declare style edicts for the entire world, like the following from the Python Style Guide, is just ridiculous:

```
Imports should usually be on separate lines, e.g.:

    Yes: import os
         import sys

    No:  import sys, os
```

Okay, so you have an arbitrary idea of what a pretty program looks like, but wouldn't you rather spend that time and energy on actually understanding and writing a good program? Besides, if there were truly a good reason for the first option, why wouldn't the language designer just disallow it in the syntax? Of course, programmers get away with it because they use automated tools to apply style guides automatically. It's much harder to do that in math, where the worst offenses are not resolvable (or discoverable!) from syntax alone. Still, I don't doubt there could be some progress made in automating some aspects of a mathematical style guide.

In an ideal world, a compiler would see how I use the "stdout" variable and be able to infer the semantics from a shared understanding about the behavior of standard output in basically every program ever. This would eliminate the need to declare module imports or even define stdout! That's basically how math solves the problem of overloaded operators. There is a clarifying and rigorous definition somewhere, but if you've forgotten it you can still understand the basic intent and infer appropriate meaning.

Sloppy notation. This is probably the area where mathematicians get the most flak, and where they could easily improve their communication with those aiming to learn.

Take summation notation, the $\sum$ symbol. Officially this symbol has three parts: an index variable, a maximum value for the index, and an expression being summed. So $\sum_{i=0}^{10} 2i+1$ sums the first ten positive odd integers. This is the kind of syntactical rigidity that makes one itch to write a parser.

However, this notation is so convenient that it's been overloaded to include many other syntax forms. A simple one is to replace the increment-by-one range of integers with a "all elements in this set" notation. For example, if B is a set, you can write $\sum_{b \in B} b^2$ to sum the squares of all elements of B.

But wait, there's more! It often happens that B has an implicit, or previously defined order of the elements $B = \{b_1, \ldots, b_n\}$, in which case one takes the liberty of writing $\sum_i b_i^2$ ("the sum over relevant i") with no mention of the set in the (local) syntax at all! As we saw in Chapter 2 with polynomials, one can additionally add conditions below the index to filter only desired values, or even have the *constraint* implicitly define the variable range! So you can say the following to sum all odd $b_i \in B$

$$\sum_{b_i \text{ odd}} b_i^2 + 3$$

The reason this makes any sense is because, as is often the case, the math notation often comes from speech. You're literally speaking, "over all b_i that are odd, sum the terms $b_i^2 + 3$." Equations are written to mimic conversation, not the other way around. You see it when you're in the company of mathematicians explaining things. They'll write their formulas down as they talk, and half the time they'll write them backwards! For a sum, they might write the body of the summation first, then add the sum sign and the index. Because out loud they'll be emphasizing the novel parts of the equation, filling the surrounding parts for completeness.

Finally, the things being summed need not be numbers, so long as addition is defined for those objects and it satisfies the properties addition should satisfy. In Chapter 10 we'll see a new kind of summation for vectors, and it will be clear why it's okay for us to reuse $\sum$ in that context. The summing operation needs to have properties that result in the final sum not depending on the order the operations are applied.

Another prominent example of summation notation being adapted for an expert audience is the so-called *Einstein notation*. This notation is popular in physics. In Einstein notation the $\sum$ symbol is *itself* implied from context! For example, rather than write

$$y = \sum_{k=1}^{n} a_k x^k,$$

The sum and the bounds on the indices are implied from the presence of the indices, as in

$$y = a_k x^k.$$

To my personal sensibilities this is extreme. But I can't fault proponents for the abuse when they find it genuinely useful. If it solidifies their intuition of the object of their study, it's a good thing.

Indeed, what makes all of this okay is when the missing parts are fixed throughout the discussion or clear from context. What counts as context is (tautologically) context dependent. More often than not, mathematicians will preface their abuse to prepare you for the new mental hoop. The benefit of these notational adulterations is to make the mathematics less verbose, and to sharpen the focus on the most important part: the core idea being presented. These "abuses" reduce the number of things you see, and as a consequence reduce the number of distractions from the thing you want to understand.

Chapter 6
Graphs

One will not get anywhere in graph theory by sitting in an armchair and trying to understand graphs better. Neither is it particularly necessary to read much of the literature before tackling a problem: it is of course helpful to be aware of some of the most important techniques, but the interesting problems tend to be open precisely because the established techniques cannot easily be applied.

– Tim Gowers

So far we've learned about a few major mathematical tools:

- Using sets for modeling
- Proof by contradiction, induction, and "trivial" proofs.
- Bijections for counting

In this chapter we won't learn any new tools. Instead we'll apply the tools above to study graphs. Most programmers have heard about graphs before, perhaps in the context of breadth-first and depth-first search or data structures like heaps. Instead of discussing the standard applications of graphs to computer science, we'll focus on a less familiar topic that still finds uses in computer science: graph coloring.

In addition to having interesting applications, graph coloring has important theorems one can prove using only the tools we've learned so far. The main theorem we'll prove in this chapter is that every planar graph is 5-colorable (I will explain these terms soon). So think of this chapter as a sort of checkpoint exam. If you're struggling to understand the definitions, theorems, and proofs here—and you've set your pace appropriately—then you should go back and review the previous chapters.

6.1 The Definition of a Graph

The definition of a graph is best done by picture, as in Figure 6.1. If you give me a bunch of "things" and a list of which things are "connected," and the result is a graph. As a simple example, the "things" might be airports, and two airports are "connected" if

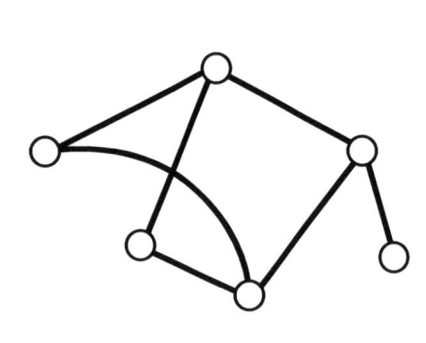

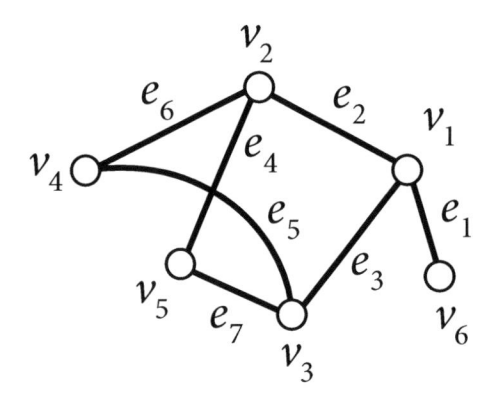

Figure 6.1: An example of a graph

Figure 6.2: A graph with labeled vertices and edges.

there is a flight between the two. Or the things are people and friends have connections. We draw the things and connections using dots and lines to erase the application from our minds. *All we care about* is the structure of the connections.

Let's lay out the definitions, using sets as the modeling language. The "things" are called *vertices* (or often *nodes*) and the "connections" are called *edges* (or *links*). For shorthand in the definition, I'll reuse a definition from Chapter 4 for the set of all ways to choose two things from a set.

$$\binom{V}{2} = \{\{v_1, v_2\} : v_1 \in V, v_2 \in V, v_1 \neq v_2\}.$$

This is like $V \times V$, but the order of the pair does not matter.

Definition 6.1. A graph G consists of a set V of *vertices*, a set $E \subset \binom{V}{2}$ of *edges*. The entire package is denoted $G = (V, E)$.[1]

Alternatively, one can think of E as just any set, and require a function $f : E \to \binom{V}{2}$ to describe which edges connect which pairs of vertices. This view is used when one wants to define a graph in a context where the vertices are complicated (we will briefly see one from compiler design later in this chapter). Despite the definition of an edge $e \in E$ as a set of size two like $\{u, v\}$, mathematicians will sloppily write it as an ordered pair $e = (u, v)$.[2]

Here's some notation and terminology used for graphs. We always call $n = |V|$ the number of vertices and $m = |E|$ the number of edges, and for us these values will always be finite. When two vertices $u, v \in V$ are connected by an edge $e = (u, v)$ we call the

---

[1] This is not the most general definition for a graph, but we will not need graphs with self loops, weights, double edges, or direction. You'll explore some of these extensions in the exercises.
[2] I have suspicions about why this abuse is commonplace: curly braces are more cumbersome to draw than parentheses, and in the typesetting language LaTeX, typing them requires an escape character. They're also just visually harder to parse when nested.

two vertices *adjacent*, and we say that e is *incident* to u and v. We call v a *neighbor* of u and we define the *neighborhood* of a vertex $N(u)$ to be the set of all neighbors. I.e.,

$$N(u) = \{v \in V : (u,v) \in E\}$$

The size of a neighborhood (and the number of incident edges) is called the *degree* of a vertex, and the function taking a vertex v to its degree is called $\deg : V \to \mathbb{Z}$. To practice the new terms, see Figure 6.2, labeling the graph from Figure 6.1. Vertices have label 'v' and edges have lebel 'e'. Verteces v_1 v_3 are adjacent, e_2 is incident to v_1, $\deg(v_2) = 3$, and all of the neighbors of v_2 are also neighbors of v_3.

Another concept we'll need in this chapter is the concept of a connected graph. First, a *path* in a graph is a sequence of alternating vertices and edges $(v_1, e_1, v_2, e_2, \ldots v_t)$ so that each $e_i = (v_i, v_{i+1})$ connects the two vertices next to it in the list. Visually, a path is just a way to traverse through the vertices of G by following edges from vertex to vertex. In Figure 6.2, there are many different paths from v_4 to v_6, four of which do not repeat any vertices. Many authors enforce that paths do not repeat vertices by definition, and give the name "trail" of "walk" to a path which does repeat vertices.

A graph is called *connected* if there is a path from each vertex to each other vertex, and otherwise it is called *disconnected*. Equivalently, $G = (V, E)$ is connected if it is impossible to split V into two subsets X, Y with no edges between X and Y. A disconnected graph is a union of connected *components*, where the component of v is the largest connected subgraph[3] containing v. A single vertex which forms a connected component is called an *isolated* vertex.

6.2 Graph Coloring

The main object of study in this chapter is called a *coloring* of a graph $G = (V, E)$, which is an assignment of "colors" (really, numbers from $\{1, 2, \ldots, k\}$) to the vertices of G satisfying some property. We realize this officially as a function.

Definition 6.2. A *k-coloring* of a graph $G = (V, E)$ is a function $\varphi : V \to \{1, 2, \ldots, k\}$. We call an edge $e = (u, v)$ *properly colored* by a k-coloring φ if $\varphi(u) \neq \varphi(v)$, and otherwise we call that edge *improperly colored*. We call φ *proper* if it properly colors every edge. If a graph G has a proper k-coloring, we call it *k-colorable*.

By now you should know to write down examples for small n and k before moving on. Because this is a crucial definition, here is a more complicated example. The Petersen graph is shown in Figure 6.3. The Petersen graph has a distinguished status in graph theory as a sort of smallest serious unit test. Conjectures that are false tend to fail on the Petersen graph.[4] The Petersen graph is 3-colorable (find a 3-coloring!) but not 2-colorable.

---

[3] A subgraph is just a subset of edges and their corresponding vertices.
[4] Why? Part of it is that the Petersen graph is highly symmetric, which we'll see more in the exercises for Chapter 16.

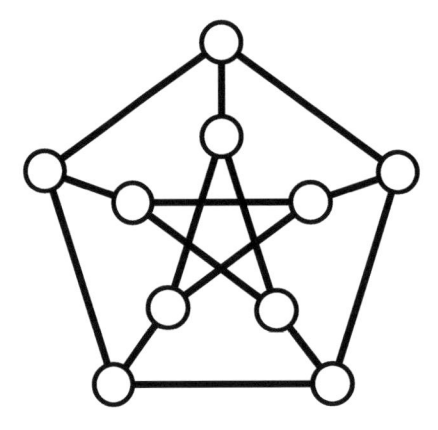

Figure 6.3: The Petersen graph.

Definition 6.3. The *chromatic number* of a graph G, denoted $\chi(G)$, is the minimum integer k for which G is k-colorable.

So by the example above, the Petersen graph has chromatic number 3. Here is a simple fact about the chromatic number.

Proposition 6.4. *If $G = (V, E)$ is a graph and d is the largest degree of a vertex $v \in V$, then $\chi(G) \leq d + 1$.*

Proof. We define a greedy algorithm for coloring a graph. Pick an arbitrary ordering $v_1, \ldots, v_n$ of the vertices of G, and then for each v_i pick the first color j which is unused by any of the neighbors of v_i. In the worst case, a vertex v of degree d will have all of its neighbors using different colors, and so it will use color $d+1$. Otherwise v could reuse one of the first d colors not used by any neighbor. So the worst-case number of colors is at most the largest degree in the graph plus one, as claimed. □

In fact, a very simple graph meets this bound and has $\chi(G) = \max_{v \in V} \deg(v) + 1$. See if you can find it. Moreover, this bound is quite loose. Consider the "star" graph which has only one vertex of degree $n - 1$, pictured in Figure 6.4. Clearly the star graph is 2-colorable, but the max degree is $n - 1$. The guarantee of the theorem is useless.

One other perspective on graph coloring I want to describe is the partition perspective. Specifically, if $G = (V, E)$ is a graph and φ is a proper k-coloring, then we can look at $\varphi^{-1}(j)$, the set of all vertices that have color j. By the fact that φ is proper, there will be no edges among these vertices. Moreover, since φ is a function, the set of all $\varphi^{-1}(j)$ form a partition[5] of the set V into "color classes," and all the edges go between the color classes. Figure 6.5 shows a picture for the Petersen graph.

---

[5] A partition of X is a set of non-overlapping (disjoint) subsets $A_i \subset X$ the union of all of them being $\cup_i A_i = X$.

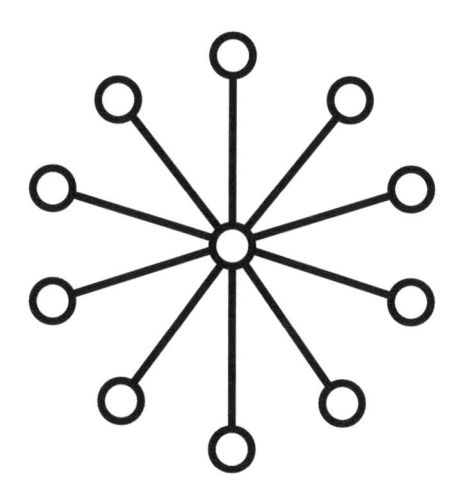

Figure 6.4: A star graph.

Figure 6.5: A coloring of the Petersen graph.

This perspective is important because one can try to properly color a graph by starting with an improper or unfinished coloring, and fiddle with it to correct the improprieties. We will do this in the main application of this chapter, coloring planar graphs. But right now we're going to take a quick detour to see why graph coloring is useful.

6.3 Register Allocation and Hardness

The wishy-washy way to motivate graph coloring is to claim that many problems can be expressed as an "anti-coordination problem," where you win when no agent in the system behaves the same as any of their neighbors. A totally made up example is radio frequencies. Radio towers pick frequencies to broadcast, but if nearby towers are broadcasting on the same frequency, they will interfere. So the vertices of the graph are towers, nearby towers are connected by an edge, and the colors are frequencies.

A more interesting and satisfying application is register allocation. That is, suppose you're writing a compiler for a programming language. Logically the programmer has no bound on the number of variables used in a program, but on the physical machine there is a constant number of registers in which to store those variables. The connection to graph coloring is beginning to reveal itself: the vertices are the logical variables and the colors are physical registers, but I haven't yet said how to connect two vertices by an edge. Intuitively, it depends on whether the logical variables "overlap" in the scope of their use. The structure of scope overlap is destined to be studied with graph theory.

To simplify things, we'll do what a compiler designer might reasonably do, and compile a program down to *almost* assembly code, where the only difference is that we allow infinitely many "virtual" registers, which we'll just call variables. So for a particular program P, there is a $n_P \in \mathbb{N}$ that is the number of distinct variable names used in the program. Each of these integers is a vertex in G.

As an illustrative example, say that the almost-compiled program looks like this, where the dollar sign denotes a variable name:

```
whileBlock:
$41 = $41 - 1
$40 = $40 + $42
$42 = $41 - $42
BranchIfZero $41 endBlock whileBlock

endBlock:
$43 = $41 + $40
```

In this example variables 41 and 42 cannot share a physical register. They have different values and are used in the same line to compute a difference. Call a variable *live* at a statement in the code if it's value is used after the end of that statement. Thinking of it in reverse: a variable is *dead* in all of the lines of code between when it was last read and when it is next written to. Whenever a variable is dead we know it's safe to reuse its physical register (storing the value of the dead variable in memory).

Now we can define the edges. Two variables $i and $j "interfere," and hence we add the edge (i, j) to G, if they are ever live at the same time in the program. With a bit of work (uncoincidentally using graphs to do a flow analysis), one can efficiently compute the places in the code where each variable is live and construct this graph G. Then if we can compute the chromatic number of G and find an actual $\chi(G)$-coloring, we can assign physical registers to the variables according to the coloring. Without some deeper semantic analysis, this provides the most efficient possible use of our physical registers.[6]

Unfortunately, in general you should not hope to compute the chromatic number of an arbitrary graph. This problem is what's called "NP-hard," which roughly means we don't know of any provably correct that is significantly better than brute-force searching through all possible colorings, and we don't hope to find one any time soon.

Moreover, it is even NP-hard to get *any reasonable approximation* of the chromatic number of a general graph. To be more specific, we can't hope to find an algorithm which, when given a graph G with n vertices, can output a number Z with the property that $\frac{Z}{\chi(G)} < n^c$ for any $0 < c < 1$. This is an asymptotic statement, meaning a hopeful algorithm might provably work for all graphs with fewer than a thousand nodes. This may be good enough for some practical purposes.[7] But to put the numbers in perspective with an example, this theorem says that for graphs with $n = 10^5$ vertices and with $c = 1/2$, algorithms will struggle to output a number guaranteed to be between $\chi(G)$ and $100 \cdot \chi(G)$. That multiplicative factor *grows polynomially quickly* with the size of the input graph!

---

[6] In fact, it can happen that the chromatic number of G is greater than the total number of registers on the target machine. In this case you have to spill some variables into memory, and deciding which variables to send to memory is both a science and an art.

[7] If you had to compute the chromatic number of a graph in a practical setting, you'd probably write it as a so-called *integer linear program* throw an industry-strength solver at it. As they say, NP-hard problems are hard in theory but easy in practice.

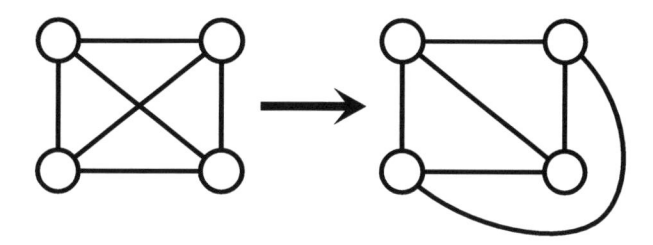

Figure 6.6: An example of a planar graph which can be drawn with no edges crossing.

But I digress. The takeaway is that coloring is a hard problem. This is a sad result for people who really want to color their graphs, but there are other ways to attack the problem. You can assume that your graph has some nice structure. This is what we'll do in the next section, and there it turns out that the chromatic number will always be at most 4. Alternatively, you could assume that you know your graph's chromatic number, and try to color it without introducing too many improperly colored edges. We'll see this approach in Section 6.6.

6.4 Planarity and the Euler Characteristic

The condition we'll impose on a graph to make coloring easier is called planarity. A graph $G = (V, E)$ is called *planar* if one can draw it on a plane in such a way that no edges cross. Figure 6.6 contains an example.

Here's a little exercise: come up with an example of a graph which is not planar. Don't be surprised if you're struggling to prove that a given graph is not planar. You personally failing to draw a specific graph without edges crossing is not a *proof* that it is impossible to do so. There is a nice rule that characterizes planar graphs, but it is not trivial. See the chapter exercises for more.

Now that you've tried the exercise: Figure 6.7 depicts two important graphs that are not planar. The left one is called the *complete graph* on 5 vertices, denoted K_5. The word "complete" here just means that all possible edges between vertices are present. The second graph is called the *complete bipartite graph* $K_{3,3}$. "Bipartite" means "two parts," and the completeness refers to all possible edges going between the two parts. The subscript of $K_{a,b}$ for $a, b \in \mathbb{N}$ means there are a vertices in one part and b in the other.

We defined planar graphs informally in terms of drawings in the plane, which doesn't use sets, functions, or anything you've come to expect. Indeed, the hand-wavy definition is the one that belongs in your head, but the *official* definition of a planar graph is one which has an *embedding* into $\mathbb{R}^2$. The problem is that defining an embedding requires opening a big can of worms, because it applies to spaces more general than a graph. We'll give you a taste in the chapter notes.

One feature about planar graphs is that when you draw a planar graph in such a way that no edges cross, you get a division of $\mathbb{R}^2$ into distinct regions called "faces." Figure 6.8

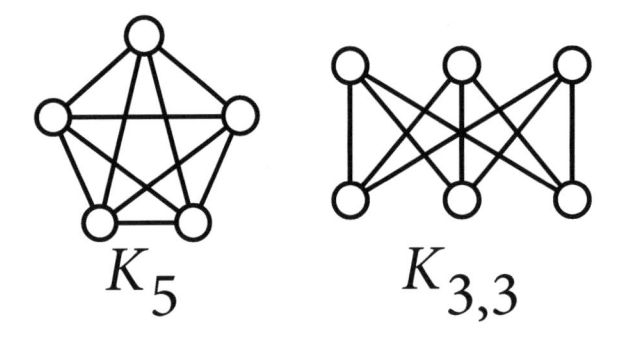

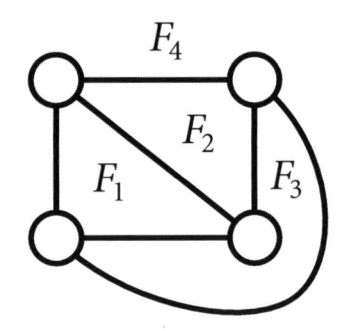

Figure 6.7: K_5 and $K_{3,3}$, two graphs which are not planar.

Figure 6.8: Faces of a planar graph.

shows a graph with four faces, because I'm calling the "outside" of the drawing also a face. If we call f the number of faces, and remember n is the number of vertices and m is the number of edges, then we can notice[8] a nice little pattern: $n - m + f = 2$.

The amazing fact is that this equation does not depend on how you draw the graph! So long as your drawing has no crossing edges, the value $n - m + f$ will always be 2. We can prove it quite simply with induction.

Theorem 6.5. *For any connected planar graph $G = (V, E)$ and any drawing of G in the plane $\mathbb{R}^2$ defining a set F of faces, the quantity $|V| - |E| + |F| = 2$.*

Proof. We proceed by induction on the total number of vertices and edges. The base case is a single isolated vertex, for which $|V| = 1$, $|E| = 0$, and $|F| = 1$, so the theorem works out.

Now suppose we have a graph G for which the theorem holds, i.e. $|V| - |E| + |F| = 2$, and we will make it larger and show that the theorem still holds. In particular, we will do induction on the quantity $|V| + |E|$. There are two cases: either we add a new edge connecting two existing vertices, or we add a new edge connected to a new vertex (which now has degree 1).

In the first case, $|V|$ is unchanged, $|E|$ increases by 1, and $|F|$ also increases by one because the new edge cuts an existing face into two pieces. So

$$|V| - (|E| + 1) + (|F| + 1) = |V| - |E| + |F| = 2$$

Notice how it does not matter how we drew the edge, so long as it doesn't cross any other edges to create more than one additional face. The second case is similar, except adding an edge connected to a new vertex does not create any new faces. Convince yourself that any vertex involved in a path that encloses a face has to have degree at least

---

[8] Why anyone would have reason to analyze this quantity is a historical curiosity; it was discovered by Euler for certain geometric shapes in three dimensions called convex polyhedra. See the following for more: http://mathoverflow.net/q/154498/6429

two. So again we get that for the new graph $|V| + 1 - (|E| + 1) + |F| = 2$. This finishes the inductive step.

Finally, it should be clear that every connected graph (regardless of whether it's planar) can be built up by a sequence of adding edges by these two cases. This completes the proof.

□

This is a surprising fact. We have some measurement derived from a drawing of a graph that *doesn't depend* on the choices made to draw it! This is called an *invariant*, and we'll discuss invariants more in Chapter 10 when we study linear algebra, and Chapter 16 when we study geometry. For now it will remain a deep mathematical curiosity. Lastly, note that the requirement the graph is connected is crucial for the theorem to hold, since a graph with n vertices and no edges has $|V| - |E| + |F| = n + 1$.

On to the main theorem!

6.5 Application: the Five Color Theorem

Here is an amazing theorem about planar graphs.

Theorem 6.6. *(The four color theorem)*
Every planar graph can be colored with 4 colors.

This was proved by Kenneth Appel and Wolfang Haken in 1976 after being open for over a hundred years. You may have heard of it because of its notoriety: it was the first major theorem to be proved with substantial aid from a computer. Unfortunately the proof is very long and difficult (on the order of 400 pages of text!). Luckily for us there is a much easier theorem to prove.

Theorem 6.7. *(The five color theorem)*
Every planar graph can be colored with 5 colors.

If you're like me and frequently make off-by-one errors, then the five color theorem is just as good as the four color theorem! In order to prove it we need three short lemmas.

Lemma 6.8. *If G is a graph with m edges, then $2m = \sum_{v \in V} \deg(v)$.*

Proof. The important observation is that the degree of a vertex is just the number of edges incident to it, and every edge is incident to exactly two vertices.

This is where the proof would usually end. As a variation on a theme, you can (and should) think of this as constructing a clever bijection like we did in Chapter 4, but it's difficult to clearly define a domain and codomain. Let me try: the domain consists of "edge stubs" sticking out from each vertex, and the codomain is the set of edges E. We're mapping each edge stub to the edge that contains that stub. This map is a surjection and a double cover of E, and the size of the domain is exactly $\sum_{v \in V} \deg(v)$.

□

Lemma 6.9. *If a planar graph G has m edges and f faces, then $2m \geq 3f$, i.e. $f \leq (2/3)m$.*

Proof. Pick your favorite embedding (drawing) of G in the plane. We'll use a similar counting argument as in Lemma 6.8: for any planar drawing, every face is enclosed by at least three edges, and every edge touches at most two faces.[9] Hence $3f$ counts each edge at most twice, while $2m$ counts each face at least three times. □

You should do what I did for Lemma 6.8 and think about how to express this as an injection from one set to another. The last lemma is the key to the five color theorem.

Lemma 6.10. *Every planar graph has a vertex of degree 5 or less.*

Proof. Suppose to the contrary that every vertex of $G = (V, E)$ has degree 6 or more. Substituting the inequality relating edges and faces from Lemma 6.9 into the Euler characteristic equation gives

$$2 = |V| - |E| + |F| \leq |V| - |E| + (2/3)|E|$$

Rearranging terms to solve for $|E|$ gives $|E| \leq 3|V| - 6$. Now we want to use the Lemma 6.8 so we multiply by two to get $2|E| \leq 6|V| - 12$. Since $2|E|$ is the sum of the degrees, and each vertex has degree at least six, $2|E|$ has to count something *at least* as large as $6|V|$. Adding this to the above inequality gives

$$6|V| \leq 2|E| \leq 6|V| - 12,$$

which is a contradiction. □

As a quick side note that we'll need in the next theorem, along the way to proving Lemma 6.10 we get a bonus fact: the complete graph K_5 is not planar. This is because we proved that all planar graphs satisfy $|E| \leq 3|V| - 6$, and for K_5, $|E| = 10 > 15 - 6$. This argument doesn't work for showing $K_{3,3}$ is not planar, but if you're willing to do a bit extra work (and take advantage of the fact that $K_{3,3}$ has no cycles of length 3), then you can improve the bound from Lemma 6.10 to work. In particular, because K_5 is not planar, no planar graph can contain K_5 as a subgraph.

Now we can prove the five color theorem.

Proof. By induction on $|V|$. For the base case, every graph which has 5 or fewer vertices is 5-colorable by using a different color for each vertex.

Now let $|V| \geq 6$. By Lemma 6.10, G has a vertex v of degree at most 5. If we remove v from G then the inductive hypothesis guarantees us a 5-coloring. So we want to extend or modify this coloring and get a good color for v, and this will finish the proof. When v

---

[9] An edge incident to a vertex of degree 1 will touch the "outside" face twice, but this only counts as one face.

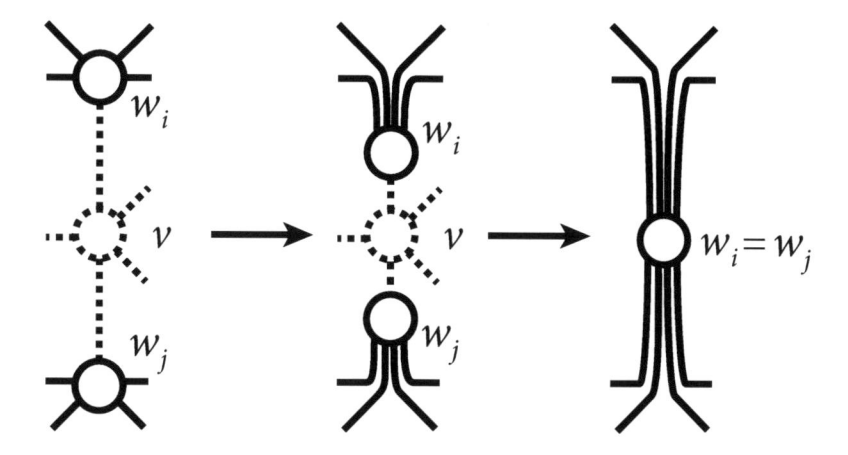

Figure 6.9: The "strands of a spider web" image guide the proof that G' is planar.

has degree at most 4, choose one of the unused colors among v's neighbors. Otherwise v has degree exactly 5, and we have to be more clever.

Call v's five neighbors w_1, w_2, w_3, w_4, w_5. Because K_5 is not planar and G is, these five neighbors can't form K_5. In particular there must be some i, j for which w_i and w_j are not adjacent. We can form a graph G' ("G prime"[10]) by *merging* these two vertices, i.e., delete w_i, w_j and add a new vertex x which is adjacent to all the vertices in $N(w_i) \cap N(w_j)$. I claim that if G' is planar then we're done: G' has $|V| - 2$ vertices and so it has a 5-coloring by the inductive hypothesis, and we can use that 5-coloring to color most of G (everything except w_i, w_j, and v). Then use the color assigned to x for both w_i and w_j; they had no edge between them in G, so this is okay. These choices ensure the neighbors of v use only 4 of the 5 colors, so finally pick the unused color for v. This produces a proper coloring of G.

So why is G' planar? To argue this, we have to show that for any planar drawing of G, removing v leaves w_i and w_j in the same face. This is equivalent to being able to trace a curve in the plane from w_i to w_j without hitting any other edges, since we could then "drag" w_i along that curve to w_j and "lengthen" the edges incident to w_i as we go. The picture in my head is like the strands of a spider web, shown in Figure 6.9.

The key is that G is planar and that v has all of the w's as neighbors. If we want to merge w_i to w_j, we can use the curve already traced by the edges from w_i to v and from v to w_j. By planarity this is guaranteed not to cross any of the other edges of G, and hence of G'. To say it a different way, if we took the drawing above and continued drawing G', and the result required an edge to cross one of the edges above, then it would have crossed through one of the edges going from v to w_i or v to w_j!

This proves G' is planar, which completes the proof.

---

[10] The tick is called the "prime" symbol, and it is used to denote that two things are closely related, usually that the prime'd thing is a minor variation on the un-primed thing. So using G' here is a reminder to the reader that G' was constructed from G.

That proof neatly translates into a recursive algorithm for 5-coloring a planar graph. We'll finish this section with Python code implementing it. In order to avoid the toil of writing custom data structures for graphs, we'll use a Python library called `igraph` to handle our data representation. As a *very* quick introduction, one can create graphs in `igraph` as follows.

```
import igraph
G = igraph.Graph(n=10)
G.add_edges([(0,1), (1,2), (4,5)])

G.vs # a list-like sequence of vertices
G.es # a list-like sequence of edges
```

For example, given a graph and a list of nodes in the graph, one might use the following function to find two nodes which are not adjacent.

```
from itertools import combinations

def find_two_nonadjacent(graph, nodes):
    for x, y in combinations(nodes, 2):
        if not graph.are_connected(x, y):
            return x, y
```

Also, the vertices of an `igraph` graph can have arbitrary "attributes" that are assigned like dictionary indexing. So if I want to assign colors to the vertices, I can literally do that. For example, this is the base case of our induction: trivially color each vertex of a ≤ 5 vertex graph with all different colors.

```
colors = list(range(5))

def planar_five_color(graph):
    n = len(graph.vs)
    if n <= 5:
        graph.vs['color'] = colors[:n]
        return graph

    ...
```

The `igraph` library overloads the assignment operator to allow for entry-wise assignments by assigning one list to another. So in the statement `G.vs['color'] = colors[:n]`, the nodes of G are being assigned the first n colors in the list of colors.

The rest of the `planar_five_color` function involves finding the vertices of the right degree, forming the graph G' to recursively color, and keeping track of which vertices were modified to make G' so you can use its coloring to color G.

Here is the part where we find vertices of the right degree and do bookkeeping:

```
deg_at_most5_nodes = graph.vs.select(_degree_le=5)
deg_at_most4_nodes = deg_at_most5_nodes.select(_degree_le=4)
deg5_nodes = deg_at_most5_nodes.select(_degree_eq=5)

g_prime = graph.copy()
g_prime.vs['old_index'] = list(range(n))
```

The `select` functions are `igraph`-specific: they allow one to filter a vertex list by various built-in predicates, such as whether the degree of the vertex is equal to 5. The `old_index` attribute keeps track of which vertex in G' corresponded to which vertex in G, since when you modify the vertex set of an `igraph` the locations of the vertices within the data structure change (which changes the index in the list of all vertices).

Next we have the part where we construct G'. This is where the two cases in the proof show up.

```
if len(deg_at_most4_nodes) > 0:
    v = deg_at_most4_nodes[0]
    g_prime.delete_vertices(v.index)
else:
    v = deg5_nodes[0]
    neighbor_indices = [x['old_index'] for x in g_prime.vs[v.index].neighbors()]

    g_prime.delete_vertices(v.index)
    neighbors_in_g_prime = g_prime.vs.select(old_index_in=neighbor_indices)

    w1, w2 = find_two_nonadjacent(g_prime, neighbors_in_g_prime)
    merge_two(g_prime, w1, w2)
```

We implemented a function called `merge_two` that merges two vertices, but the implementation is technical and not interesting. The official `igraph` function we used is called `contract_vertices`. The remainder of the function executes the recursive call, and then copies the coloring back to G, computing the first unused color with which to color the originally deleted vertex v.

```
colored_g_prime = planar_five_color(g_prime)

for w in colored_g_prime.vs:
    # subset selection handles the merged w1, w2 with one assignment
    graph.vs[w['old_index']]['color'] = w['color']

neighbor_colors = set(w['color'] for w in v.neighbors())
v['color'] = [j for j in colors if j not in neighbor_colors][0]
return graph
```

The entire program is in the Github repository for this book.[11] The second case of the algorithm is not trivial to test. One needs to come up with a graph which is planar, and hence has *some* vertex of degree 5, but has no vertices of degree 4 or less. Indeed, there is

---

[11] See `pimbook.org`.

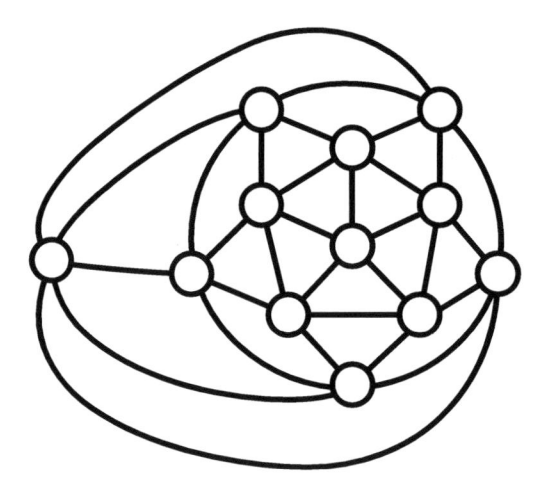

Figure 6.10: A planar graph which is 5-regular.

a planar graph in which every vertex has degree 5. Figure 6.10 shows one that I included as a unit test in the repository.

6.6 Approximate Coloring

Earlier I remarked that coloring is probably too hard for algorithms to solve in the worst case. To get around the problem we added the planarity constraint. Though a practical coloring algorithm would likely use an industry standard optimization problem solver to approximately color graphs, let's try something different to see the theory around graph coloring. Let's say that we're given a graph and promised it can be colored with 3 colors, and let's try to find a coloring that uses some larger number of colors.[12]

The first algorithm of this kind colors a 3-colorable graph with $4\sqrt{n}$ colors, where $n = |V|$. To make the numbers concrete, for a 3-colorable graph with 1000 vertices, this algorithm will use no more than 127 colors. Sounds pretty rotten, but the algorithm is quite simple. As long as there is an uncolored vertex v with degree at least $\sqrt{n}$, pick three new colors. Use one for v, and the other two to color $N(v)$. Then remove all these vertices from the graph and repeat. If there are no vertices of degree $\sqrt{n}$, then use the greedy algorithm to color the remaining graph.

Theorem 6.11. *This algorithm colors any 3-colorable graph using at most $4\sqrt{n}$ colors.*

Proof. Let G be a 3-colorable graph. For the first case, where there is a vertex v of degree $\geq \sqrt{n}$, we have to prove that the neighborhood $N(v)$ can be colored with two colors. But this follows from the assumption that G is 3-colorable: in any 3-coloring of G, v uses a color that none of its neighbors may use. Only two colors remain.

---

[12] Ideally we might hope to color a 3-colorable graph with 4 colors, but this was shown to be NP-hard as well. See http://dl.acm.org/citation.cfm?id=793420.

If there are no vertices of degree $\sqrt{n}$, then the maximum degree of a vertex is $\leq \sqrt{n}-1$, and we proved in Proposition 6.4 that the greedy algorithm will use no more than $\sqrt{n}$ colors on this graph.

Now we have to count how many colors get used total. The first case can only happen $\sqrt{n}$ times, because each time we color v and its neighbors, we remove those $\sqrt{n}+1 \geq \sqrt{n}$ vertices from G ($\sqrt{n} \cdot \sqrt{n} = n$). Since we add 3 new colors in each step, this part uses at most $3\sqrt{n}$ colors. The greedy algorithm uses at most $\sqrt{n}$ colors, so in total we get $4\sqrt{n}$, as desired.

$\square$

One might naturally ask whether we can improve $\sqrt{n}$ to something like $\log(n)$, or even some very large constant. This is actually an open question. Recent breakthroughs[13] have got the number of colors down to roughly $n^{0.2}$ colors. For reference, a thousand-node 3-colorable graph would have $n^{0.2} \approx 4$. That's quite an improvement over 127 colors given by the $4\sqrt{n}$ bound.

I should make a clarification here: the open problem is on the existence of an algorithm which is guaranteed to achieve some number of colors (depending on the size of the graph) *no matter what the graph is*. As a programmer you are probably somewhat familiar with this idea that one often measures an algorithm by its worst-case guarantees, but the point is important enough to emphasize. So when I say a problem is "possible" or "impossible" to solve, I mean that there exists (or does not exist, respectively) an efficient algorithm that achieves the desired worst-case guarantee on all inputs. In particular, there is no evidence for either claim that it is possible or impossible to color a 3-colorable graph with $\log(n)$ colors (or anything close to that order of magnitude, like $(\log(n))^{10}$). A ripe problem indeed.

6.7 Cultural Review

1. Invariants are measurements intrinsic to a concept, which don't depend on the choices made for some particular representation of that concept.

2. Sometimes if you want to come up with the right rigorous definition for an intuitive concept (like a planar graph), you need to develop a much more general framework for that concept. But in the mean time, you can still do mathematics with the informal notion.

3. Every conjecture about graphs must be tested on the Petersen graph.

---

[13] Using a technique called *semidefinite programming*.

6.8 Exercises

6.1 Write down examples for the following definitions. A graph is a *tree* if it contains no cycles. Two graphs G, H are *isomorphic* if they differ only by relabeling their vertices. That is, if $G = (V, E)$ and $H = (V', E')$, then G and H are isomorphic if there is a bijection $f : V \to V'$ with the property that $(i, j) \in E$ if and only if $(f(i), f(j)) \in E'$. Given a subset of vertices $S \subset V$ of a graph $G = (V, E)$, the *induced subgraph on S* is the subgraph consisting of all edges with both endpoints in S. Given a vertex v of degree 2, one can *contract* it by removing it and "connecting its two edges," i.e., the two edges $(v, w), (v, u)$ become (w, u). Likewise, one can contract an edge by merging its endpoint vertices, or *subdivide* an edge by adding a vertex of degree two in the middle of an edge. If H can be obtained as a subgraph of G after some sequence of contractions and subdivisions, it is called a *minor* of G.

6.2 Look up the statement of Wagner's theorem, which characterizes planar graphs in terms of contractions and the two graphs $K_{3,3}$ and K_5. Find a proof you can understand.

6.3 Here's a simple way to make examples of planar graphs: draw some non-overlapping circles of various sizes on a piece of paper, call the circles vertices, and put an edge between any two circles that touch each other. Clearly the result is going to be a planar graph, but an interesting question is whether *every* planar graph can be made with this method. Amazingly the answer is yes! This is called Koebe's theorem. It is a relatively difficult theorem to prove for the intended reader of this book, but as a consequence it implies Fáry's theorem. Fáry's theorem states that every planar graph can be drawn so that the edges are all straight lines. Look up a proof of Fáry's theorem that uses Koebe's theorem as a starting point, and rewrite it in your own words.

6.4 Given a graph G, the *chromatic polynomial* of G, denoted $P_G(x)$, is the unique polynomial which, when evaluated at an integer $k \geq 0$, computes the number of proper colorings of G with k colors. Compute the chromatic polynomial for a path on n vertices, a cycle on n vertices, and the complete graph on n vertices. Look up the chromatic polynomial for the Petersen graph.

6.5 Look up a recursive definition of the chromatic polynomial of a graph in terms of edge contractions, and write a program that computes the chromatic polynomial (for small graphs). Think about a heuristic that can be used to speed up the algorithm by cleverly choosing an edge to contract.

6.6 In the chapter I remarked that the Euler characteristic is a special quantity because it is an invariant. Look up a source that explains why the Euler characteristic is special.

6.7 Find a simple property that distinguishes 2-colorable graphs from graphs that are not 2-colorable. Write a program which, when given a graph as input, determines if it is 2-colorable and outputs a coloring if it is.

6.8 Implement the algorithm presented in the chapter to $(4\sqrt{n})$-color a 3-colorable graph. Use the 2-coloring algorithm from the previous problem as a subroutine.

6.9 A *directed graph* is a graph in which edges are oriented (i.e., they're ordered pairs instead of unordered pairs). The endpoints of an edge $e = (u, v)$ are distinguished as the *source* u and the *target* v. A directed graph gives rise to natural *directed paths*, which is like a normal path, but you can only follow edges from source to target. A graph is called *strongly connected* if every pair of vertices is connected by a directed path. Write a program that determines if a given directed graph is strongly connected.

6.10 A *directed acyclic graph* (DAG) is a directed graph which has no directed cycles (paths that start and end at the same vertex). DAGs are commonly used to represent dependencies in software systems. Often, one needs to *resolve* dependencies by evaluating them in order so that no vertex is evaluated before all of its dependencies have been evaluated. One often solves this problem by sorting the vertices using what's called a "topological" sort, which guarantees every vertex occurs before any downstream dependency. Write a program that produces a topological sort of a given DAG.

6.11 A *weighted* graph is a graph G for which each edge is assigned a number $w_e \in \mathbb{R}$. Weights on edges often represent *capacities*, such as the capacity of traffic flow in a road network. Look up a description of the maximum flow problem in directed, weighted graphs, and the Ford-Fulkerson algorithm which solves it. Specifically, observe how the maximum flow problem is modeled using a graph. Find real-world problems that are solved via a related max flow problem.

6.12 A *hypergraph* generalizes the size of an edge to contain more than two vertices. Hypergraphs are also called *set systems* or *families of sets*. Edges of a hypergraph are called *hyperedges*, and a $k-$*uniform* hypergraph is one in which all of its hyperedges have size k. Look up a proof of the Erdős-Ko-Rado theorem: let G be a k-uniform hypergraph with $n \geq 2k$ vertices, in which every pair of hyperedges shares a vertex in common. Then G has at most $\binom{n-1}{k-1}$ hyperedges in total. Find a construction that achieves this bound exactly when $n > 2k$.

6.9 Chapter Notes

Some Topology and the Rigorous Definition of an Embedding

The reason a planar graph is so hard to define rigorously is because the right definition of what it means to "draw" one thing inside another is deep and deserves to be defined in general. And such a definition requires some amount of topology, the subfield of mathematics that deals with the intrinsic shape of space without necessarily having the ability to measure distances or angles.

If you *really* pressed me to define a planar graph without appealing to topology I could do it with a tiny bit of calculus. Here it goes.

Definition 6.12. An *embedding* of a graph $G = (V, E)$ in the plane is a set of continuous functions $f_e : [0, 1] \to \mathbb{R}^2$ for each edge $e \in E$ mapping the unit interval to the plane with the following properties:

- Every f_e is injective.

- There are no two f_{e_1}, f_{e_2} and values $0 < t_1, t_2 < 1$ for which $f_{e_1}(t_1) = f_{e_2}(t_2)$, i.e., the images of f_{e_1} and f_{e_2} do not intersect except possibly at their endpoints.

- Whenever there are two edges (u, v) and (u, w), the corresponding functions must intersect at one endpoint, and these intersections must be consistent across all the vertices. I.e., every $u \in V$ corresponds to a point $x_u \in \mathbb{R}^2$ such that for every edge (u, v) incident to u, either $f_{(u,v)}(0) = x_u$ or $f_{(u,v)}(1) = x_u$.

Disgusting! Why did you make me do that?

The problem is that the definition is full of a bunch of "except" and special cases (like that the endpoint could either be zero or one). This makes for ugly mathematics, and the mathematical perspective is to spend a little bit more time understanding exactly what we want from this definition. We are humans, after all, who are inventing this mathematics so that we can explain our ideas easily to others and appreciate the beautiful proofs and algorithms. Keeping track of such edge cases is dreary.

We really want to define an embedding as a single function f whose codomain is $\mathbb{R}^2$. And because we said we don't want any of the edges to cross each other in the plane, we probably want f to be injective. Finally, because the drawing has to be a sensible *drawing*, we need f to be continuous. Recall from calculus that a continuous function intuitively maps points that are "close together" in the domain to points that remain close together in the codomain. Without continuity, a "drawing" could break edges into disjoint pieces and there would be nothing but madness!

The real question is: what is the *domain* of this function? It can't be G as a set because we don't have a notion of "closeness" for pairs of vertices, and we really want to think of an edge as a line-like thing.

The trick is to start imagining *abstract spaces* that are not sitting in any ambient geometric space. This is where the formalisms of *topology* really come in handy, but unfortunately a satisfying overview of the basic definitions of topology is beyond the scope of this book. It suffices for our purposes to understand two concepts:

One can take the *disjoint union* of two abstract spaces and get another abstract space in which the points comprising the two pieces are different. In other words, we can take lots of different copies of the same space (in our case $[0, 1]$), their disjoint union is like a bunch of lines, but we aren't presuming any way to compare the different pieces.

The second idea is that one can *identify* two points in an abstract space. Intuitively, one can "glue together" two points and maintain the rest of the space unhindered. For us, if a copy of $[0, 1]$ represents an edge, then we'll want two edges incident to the same vertex to have one of their two endpoints identified.[14]

So putting these two ideas together, the abstract space X_G corresponding to a graph G is the disjoint union of copies of $[0, 1]$ for each edge, with endpoints identified when two edges intersect at a vertex. Then we can define a function $f : X_G \to \mathbb{R}^2$, enforce it to be injective (it's just a function between two sets), and call it continuous if points that are close in X_G, using the natural distance for points in the interval $[0, 1]$, get sent to points that are close in $f(X_G)$. How do I measure distance between two points $a, b \in X_G$ that might be on *different* edges? Well a, b are either vertices or on some copy of $[0, 1]$, so I can find a path in the graph G, that gets from one edge to another (if not, then the distance can be called infinite). Then I could measure the length of each full edge on this path, and add up the partial edges required to get from a or b to the desired endpoint of the edge they're in.

This is a very fancy way to say that I can impose the same geometry that was on $[0, 1]$ onto the different pieces of X_G and patch them together. But once you get comfortable with that idea, you have a natural way to define an embedding of any abstract space into any other abstract space: a continuous injective function!

For more mathematics like this, I suggest you pick up a book on topology. Unfortunately I haven't yet found one that I like particularly better than any other. Most books tend to be terse and contain few pictures (which is the opposite of how topology is done!). Topology also aims to generalize much of calculus, so waiting until after Chapter 14 might be prudent.

---

[14] This foreshadows a topic in a later chapter called the *equivalence relation*, which formalizes how to identify points in a consistent way.

Chapter 7
The Many Subcultures of Mathematics

Some people may sit back and say, "I want to solve this problem" and they sit down and say, "How do I solve this problem?" I don't. I just move around in the mathematical waters, thinking about things, being curious, interested, talking to people, stirring up ideas; things emerge and I follow them up. Or I see something which connects up with something else I know about, and I try to put them together and things develop. I have practically never started off with any idea of what I'm going to be doing or where it's going to go. I'm interested in mathematics; I talk, I learn, I discuss and then interesting questions simply emerge. I have never started off with a particular goal, except the goal of understanding mathematics.

– Sir Michael Atiyah

A mathematician is a machine for turning coffee into theorems.

– Alfréd Rényi

There is a fascinating bit of folk lore, which as far as I know originated with a 2010 blog post of Ben Tilly, that you can tell what type of mathematician you are by how you eat corn on the cob. It turns out there are multiple ways to eat corn, and they are roughly grouped as "eat in rows like a typewriter, left to right," and "eat in a spiral, teeth scraping the corn into your mouth."

The corresponding two types of mathematicians are roughly grouped as *algebraists* and *analysts*. An algebraist, as we'll see in Chapters 10, 12, and 16, supposedly prefers orderliness and working with the inherent structure of the corn cob. Analysis, the topic of Chapters 8, 14, and 15, alternatively prioritizes efficiency, approximation, and getting the job done. One's underlying preference apparently explains both the choice of a mathematical domain of study, and the less conscious choice of how to eat corn.

According to Tilly, who surveyed 40-ish mathematicians and received countless more self-selected responses via the internet, corn eating predicts mathematical preference with surprising accuracy. Since his post, this observation has become a bit of folk lore that reinforces the idea that mathematics has many subcultures organized around preference and character.

One of the more prominent distinctions is the concept described by mathematician Tim Gowers and others, between mathematicians who prioritize *problem solving* versus those who prioritize *theory building*. As the quotes at the beginning of the chapter emphasize, these are very different styles of doing mathematics. Gowers defines them via example in a 2000 essay:

If you are unsure to which class you belong, then consider the following two statements.

1. *The point of solving problems is to understand mathematics better.*
2. *The point of understanding mathematics is to become better able to solve problems.*

Most mathematicians would say that there is truth in both (1) and (2). Not all problems are equally interesting, and one way of distinguishing the more interesting ones is to demonstrate that they improve our understanding of mathematics as a whole. Equally, if somebody spends many years struggling to understand a difficult area of mathematics, but does not actually do anything with this understanding, then why should anybody else care?

The Hungarian mathematician Paul Erdős was a pillar of the problem solving camp. Though this short essay could not possibly do justice to his outlandish life story, I will try to summarize. Erdős is the most prolific mathematician in history, by count of papers published (over 1500). He was able to do this because he renounced every aspect of life beyond mathematics. He had no home, and lived out of a suitcase while traveling from university to university. At each stop, he would show up, knock on the department chair's office door, and be provided housing and food by an attendant professor. In the subsequent weeks, Erdős and his host would work on problems and usually publish a paper or two, until such time as Erdős decided to move on to his next host. As Erdős said, "Another roof, another proof." He never married and had no children.

Erdős would often do bizarre things like wake up his host in the middle of the night, exclaiming, "My mind is open," meaning he was ready to do mathematics. He was a serious user of methamphetamines, and since he had no possessions or money, it fell to his hosts to procure his drugs. Despite being an atheist, he called God the "Supreme Fascist." He also claimed God kept a Book of the most beautiful proofs of every theorem. He didn't believe in God, but he believed in the Book.

Erdős's hosts tolerated his idiosyncratic behavior because his presence was a boon to one's career. Mathematicians jumped at the chance to work with Erdős, and in turn they started to track their so-called Erdős number. In the graph whose vertices are people and whose edges are coauthorship, your Erdős number tracks the length of the shortest path from you to Erdős.[1]

---

[1] You didn't ask, but my Erdős number is three, by way of György Turán → Endre Szemerédi (and others) → Erdős.

His work focused on problems in combinatorics, number theory, graph theory, and incidence geometry (statements about configurations of points and lines), the sort of counting arguments that we saw in Chapters 4 and 6—though much more sophisticated and interesting. As he spread his ideas from university to university, he both gave combinatorics credibility as a field of study, and also established its reputation as a field that prioritizes problem solving over theory building. To Erdős, mathematics was "conjecture and proof."

Indeed, as Tim Gowers writes, graph theory tends not to benefit from extensive theory-building.

> *At the other end of the spectrum is, for example, graph theory, where the basic object, a graph, can be immediately comprehended. One will not get anywhere in graph theory by sitting in an armchair and trying to understand graphs better. Neither is it particularly necessary to read much of the literature before tackling a problem: it is of course helpful to be aware of some of the most important techniques, but the interesting problems tend to be open precisely because the established techniques cannot easily be applied.*

Michael Atiyah is Gowers's example of a theory builder. Theory builders focus on the conceptual unity of mathematics, and on connecting disparate subjects and identifying their commonalities. Atiyah even argues against my claims in this book, that proof is not necessarily central to mathematics. From Atiyah's essay, "Advice to a Young Mathematician."

> *It is a mistake to identify research in mathematics with the process of producing proofs. In fact, one could say that all the really creative aspects of mathematical research precede the proof stage. To take the metaphor of the "stage" further, you have to start with the idea, develop the plot, write the dialogue, and provide the theatrical instructions. The actual production can be viewed as the "proof": the implementation of an idea.*
>
> *In mathematics, ideas and concepts come first, then come questions and problems. At this stage the search for solutions begins, one looks for a method or strategy. Once you have convinced yourself that the problem has been well-posed, and that you have the right tools for the job, you then begin to think hard about the technicalities of the proof.*
>
> *Before long you may realize, perhaps by finding counterexamples, that the problem was incorrectly formulated. Sometimes there is a gap between the initial intuitive idea and its formalization. You left out some hidden assumption, you overlooked some technical detail, you tried to be too general. You then have to go back and refine your formalization of the problem. It would be an unfair exaggeration to say that mathematicians rig their questions so that they can answer them, but there is undoubtedly a grain of truth in the statement. The art in good mathematics, and mathematics is an art, is to identify and tackle problems that are both interesting and solvable.*
>
> *Proof is the end product of a long interaction between creative imagination and critical reasoning.*

I interpret this in more of a metaphysical sense than a literal sense; one needs to know what questions are worth asking before one can provide a proof answering them. For whatever reason, Atiyah doesn't consider the validations or refutations of these initial ideas as "proofs" in the formal sense.

One person who might be said to be the antithesis to Paul Erdős is the French mathematician Alexander Grothendieck. He also lived a curiously eccentric lifestyle involving radical anti-military politics and an eventual self-exile to a small village in Southern France. Grothendieck declined various prizes for his life's work, and decried the mathematical establishment as being obsessed by status to the point of intellectual bankruptcy. Toward the end of his life he also turned to mysticism and spiritualism, almost starving himself to death via unusual diets and fasting.

Grothendieck's work was a complete rebuilding of the foundations of the subfield of *algebraic geometry* in terms of category theory. These developments concurrently reshaped the foundations of adjacent and burgeoning fields of cohomology theory, algebraic topology, and representation theory. His work also led to the resolution of a number of high-profile conjectures, and important generalizations of famous theorems.

In particular, his theory elucidated the role of category theory in connecting disparate fields of mathematics together via *universality*. In brief, universality is a uniqueness property of a particular pattern or structure that occurs within a subfield of mathematics. For example, the product of two sets has a universal property, and it is the same property as the product of vector spaces (Chapter 10) as well as groups (Chapter 16). Noticing these similarities allows one to formalize a "product" in a domain-independent way, and then prove theorems about it that apply to all relevant domains at once! Grothendieck's attitude takes theory-building to the extreme.

Mathematicians David Mumford and John Tate wrote about Grothendieck,

Although mathematics became more and more abstract and general throughout the 20th century, it was Alexander Grothendieck who was the greatest master of this trend. His unique skill was to eliminate all unnecessary hypotheses and burrow into an area so deeply that its inner patterns on the most abstract level revealed themselves—and then, like a magician, show how the solution of old problems fell out in straightforward ways now that their real nature had been revealed.

Grothendieck's ideas were to find out what theorems are important, and then rewrite the basic definitions of mathematics until those theorems become completely trivial. In his mind, a theory is powerful only insofar as what it makes obvious. A radical conviction indeed!

Subcultures and styles go beyond theory-building/problem-solving and algebra/analysis, deep into subfields of mathematics. Even those working entirely within geometry having specific styles. Henri Poincaré remarks in his essay, "Intuition and Logic in Mathematics,"

Among the German geometers of this century, two names above all are illustrious, those of the two scientists who have founded the general theory of functions, Weierstrass and

Riemann. Weierstrass leads everything back to the consideration of series and their analytic transformations; to express it better, he reduces analysis to a sort of prolongation of arithmetic; you may turn through all his books without finding a figure. Riemann, on the contrary, at once calls geometry to his aid; each of his conceptions is an image that no one can forget, once he has caught its meaning.

We'll see the two sides of this analytic/geometric coin in the forthcoming chapters: the view that geometric ideas should be studied using series is how we will approach Calculus in Chapter 8 (and to a lesser extent Chapter 14), while the geometric view is the heart of the study of hyperbolic geometry in Chapter 16. These could have easily been swapped, with geometric ideas founding calculus and analytic ideas underlying hyperbolic geometry.

As with most "classifications" of things, the problem-solving and theory-building groups, along with the algebra/analysis divide, are neither wholly distinct nor discrete. Styles fall along a spectrum, depending on the occasion and whether one has had breakfast. Whether Poincaré, Mumford, Atiyah, or Tilly, the mathematical universe is as varied in attitudes and preferences as any other community, and mathematics reaps the benefits of diversity.

For the record, I eat corn like a typewriter, and I do prefer algebra. Although, much of my mathematical research involved analysis-style arguments, and I have come to appreciate the beauty of a good bound. Maybe next time I'm in a rush I'll try scraping that corn.

Chapter 8
Calculus with One Variable

The derivative can be thought of as infinitesimal, symbolic, logical, geometric, a rate, an approximation, microscopic.

This is a list of different ways of thinking about or conceiving of the derivative, rather than a list of different logical definitions. Unless great efforts are made to maintain the tone and flavor of the original human insights, the differences start to evaporate as soon as the mental concepts are translated into precise, formal and explicit definitions.

I can remember absorbing each of these concepts as something new and interesting, and spending a good deal of mental time and effort digesting and practicing with each, reconciling it with the others. I also remember coming back to revisit these different concepts later with added meaning and understanding.

– William Thurston

Calculus is a difficult subject to introduce. It has a hundred different motivating angles, a thousand books you could read, and millions of applications. You can start with basic physics, where position is a function, and derivatives are velocity and acceleration, and work your way to Newtonian mechanics. You could aim for systems of differential equations and numerical simulations, tread the probability path and dabble in measure theory, or take a purely mathematical approach. Your ultimate goal might be machine learning, weather modeling, the frontiers of theoretical physics, economics, or operations research and optimization. These all rely on the fundamental idea of calculus: that progressively better approximations ultimately produce the truth.

Luckily, as a programmer you're familiar with the existence of these fantastic applications. You may have seen and played with programmed physics models before, or programmed a sprite jumping on a screen. You're probably aware at least in a vague sense that many widely-used algorithms involve calculus. This makes the job of learning calculus much easier, because I don't have to convince you it's worth learning.

Much of the mastery of calculus (and any subject!) comes with practice. Even so, in this chapter and the next we can survey most of the important features of a more complete calculus course *and* do a bit of machine learning at the end. This chapter will be about calculus for functions with one input, while Chapter 14 will cover functions with many inputs.

If you've seen a lot of calculus before, you can probably tell that I don't regard it as reverently as most other authors. While I can appreciate its place in history and its applications to physics and everything else, my esteem for calculus is essentially limited to "It's a great tool for computation." I avoid nonsense rhetoric about calculus like a plague ("With calculus you can hold infinity in the palm of your hand!"). I'd much rather use it to do something useful and draw divine inspiration from other areas of math. But that's a personal preference.

Besides calculus, in this chapter we'll dive into more detail about the process of *designing* a good mathematical definition. In doing this we'll introduce the idea of a *quantifier*, which is the basis for compound (recursive) conditions and claims. We'll also come to understand the idea of *well-definition* in mathematics, which is how a mathematician proves (or asserts) that the definition of a concept doesn't depend on certain irrelevant details in its construction. Finally, we'll level up our proof skills by using multiple definitions in conjunction to prove theorems. The application for this chapter is an analysis of the classic Newton's method for finding roots of functions.

8.1   Lines and Curves

Let's start with something we know well. If you give me a line in the plane, with tick marks forming integer coordinates like in Figure 8.1, then I can tell you how "steep" the line is. That is, I can assign a number to the line, and larger numbers correspond to steeper lines while smaller numbers correspond to more gradual lines. Also recall that the picture with coordinate axes is just one representation of the line, while another might be as a set of points $\{(x, y) \in \mathbb{R}^2 : 2x + 3y = 4\}$. How we choose to draw the line isn't as important as the set-with-equation definition, but a good drawing swiftly reveals qualitative facts about the line (such as whether it's "steepness" goes up or down).

Assigning a steepness number is easy, something most students do when they're 11 or 12 years old. Just pick two different points on the line, *any two*, call them $(x_1, y_1), (x_2, y_2)$, and then call the *slope* of the line

$$\text{slope}(L) = \frac{y_2 - y_1}{x_2 - x_1}.$$

The difference in the y's correspond to a vertical change, while the difference in x's correspond to a horizontal change. The slope is an invariant of the line because any choice of two points because any for any two choices of points you can draw a right triangle (Figure 8.2), and all of the triangles drawn this way are similar (i.e., they have the same angles at all vertices). In Figure 8.2, the slope between A and B is the same as between C and D because if I move point B to D the ratios stay the same (similar triangles), likewise for A to C.

Lines and other simple functions often represent the 1-dimensional position of an object over time, while the steepness—the ratio of the change in position to the change in time—is the velocity of that object.

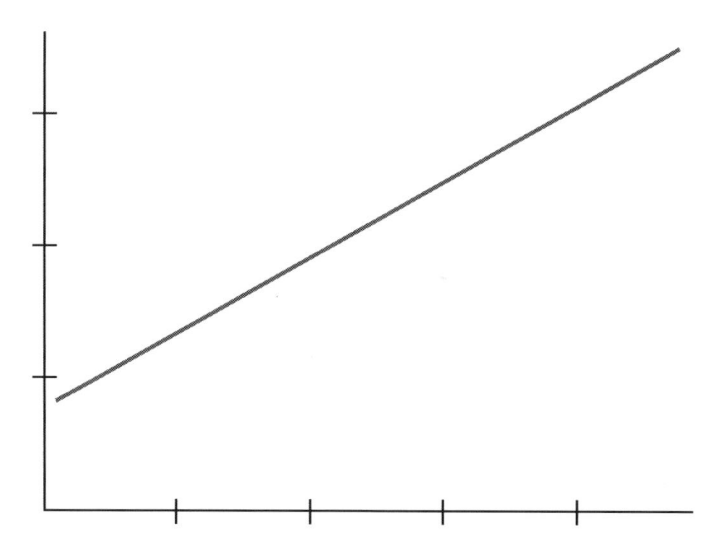

Figure 8.1: A line in the plane.

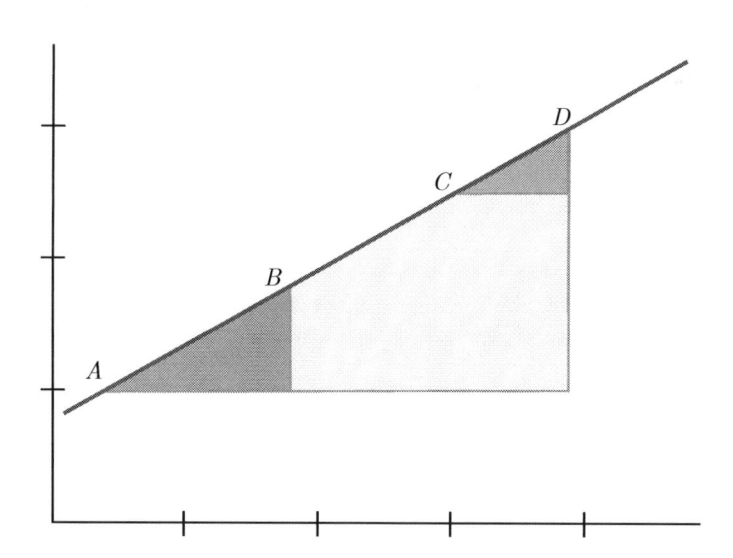

Figure 8.2: Slope is consistent no matter where you measure, because the triangles are all similar.

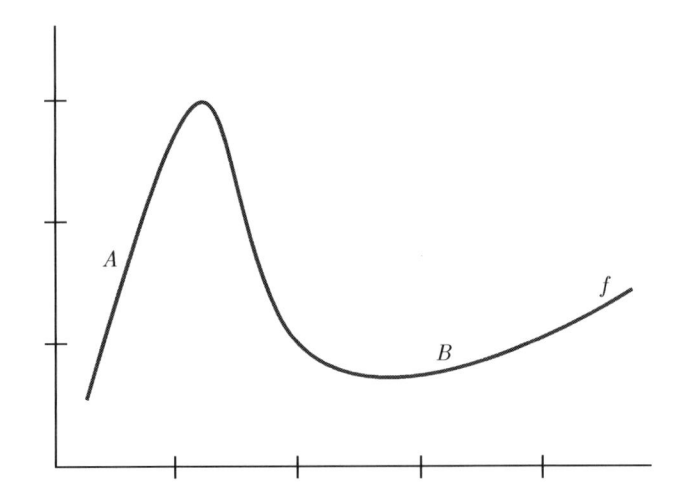

Figure 8.3: For a general curve, steepness depends on where you measure.

Before graduating from lines, let me point out that not all lines are functions from the x coordinate to the y coordinate.[1] If you pick a line which is a function $f : \mathbb{R} \to \mathbb{R}$, then the formula for the slope can be written as

$$\text{slope}(f) = \frac{f(x_2) - f(x_1)}{x_2 - x_1}.$$

This makes it clear that the slope imposes an orientation on the line, that the x coordinate is "horizontal" while the y coordinate is "vertical." This is an arbitrary choice of perspective, albeit the standard one.

Now say we have a function $f(x)$ that isn't a line. It's curved, and it has some complicated formula we won't write down. The curve in Figure 8.3 is steeper at some places (e.g., A) and less steep at others (B). Despite the self-evident *fact* that the line is steep at A and gradual at B, if we were pressed to say precisely and consistently *how* the two steepnesses compare, we'd be at a loss. This is because the picture only tells us qualitative information, and we have to leave the picture behind to get useful quantitative data.

To motivate an exact answer, let's approximate steepness using tools we know. Focus on the point labeled A, and call it $A = (x, f(x))$. After a moment of thought, the idea naturally occurs to draw a line between $(x, f(x))$, and a nearby point $(x', f(x'))$, and have our approximation be the slope of that line, as in Figure 8.4.

$$\text{steepness at } A \approx \frac{f(x') - f(x)}{x' - x}$$

As a reminder, we adorn a variable with the tick $'$ (called a "prime") to denote a slight

---

[1] such as $\{(x, y) : x = 1\}$

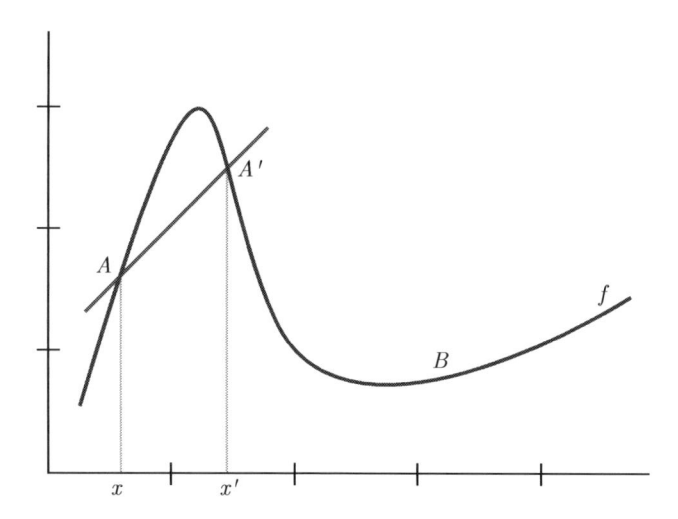

Figure 8.4: We can use the slope of a line as a proxy for the corresponding "steepness" measurement on a curve.

difference. So x, and x' play similar roles, but x' is slightly different from x in some way.[2] We also use the $\approx$ symbol as a stand-in for the phrase "is approximately." I also went back to using the word "steepness" instead of slope because we're using the slope of a line to reason about this new kind of steepness.

My choice of x' isn't *that* close to x, but I chose it to illustrate a point. The approximation isn't perfect, but it's still good enough to concretely distinguish it from my approximation of a similarly bad approximation of the steepness of f at B, as shown by Figure 8.5. Concrete numbers for the slopes of these two lines suggest that f is twice as steep at A as at B. Our brains still nag us to be more precise. Otherwise, how could we be certain we aren't fooling ourselves with inadequate picture-drawing skills? To that effort, let's try to improve our estimate.

Once blessed with the idea of approximating the steepness of f at A by drawing a line from x to some other x', we neurotically yearn to move x' closer to x. We could move x' halfway closer to x, call this new point x_1, and update our slope approximation, as in Figure 8.6.

$$\text{steepness at } A \approx \frac{f(x_1) - f(x)}{x_1 - x}$$

Our yearnings are destined for iteration. Do it again, and again, getting $\frac{f(x_2)-f(x)}{x_2-x}$ and $\frac{f(x_3)-f(x)}{x_3-x}$, and so on. With each step the line approximation gets better and better, closer and closer to our brain's intuitive picture of the steepness at A.

---

[2] It's a shame that the tick symbol is also used in calculus to denote the derivative of a function, but this will be a good opportunity to practice disambiguating notation using context. We'll get to that shortly.

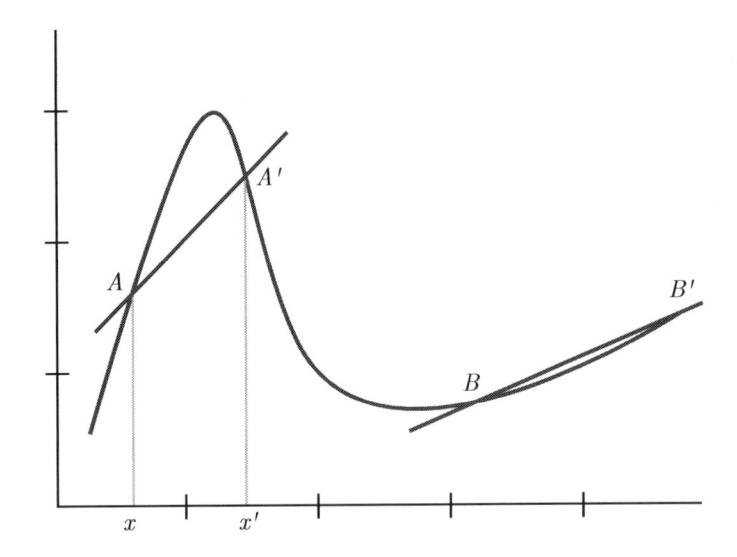

Figure 8.5: Two different lines show how the approximation can be better or worse, depending on where it is.

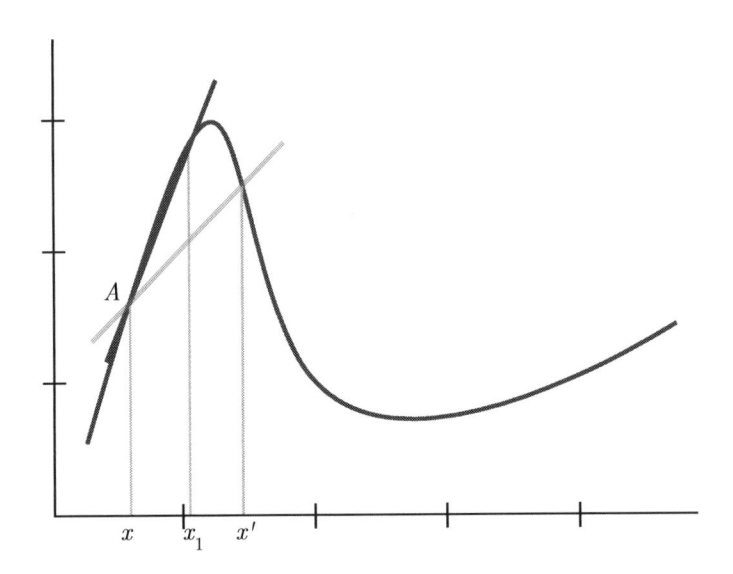

Figure 8.6: Moving x' halfway closer to x improves the approximation.

How do we reason about the "end" of this process? We get a number at every step. If we were to run this loop forever, would these approximate numbers would approach some concrete number? If so, we could reasonably call that number the "true" steepness of f at A.

That is exactly what limits do. Limits are a computational machinery that allows one to say "this sequence of increasingly good approximations would, if followed forever, end up at a specific value." The limit of this particular line-approximation-scheme is called the *derivative*. We'll return to derivatives in a bit. Note in particular that whether this "limiting process" works shouldn't depend on how we move x' closer to x. A good definition should work so long as x' approaches x *somehow*.

8.2 Limits

In the last section we saw a strong motivation for inventing limits, and an intuitive understanding for what a limit *should* look like. It's the "end result" of iteratively improving an approximation forever. You have some quantity a_n indexed by a positive integer n, and as n grows a_n eventually gets closer and closer to some target. For example, if $a_n = 1 - 1/n$, the numbers in the sequence $0, \frac{1}{2}, \frac{2}{3}, \frac{3}{4}, \frac{4}{5}, \ldots$ seem to approach 1.

But we need a *definition*. A definition is like the implementation of a program spec. From a specification standpoint, you care mostly about how one intends to use an interface. When actually writing the program you have to worry about people misusing your code, intentionally or not. You have to anticipate and defend against the edge case inputs which are syntactically allowed but semantically unnatural. Anyone who has spent time designing a software library has spent hours upon hours thinking about:

- How to organize code to handle all inputs generically and elegantly.

- How to reduce cognitive load by maintaining conceptual consistency.

- How to avoid writing a mess of extra code just to handle edge cases.

And ideally a library author wants to meet all of these criteria at once! We have the same problem in mathematics.

Most concepts in math—in this case limits—usually make intuitive sense in the overwhelming majority of cases you encounter in real life. However, 99% of the work in making the math rigorous is converting the concept into concrete definitions that can handle pathological counterexamples. By *pathological*, I mean examples that are mathematically valid, but which nobody would ever encounter in the wild.[3] The best pathological examples are edge cases on steroids, and some mathematicians gain fame for constructing

---
[3] This is relative, of course. Once upon a time complex numbers like $1 + i$ were thought to be pathological, but now they're standard.

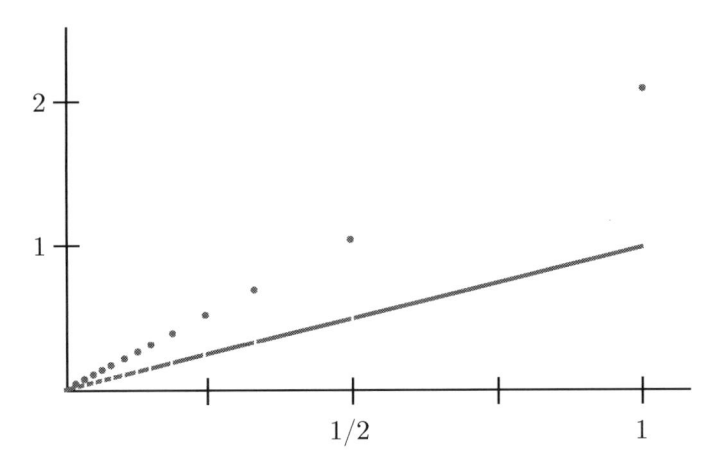

Figure 8.7: This pathological function admits two different possibilities for the derivative depending on the sequence of approach.

particularly vexing pathological examples. They're the penetration testers of mathematics.

Indeed, much like a program, once a mathematical definition is written down it must be judged on its own merits. It must behave properly under any "input" (being applied to any mathematical object). Best practices also suggest definitions reduce cognitive load and avoid too many special cases. Achieving the right balance is a serious challenge.

An unfortunate consequence of all this is that math books start with the final definition—the end result of this arduous design process—followed by many pages of theorems and proofs explaining why it doesn't succumb to edge cases. Calculus is no different, and in fact most of how Isaac Newton and Gottfried Leibniz originally did calculus was in this informal, intuitive setting, without much rigor at all. It was a less famous mathematician, Karl Weierstrass, who is considered to have finally "set calculus straight" (though it was really a team effort over decades). Modern calculus textbooks are a strange mix. They want to capture the informality of Leibniz, feel obliged to Weierstrass's rigor, but can't commit fully due to a lack of proof-reading skills. Alas, it's hard to imagine a better way. Only mathematicians enjoy the elaborate tour of blunders and false starts that historically sculpted a modern definition. One could hardly cajole the average student to care, or even the brightest student who ultimately wants to apply mathematics to the problems of their choosing.

To my delight, you're still reading. My goal for the rest of the chapter is to whet your appetite for definition crafting. Let's continue with the "steepness of a function" as our prototypical example of a limit. Here's one of those pathological examples that makes limits hard. I'm going to define a non-curve and not-even-connected function $f : \mathbb{R} \to \mathbb{R}$ as follows: if x is $1/k$ for some integer k, then $f(x) = 2x$, otherwise $f(x) = x$. Figure 8.7 sketches f.

Now we can ask: what's the steepness of f at $x = 0$? We pick some starting x_1, compute the slope, pick an x_2, compute the slope, and keep going until we see convergence.

But I dastardly chose f in such a way that the limit changes depending on how you pick the sequence $x_1, x_2, \ldots$. In fact, if you pick $x_k = 1/k$, every slope in the sequence is 2, implying the limit is 2. There isn't even an approximation because the values in the sequence are constant. But if you choose $x_k = \frac{1}{k+0.5}$, the slopes are always 1. So should the limit be 1 or 2? Neither?

This will be the last pathological example I inflict upon you,[4] but it emphasizes an important point. However we choose to define limits, it can't depend on the arbitrary choice of which points you choose in the sequence. It should be a definition like "no matter how your values approach the limit, the limit is the same." The generic mathematical term for this is that the limit should be *well-defined*.

With that thought, let's start with the limit of a sequence of numbers, which will be used to define limits for functions. Since sequences of numbers can have repetition, we won't use set notation (though some authors do). Instead we'll use a comma notation $x_1, x_2, \ldots$ which the strongly-typed programmer can think of as the output of an iterator which never terminates, or a tuple/array of infinite length $(x_1, x_2, \ldots)$. The ε character is a lower-case Greek epsilon, contextually used across mathematics as an arbitrarily small positive real number.

Definition 8.1. Let $x_1, x_2, \ldots$ be a sequence of real numbers, one x_n for each $n \in \mathbb{N}$, and let $L \in \mathbb{R}$ be fixed. We say that x_n *converges to* L if for every threshold $\varepsilon > 0$, there is a corresponding $k \in \mathbb{N}$ so that all the x_n after x_k are within distance ε of L. We also equivalently say the *limit* of x_n is L.

This is the first time we've encountered a definition that relies heavily on alternating quantifiers (for every..., there is...), so let's discuss it in detail. A statement like "for every FOO there is a BAR," means there's a functional relationship. If you give me a FOO as input, I can produce a BAR with the desired property as output.[5] Interpreting this for Definition 8.1, the input is a real number threshold $\varepsilon > 0$, and the output is an integer k with a special property. So the relationship is:

```
int sequence_index_from_threshold(float epsilon) {
    // compute k depending on epsilon
    return k;
}
```

The special property of k is that all the sequence elements after k are close to L, in fact as close as the input ε specified.

As a simple non-pathological example, let's take the sequence $x_n = 1 - \frac{1}{n}$. This is the sequence $0, \frac{1}{2}, \frac{2}{3}, \frac{3}{4}, \frac{4}{5}, \ldots$. Our intuition tells us that the limit should be $L = 1$, so let's prove it strictly by the letter of the definition.

---

[4] If you want more, check out the book "Counterexamples in Calculus."
[5] It isn't strictly true in math that there's always a functional relationship that you can compute. Sometimes you can prove a thing exists without knowing how to compute it. But in most important cases you can compute, and it makes the explanation here simpler.

First let's see a concrete example of the threshold-to-sequence-index functional relationship. If you require $\varepsilon = 1/4$, I need to find an index after which all x_n are within $1/4$ of 1. I.e., all these x_n's should satisfy $1 - 1/4 < x_n < 1 + 1/4$. Another way to write this is with the absolute value: $|x_n - 1| < 1/4$. Since we already see that $3/4$, also known as $1 - 1/4$, is one of the sequence elements, it should be easy to guess that everything starting at $k = 5$ will be close enough to 1. Indeed, we can do the algebra that if $n > 4$,

$$|x_n - 1| = \left|\left(1 - \frac{1}{n}\right) - 1\right| = \left|-\frac{1}{n}\right| = \frac{1}{n},$$

and $1/n < 1/4$ when $n > 4$.

Now let $\varepsilon \geq 0$ be unknown, but fixed. We can do the same algebra as above. How large of an index k do we need to ensure $|x_n - 1| < \varepsilon$ for all $n > k$? In other words, can I write ε in terms of n so that all of the above equations and inequalities are still true when I replace $1/4$ with ε?

Above we showed that $|x_n - 1| \leq 1/n$, so to ensure that $1/n < \varepsilon$ we can rearrange to get $n > 1/\varepsilon$. Picking any index k bigger than that will work.[6] Since ε is fixed, just pick k to be the integer that immediately follows $\frac{1}{\varepsilon}$ (the "ceiling" of $1/\varepsilon$). This formally proves that 1 is the limit of the sequence $x_n = 1 - \frac{1}{n}$.

Let me restate all of this as a theorem with a proof as you might see in a book.

Theorem 8.2. *The limit of the sequence $x_n = 1 - \frac{1}{n}$ is 1.*

Proof. Let $\varepsilon > 0$ be fixed. Pick any integer $k > 1/\varepsilon$. We will show that $|x_n - 1| < \varepsilon$ for all $n \geq k$. Indeed,

$$|x_n - 1| = \left|\left(1 - \frac{1}{n}\right) - 1\right| = \left|-\frac{1}{n}\right| = \frac{1}{n},$$

and because $n \geq k > 1/\varepsilon$, we have $1/n \leq 1/k < \varepsilon$. □

You can think of this ε-to-k process as a game. A skeptical contender doesn't believe x_n converges to L, and challenges you to find the tail of the sequence that stays within $\varepsilon = 1/2$ of L. You provide such a k, but the contender isn't happy and re-ups the challenge using $\varepsilon = 1/100$. You comply with a bigger k. The contender retorts with $\varepsilon = (1/2)^{99}$. Unfazed, you still produce a working k.

If there's any way for the contender to stump you in this game, then x_n doesn't converge to L. But if you can always produce a good k no matter what, the sequence converges to L.

As a notational side note, the phrase "for every x there is a y" can be long and annoying to write all the time. It also makes it difficult to look at the *syntactic* structure of statements like this, since language tends to vary across the world and it can be unclear what depends on what. This is exacerbated by slightly ambiguous words like "each"

---

[6] The fraction $1/\varepsilon$ shouldn't be scary because, looking again at Definition 8.1, we require $\varepsilon > 0$, so we'll never divide by zero.

and "unless." Mathematicians designed an unambiguous notation for this situation called *quantifiers*. We briefly introduced quantifiers in Chapter 4, and promised we wouldn't use them in this book. However, standard textbook definitions in analysis often use the symbols heavily, so this digression helps put what you might see elsewhere in context.

The first quantifier is the symbol $\forall$, which means "for all" (the upside-down A stands for All). The second is $\exists$, which stands for "there exists" (the backwards E in "Exists"). Quantifiers may appear in any order. If I claim

$$\exists x \in \mathbb{R}, \forall y \in \mathbb{R}, x + y = 3,$$

I'm saying I can come up with a real number x, such that no matter which y you produce, it's true that $x + y = 3$. Obviously no such x exists, so the statement is false. Note the statement changes if the order of the quantifiers is reversed: for every y, there is indeed an x for which $x + y = 3$, it's $x = 3 - y$.

If I were to state the definition of the limit in its briefest form, I might say:

$$x_n \text{ converges to } L \text{ if: } \forall \varepsilon > 0, \exists k > 0, \forall n > k, |x_n - L| < \varepsilon.$$

We've just packed the math like sardines in a tin box. That being said (and now we're really digressing), some situations benefit from writing logical statements in this form. Particularly in the realm of formal logic, it turns out that as you add more "alternating" quantifiers ($\forall x \exists y \forall z$), you get progressively more expressive power. In theoretical computer science this is formalized by the so-called *polynomial hierarchy*, which conjecturally asserts that the computational cost of deciding the truth of generic logical statements increases dramatically with the number of alternating quantifiers. That's why one might believe factoring integers ($\exists a, \exists b, ab = n$) is easier than deciding if one can force a win in a two player game like chess (there exists a move for me, such that for every move for my opponent, there exists a move for me, such that..., such that I have a winning move).

Back to limits. The definition of a limit allows a sequence to have no limit, like the sequence $0, 1, 0, 1, 0, \ldots$, which isn't pathological at all. For this sequence you can't even satisfy the limit definition with $\varepsilon = 1/3$ (no matter what you think the limit L might be!). This fits with our intuition that an alternating $(0, 1, 0, 1, \ldots)$ sequence doesn't "get closer and closer" to anything. So now we can add to our definition.

Definition 8.3. Let x_n be a sequence of real numbers. If there is an L satisfying the definition of the limit for x_n, we say that x_n *converges*. Otherwise, we say it *does not converge*.

Sometimes we abbreviate the claim that x_n converges to L by the notation $\lim_{n \to \infty} x_n = L$, and sometimes even more compactly as $x_n \to L$. In this setting, the symbol ∞ doesn't have any concrete mathematical meaning by itself, it's just notation to remind us that we're talking about n's that get arbitrarily large.

Now we're ready to define the limit of a function.

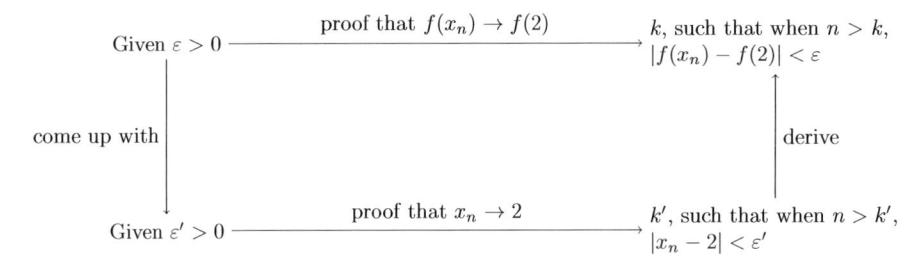

```
                    proof that f(xₙ) → f(2)        k, such that when n > k,
Given ε > 0  ─────────────────────────────────▶   |f(xₙ) − f(2)| < ε

   │                                                    ▲
come up with                                          derive
   │                                                    │
   ▼            proof that xₙ → 2                  k', such that when n > k',
Given ε' > 0 ─────────────────────────────────▶   |xₙ − 2| < ε'
```

Figure 8.8: Starting in the top left corner, we want to deduce the top right corner. We do this by taking the longer route down and around.

Definition 8.4. Let $f : \mathbb{R} \to \mathbb{R}$ be a function. Let c and L be real numbers. We say that $\lim_{x \to c} f(x) = L$ if for every sequence x_n that converges to c (and for which $x_n \neq c$ for all n), the sequence $f(x_n)$ converges to L.

The notation $f(x_n)$ is shorthand for a sequence $y_n = f(x_n)$. In this context we're implicitly "mapping" f across the sequence x_n as one would say in functional programming, or alternatively we're "vectorizing" f. The notation $x \to c$ is used to signify that x_n is a sequence converging to c, and the value of x is used in the expression inside the limit.

Let's do another simple example: compute $\lim_{x \to 2} x^2 - 1$. We prove it directly. Given any sequence x_n for which $x_n \to 2$, we must prove that $f(x_n) \to L$ for a specific L. Most often $L = f(c)$, which in this example is $f(2) = 3$.

Proposition 8.5.
$$\lim_{x \to 2} x^2 - 1 = 3.$$

Proof. Let $\varepsilon > 0$ be an arbitrarily small threshold required by the definition of $f(x_n) \to 3$. In this proof we'll actually need ε to be small enough (say, less than $1/5$). What we're going to do is use the proof of the fact that $x_n \to 2$ as a *subroutine* for some special ε' that we choose, and use the index we get as output to prove that $f(x_n) \to 3$.

Figure 8.8 contains a diagram to illustrate the gymnastics. The top row is the theorem we want to prove, with the input on the left and the desired output on the right. Likewise, the bottom row is the black box subroutine for $x_n \to 2$. So given that first $\varepsilon > 0$ that we don't get to pick, we choose a threshold ε' to use for $x_n \to 2$. Picking a useful ε' is the tricky part of these kinds of proofs, and I'll be momentarily opaque and choose $\varepsilon' = \varepsilon/5$.

So the output of our subroutine for $x_n \to 2$ gives us an index k' after which all x_n are within $\varepsilon/5$ of 2. Now we'll use that same index $k' = k$ for $f(x_n) \to 3$. All we need to show is that $|f(x_n) - 3| < \varepsilon$ for $n > k$. To that effect, a little algebra:

$$f(x_n) - 3 = x_n^2 - 4 = (x_n + 2)(x_n - 2)$$

We know that $x_n - 2$ is less than ε. Moreover, if you take a number that's very close to 2, and you add 2 to that number, it must be close to 4. At the very least, it won't be *way* bigger than 4. In symbols, since we required $\varepsilon < 1/5$ then it must be the case that $|(x_n + 2)| < 5$.[7]

Putting these two facts together gives us

$$|f(x_n) - 3| = |x_n + 2| \cdot |x_n - 2| < 5 \cdot \frac{\varepsilon}{5} < \varepsilon.$$

Which proves that $f(x_n) \to 3$.

□

All of this was a formal way of saying that to compute $\lim_{x \to 2} x^2 - 1$, you may "plug in" 2 to the expression $x^2 - 1$. Indeed, in almost all cases where the expression inside the limit is *defined* at the limiting input (in this case $x = 2$), you can do that. But there are non-pathological functions with useful limits (not just the derivative) for which you can't simply "plug the value in." See the exercises for a famous example. To reiterate from earlier, all of this hefty calculus machinery was invented to deal with those difficult functions.

As we saw with our pathological "two lines" example from Figure 8.7, not every sequence has a limit. For the "two lines" $f(x)$, we computed the slope as $\frac{f(x_n) - f(0)}{x_n - 0}$ where x_n was part of a sequence tending to zero. I.e., we informally computed the limit $\lim_{x \to 0} \frac{f(x) - f(0)}{x - 0}$. But then we found two sequences a_n, b_n that both converge to zero, but their vectorized slope-sequences $\frac{f(a_n - x)}{a_n - x}, \frac{f(b_n - x)}{b_n - x}$ gave different slope values. As a consequence, the limit does not exist, corroborating our intuition. So we've seen that this definition of the limit passes a litmus test: good functions have limits, and bad functions do not.

8.3 The Derivative

Now we define the derivative, which formalizes the steepness of a function $f(x)$ at a given input $x = c$.

Definition 8.6. Let $f : \mathbb{R} \to \mathbb{R}$ be a function. Let $c \in \mathbb{R}$. The *derivative* of f at c, if it exists, is the limit

$$\lim_{x \to c} \frac{f(x) - f(c)}{x - c}$$

This value is denoted $f'(c)$.[8] In the limit, sequences $x \to c$ are taken so that $x_n \neq c$ to avoid division by zero.

---

[7] Even if we didn't require $\varepsilon < 1/5$, we can always choose a k at *least* as large as when we do impose this restriction, even if it's larger than k'. This is a sleight-of-hand that allows us to add extra assumptions that simplify a computation, and it's often paired with the phrase "without loss of generality" to signal what's going on.

[8] Here is where the prime ' is being used to denote the derivative.

Let's compute an example, the derivative of $f(x) = x^2 - 6x + 1$ at $c = 3$. A priori (without looking at a plot of the function) we might have no clue whether the derivative is even positive or negative at 3. By definition, it's:

$$f'(3) = \lim_{x \to 3} \frac{f(x) - f(3)}{x - 3}$$

$$= \lim_{x \to 3} \frac{x^2 - 6x + 9}{x - 3}$$

$$= \lim_{x \to 3} \frac{(x - 3)(x - 3)}{x - 3}$$

We can now simplify $(x - 3)/(x - 3) = 1$. Indeed, recalling the definition of the limit, the expression $\frac{(x-3)(x-3)}{x-3}$ is evaluated at the entries of a sequence x_n that for which $x_n \neq 3$. Hence, we never divide zero by zero and may simplify.

$$f'(3) = \lim_{x \to 3} \frac{(x - 3)(x - 3)}{x - 3}$$

$$= \lim_{x \to 3} x - 3$$

$$= 0$$

This was a nice exercise, but it's tedious to compute derivatives over and over again for every input. It would be much more efficient to instead compute a compact representation of the derivative at all possible points. That is, we want a process which, when given a function $f : \mathbb{R} \to \mathbb{R}$ as input, produces another function $g : \mathbb{R} \to \mathbb{R}$ as output, such that $g(c) = f'(c)$ for every c. While computing the limit may be tedious, our representation of g should make subsequent derivative calculations as computationally easy as evaluating f.

If you ask a mathematician how to come up with such a g, you'd probably receive the reply, "You just do it." This means we can calculate directly from the definition. If, for example, $f(x) = x^2$,

$$f'(c) = \lim_{x \to c} \frac{f(x) - f(c)}{x - c}$$

$$= \lim_{x \to c} \frac{x^2 - c^2}{x - c}$$

$$= \lim_{x \to c} \frac{(x - c)(x + c)}{x - c}$$

$$= \lim_{x \to c} x + c$$

$$= 2c$$

Forever after, we may plug in the desired value of c to get the derivative at c. Most mathematicians don't switch variables, so they'd call the derivative function $f'(x)$ instead of $f'(c)$. This has the added advantage of displaying patterns in derivative computations.

For example, if you compute the derivative of x^4, you get $4x^3$, and the derivative of x^8 is $8x^7$, suggesting the correct rule that the derivative of x^n is nx^{n-1} (for a positive integer n). Here, the notation makes this pattern clear in a way that pictures do not. In fact, if you want to prove this, the following theorem makes the limit calculation less painful.

Theorem 8.7. *For any real numbers x, c and any positive integer n,*

$$x^n - c^n = (x-c)(x^{n-1} + x^{n-2}c + x^{n-3}c^2 + \cdots + xc^{n-2} + c^{n-1}).$$

I'll call the sum $(x^{n-1} + x^{n-2}c + \cdots + c^{n-1})$ "the ugly sum."

Proof. Start to multiply the right-hand side and notice that each term, except the first and last, pair off and sum to zero. In particular, you get

$$x^n + [-c \cdot x^{n-1} + x \cdot x^{n-2}c]$$
$$+ [-c \cdot x^{n-2}c + x \cdot x^{n-3}c^2]$$
$$\vdots$$
$$+ [-c \cdot xc^{n-2} + x \cdot c^{n-1}]$$
$$+ (-c \cdot c^{n-1})$$

Each of the square-bracketed terms is zero and can be removed.

$\square$

Tenderly applying Theorem 8.7 while computing the derivative of $f(x) = x^n$ reveals that in the limit defining $f'(x)$ you can cancel two $(x - c)$ terms, as in our previous examples, leaving just the ugly sum. Plugging $x = c$ in to the ugly sum gives nc^{n-1}.

Theorem 8.8. *For every integer $n \geq 0$, the derivative of x^n is nx^{n-1}.*

At this point in a standard calculus course, a student would spend a few weeks (or months) learning:

1. The derivatives of particular "elementary" functions, such as polynomials, $\sin(x)$, e^x, and $\log x$.

2. When given two functions f, g whose derivatives you know separately, how to compute the derivative of an elementary combination of f and g, such as $f + 3g$ and $f(g(x))$.

3. How to use special values of the derivative (such as zero) to find maxima and minima of various functions, such as maximizing profit from selling a widget subject to costs for creating certain variations of that widget.

4. Assorted nonsense like the derivative of the inverse cosine function.[9]

Because this book can only give you a taste of calculus, and because we're rushing to an interesting application, we'll skip most of this in favor of stating (what I believe is) the most important facts for applications.

Let F be the set of all functions $\mathbb{R} \to \mathbb{R}$ that have derivatives. Let $D : F \to F$ be the function that takes as input a function f and produces as output its derivative f'.

Theorem 8.9. *D is a linear function. Meaning $D(f + g) = D(f) + D(g) = f' + g'$, and $D(cf) = cD(f) = cf'$ for any $c \in \mathbb{R}$.*

As a functon, "cf" is the function that takes as input x and produces as output $c \cdot f(x)$. Likewise, $f + g$ takes as input x and produces as output $f(x) + g(x)$.

As a quick aside, I hate writing sentences like "the function that on input x produces as output $c \cdot f(x)$." Instead I like to use the mathematical analogue of "anonymous function" notation, using the $\mapsto$ symbol. So I can instead say "cf is defined by $x \mapsto c \cdot f(x)$," or even "D is the function $f \mapsto f'$." When you're reading this out loud, $\mapsto$ is pronounced "maps to."

This derivative-computing function D is also often written as $\frac{d}{dx}$, but this causes inconsistent notation like $\frac{d}{dx}(f)$ versus $\frac{df}{dx}$ and forces one to choose a variable name x. In my opinion, this notation exists for *bad* reasons: backwards compatibility with legacy math, and trying to trick you into thinking that derivatives are fractions so you'll guess the forthcoming chain rule. But it is too widespread to avoid.

Theorem 8.9 immediately lets us compute the derivative of any polynomial, because we can use Theorem 8.8 to compute the derivatives of each term and add them up. E.g., the derivative of $3 + 2x - 5x^3$ is $2 - 15x^2$. Quick spot check exercise: using intuition, reason that a constant function like $f(x) = 3$ has derivative $f'(x) = 0$. If your intuition fails you, use the definition of the limit to compute it.

The other crucial fact, which we'll use later, is the chain rule.

Theorem 8.10 (The chain rule). *Let $f, g : \mathbb{R} \to \mathbb{R}$ be two functions which have derivatives. Then the derivative of $f(g(x))$ is $f'(g(x))g'(x)$.*

In the chapter exercises you'll look up a proof of this theorem. The chain rule makes it easy to compute derivatives that would require a lot of algebra to compute, such as $(x^2 - 10)^{50}$. Here f is $z \mapsto z^{50}$ and g is $x \mapsto x^2 - 10$, so the derivative is $50(x^2 - 10)^{49} \cdot (2x)$. The chain rule also lets us compute derivatives that would otherwise be completely mysterious, such as that of $\sin(e^x)$. If you're told what the derivatives of $\sin(x)$ and e^x are separately, then you can compute the derivative of the composition.

---

[9] I sneer, but if you're serious about mathematics then at some point you need to become intimately familiar with specific derivatives of elementary functions. This book is not the place for that, and I suspect many of my readers will have seen calculus at least once before, and knows how to google "derivative of arctan(x)" should they forget.

As a notational side note, let me explain the "fractions make you guess the chain rule" remark. Call $h(x) = f(g(x))$. Then if we use the fraction notation $\frac{dh}{dx}$ for the derivative of h, the standard way to write the chain rule for this would be $\frac{dh}{dx} = \frac{dh}{dg} \cdot \frac{dg}{dx}$. The "hint" of the notation is that if you're a reckless miscreant, you might jump to the conclusion that the dg's "cancel" like fractions do. Rest assured that is not how it works, but calculus students the world over are encouraged to do it this way because the resulting rule is correct. We'll return to this in Chapter 14.

Historically, symbols like dx had no concrete mathematical meaning. They were called "infinitesimals" and regarded informally as quantities infinitely smaller than any fixed value. More recently, dx was retroactively assigned a semantic meaning that allows one to work with it as the notation suggests. The formalism is beyond the scope of this book.[10]

8.4 Taylor Series

Approximation by a Line

If you got ten mathematicians in a room they'd come up with twenty different ways to motivate calculus. In this chapter we used, "generalize the slope of a line to curvy things," but here's another. One prevalent idea is to take a complicated thing and approximate it by simpler things. Without calculus, the simplest function we fully understand is a straight line. So we might ask, "Given a function $f : \mathbb{R} \to \mathbb{R}$ and a point $x \in \mathbb{R}$, what line best approximates f at x?"

If you define "best approximates" in a particular but reasonable way, the answer to this question uniquely defines the derivative. Call $L(x)$ the line approximation of f we get using the derivative of f at $x = c$. That is, $L(x) = f'(c)(x - c) + f(c)$. This is just the line passing through $(c, f(c))$ with slope $f'(c)$, often called the "tangent line" to f at c.

The definition of "best approximates" we *wish* we had is that, for any other line $K(x)$ that passes through $(c, f(c))$, $L(x)$ is *always* closer to $f(x)$ than $K(x)$. But that just isn't possible. Take our example from earlier, replotted in Figure 8.9. There, the line between A and A' is not the tangent line at A, and it is also far closer to f at A' than the tangent line would be. However, for points close to A, the tangent line is a much better approximator. If we're trying to approximate f "at" A, we care more about points closer to A than points far from A. Here's how we make this clear in the math.

Take any line $K(x)$ that is supposedly challenging the tangent line for the title of "best approximating line of f at $x = c$." Then I claim I can choose a small enough interval around c (the width of this interval depends on the features of the challenger K) so that L beats K on all points in this interval. Here's the formal theorem I'll prove momentarily.

Theorem 8.11. *Let $f : \mathbb{R} \to \mathbb{R}$ be a function and $A = (c, f(c))$ be a point on f. Let $L(x)$ be the tangent line at c, i.e. $L(x) = f'(c)(x - c) + f(c)$. Then for every line $K(x)$*

---

[10] If you are insistent on reading more about the modern formalism, look up "differential forms" and the "exterior derivative." Then you'll understand why one would opt for fractions as a simpler mechanism.

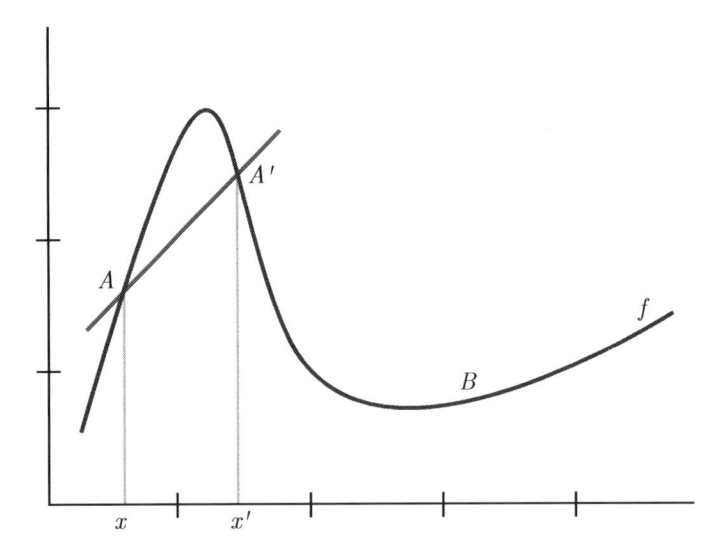

Figure 8.9: The line between A and A' does not approximate f well close to A.

passing through $(c, f(c))$, there is a sufficiently small $\varepsilon > 0$ such that if $|x - c| < \varepsilon$, then $|L(x) - f(x)| \leq |K(x) - f(x)|$.

Notation time: people often write the set of points $\{x \in \mathbb{R} : |x - c| < \varepsilon\}$ using the notation $(c - \varepsilon, c + \varepsilon)$. They also often call this an *epsilon-ball* around c. Using this, the last sentence of the theorem might read, "For all $x \in (c - \varepsilon, c + \varepsilon)$, it holds that $|L(x) - f(x)| \leq |K(x) - f(x)|$." This makes the statement clearer. Instead of saying "if this then that," you're saying what you want to say outright, that "FOO is always true in my domain of interest."

Proof. If K is a line passing through $(c, f(c))$, then it can be written in the same way as L but with a different slope. I.e., for some $m \in \mathbb{R}$, $K(x) = m(x - c) + f(c)$.

Expanding K and L according to their formulas, the theorem's conclusion requires us to choose a $\varepsilon > 0$ such that when $|x - c| < \varepsilon$ the following inequality is true.

$$|f'(x)(x - c) + f(c) - f(x)| \leq |m(x - c) + f(c) - f(x)|$$

We don't yet know this inequality is true, but we can "work backwards" by doing valid algebraic manipulations until we get to something we know is true. In particular, one might recognize the definition of the derivative hiding in there and divide by $(x - c)$ to get

$$\left| f'(x) - \frac{f(x) - f(c)}{x - c} \right| \leq \left| m - \frac{f(x) - f(c)}{x - c} \right|.$$

The fraction $\frac{f(x) - f(c)}{x - c}$, which is on both sides, is most of the definition of the derivative, missing only the limit. And $f'(x)$ is the *value* of that limit, whereas m is some other number. This should already make it pretty clear that the inequality above holds, but let's prove it formally by contradiction.

Suppose to the contrary that no matter which ε I choose, there is some x in $(c-\varepsilon, c+\varepsilon)$ that contradicts the inequality above. I would like to pick a sequence of x values going to c that violates the definition of the derivative. I will do that by picking a sequence of ε's, using the fact that the inequality above is false for *every* ε, and arriving at the sequence of x's needed for my contradiction. Let

$$(\varepsilon_1, \varepsilon_2, \varepsilon_3, \dots) = (1, 1/2, 1/3, \dots)$$

and let $x_1, x_2, x_3, \dots$ be the corresponding x's violating the inequality for each ε_i. Since each x_i is in $(c - \varepsilon_i, c + \varepsilon_i)$, it follows that $x_i \to c$, but because (by assuming the contradictory hypothesis) the inequalities are false, the sequence $\frac{f(x_i)-f(c)}{x_i-c}$ *does not* converge to $f'(x)$. The contradictory hypothesis says it's closer to m instead. This contradicts the definition of the derivative.

$\square$

We have proved that derivatives provide the best linear approximation to a function at a point for a concrete sense of "best." This perspective brings up the natural question of whether we can improve this approximation by using more complicated functions than lines. The answer is yes, and it's called the Taylor polynomial.

Taylor Polynomials

Lines are degree 1 polynomials. One thing that's nice about polynomials is that they have a *grading*. By that I informally mean, if you increase the degree of your polynomial, you can express a wider variety of functions. There is a rigorous way to state this using linear algebra (see Definition 10.9), but the gist of it is that the data defining a degree 3 polynomial is four unrelated numbers, while the data defining a degree 4 polynomial is five unrelated numbers. In principle, higher degree includes more complexity, and allows better approximations of f.

You can derive exactly how this works by following the steps of Theorem 8.11, and asking for a degree 2 polynomial *whose derivative best approximates f' close to a*. Indeed, let our candidate be the following (where below $q^* \in \mathbb{R}$ is the unknown parameter we must set to get a degree 2 polynomial).

$$p(x) = q^*(x - a)^2 + f'(a)(x - a) + f(a)$$

We can't avoid using $f'(a)$ for the coefficient of the $(x - a)$ term, because $p'(a)$ needs to be exactly $f'(a)$ and $p'(x)$ is

$$p'(x) = 2q^*(x - a) + f'(a).$$

Plugging in $x = a$ leaves only $f'(a)$. If we had used some other number R instead of $f'(a)$, then $p'(a) = R$. In the same way, in Theorem 8.11 we couldn't avoid using $f(a)$ for the constant term because the line had to pass through $(a, f(a))$.

And so if we want to optimize $p'(x)$ by choosing q^*, it's almost *exactly* the same proof as Theorem 8.11, with the different being an extra factor of 2. We'll leave it as an exercise for the reader to redo the steps, but at the end you get $q^* = \frac{f''(a)}{2}$, where f'' is the derivative of the derivative of f (the "second" derivative of f).

Two quick asides. First, the second derivative only makes sense if f has a first derivative, and as we saw not all functions have derivatives at all points. Second, adding more and more primes to denote repeated applications of the derivative operation is cumbersome. Rather, it's customary to use a parenthetical superscript notation $f^{(n)}(x)$ for the n-th derivative of f. You call a function *n-times differentiable* if it has n derivatives at every point. Finally, if f has infinitely many derivatives (i.e., it is n-times differentiable for every $n \in \mathbb{R}$), f is called *smooth*. The typical example of a smooth function is $\sin(x)$ or 2^x. A default assumption is that life is smooth, and when it's not you pay very close attention.

Our exploration has led us to the Baby Taylor Theorem.

Theorem 8.12 (The Baby Taylor Theorem). *Let $f : \mathbb{R} \to \mathbb{R}$ be a twice-differentiable function and let $(a, f(a))$ be a point on f. Then the degree 2 polynomial that best approximates f and f' simultaneously close to a is*

$$p(x) = f(a) + f'(a)(x-a) + \frac{f^{(2)}(a)}{2}(x-a)^2$$

A proof by induction, which the reader should finish (we just did the step from $n = 1$ to $n = 2$ which has all the features of the general induction), extends the Baby Taylor Theorem to the Adolescent Taylor Theorem. Note that by $n!$ we mean the factorial function $n \mapsto n \cdot (n-1) \cdot (n-2) \cdots 2 \cdot 1$ where n is a positive integer. We're not merely excited about n, though it is bittersweet to have watched n grow up so fast.

Theorem 8.13 (The Adolescent Taylor Theorem). *Let $f : \mathbb{R} \to \mathbb{R}$ be a k-times differentiable function and let $(a, f(a))$ be a point on f. Then the degree k polynomial that best approximates f and all of the k derivatives of f simultaneously close to a is*

$$f(a) + \sum_{n=1}^{k} \frac{f^{(n)}(a)}{n!}(x-a)^n$$

This is called the degree k Taylor polynomial of f at a.

Definitions are usually introduced in their most general and often-used form, in this case with a summation. It is almost always helpful to write out the first few terms to familiarize yourself with the pattern. Here are the first three.

$$f(a) + \frac{f'(a)}{1!}(x-a) + \frac{f^{(2)}(a)}{2!}(x-a)^2 + \frac{f^{(3)}(a)}{3!}(x-a)^3$$

As if possessed by the spirit of Leonhard Euler, we write down examples. Just so we can work with an example that's not already a polynomial, let $f(x) = e^x$. Recall or learn

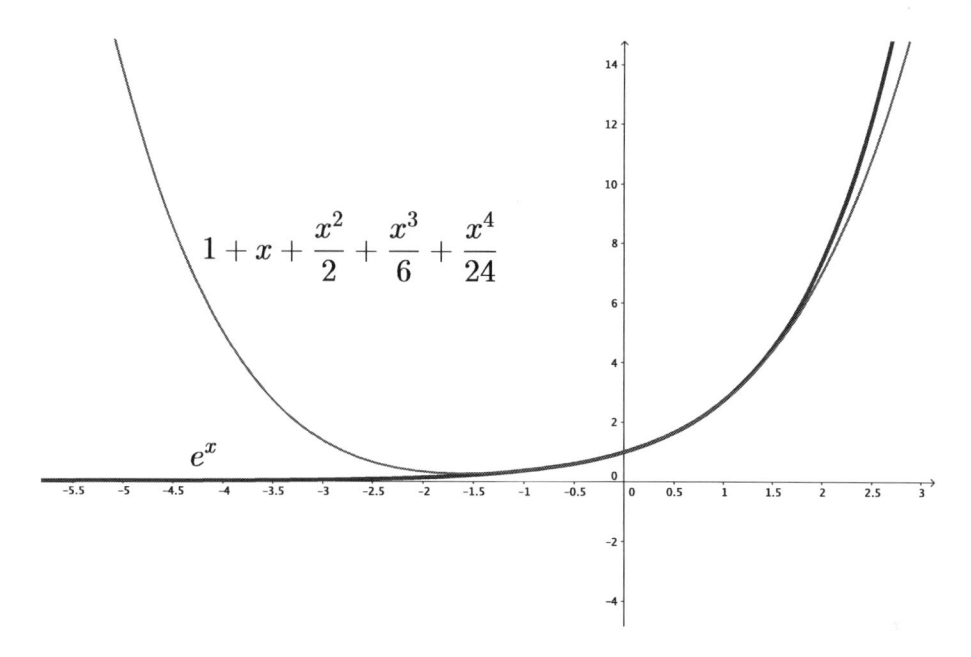

Figure 8.10: The degree 4 Taylor series approximation of $f(x) = e^x$.

now that the derivative of e^x is also e^x. In fact, the number e is uniquely defined by this property.

Then the degree 4 Taylor series for e^x at $x = 0$ is particularly simple because e^0 is 1 in every term:

$$1 + x + \frac{x^2}{2} + \frac{x^3}{6} + \frac{x^4}{24}.$$

Figure 8.10 contains a picture of e^x and its approximation by the degree 4 Taylor polynomial. The approximation is faithful to the original function, but only close to $x = 0$. Elsewhere it can be arbitrarily bad.

The Taylor polynomial is one of the most often used applications of mathematics to itself. The reason is because when you're analyzing a mathematical problem, it's easy to define functions with convoluted behavior. One example of this is in machine learning, when you analyze the probability that some event occurs. You can often write down the probability as a massive product, but can't compute it exactly. Instead, one often uses a small-degree Taylor polynomial to approximate the complicated thing at a point of interest. With knowledge of whether the Taylor polynomial is an over- or under-approximation of the truth, one can bound the complicated behavior enough to prove something useful.

Theorem 8.13 seems to show us that every function can be approximated arbitrarily well using polynomials. As useful as polynomials are, it turns out this is not entirely true. Let's say we're working with a function where the polynomial approximation *does* get progressively better at higher degrees. If you're in the proper mindset for calculus, you naturally ask what happens in the limit? If I call p_k the degree k Taylor polynomial for f at $a = 0$, how can we make sense of the expression

$$\lim_{k \to \infty} p_k(x) \ldots ?$$

Remember, we only defined what it means for a sequence of *numbers* to converge, but this is a sequence of *functions* $\mathbb{R} \to \mathbb{R}$. In order to define convergence for a sequence of functions, we need to define what it means for two functions to be "close" together, which is not easy. But suppose we did that and we can make sense of this expression, we'd hope that this limit was also *equal* to f, and least close to $x = 0$. This expression, the limit of Taylor polynomials, is called the *Taylor series* of f at that point.

Mathematics is not so kind to us here. There are certain simple functions, like the base 2 logarithm, for which the Taylor series breaks down in certain regions. In particular, if $f(x) = \log(1 + x)$ and you compute the limit at $a = 0$, the resulting function would only be equal to $f(x)$ between $x = -1$ and $x = 1$. When $x > 1$ the limit does not converge, even though $\log(1 + x)$ exists for $x > 1$. In that case, you have to compute a different Taylor series at, say, $a = 2$. The complete function is then joined together piece-wise by enough Taylor series pieces until you get the whole function. The functions which can be reconstructed in this way (and aren't sensitive to which points you choose within a region, again in the interest of well-definition) are called *analytic* functions.[11]

There are somewhat natural functions that fail to accommodate Taylor series worse than the logarithm. Let $f(x) = 2^{-1/x^2}$ when $x \neq 0$, and let $f(0) = 0$. Figure 8.11 contains a plot of this function. You will prove in Exercise 8.8 that $f^{(n)}(0) = 0$ for every $n \in \mathbb{N}$. As a consequence, all of its Taylor polynomials at $x = 0$ are the zero function, and the "limit function" should be the constant zero function.[12] In this case, the Taylor series tells you nothing about the function except its value at $x = 0$. Polynomials aren't able to express what f looks like near zero.

This highlights the shortcomings of Taylor polynomials. They're not the perfect tool for every job. It also leads us to ask why, for this mildly pathological f, the Taylor series fails so spectacularly. Complex analysis provides a satisfactory answer, but the subject is unfortunately beyond the scope of this book.

8.5 Remainders

The Adolescent Taylor Theorem tells us how to compute the best polynomial of a given degree that approximates the behavior of a function. In fact, it approximates the behavior

---

[11] There is a more rigorous way to say "not sensitive to the points you choose," which is to say that computing the Taylor series of f at every input a in the domain of f converges to f in some open set around a. Saying what an "open" set is another can of worms, but for most functions $\mathbb{R} \to \mathbb{R}$ this just means "any interval containing a." This can fail, e.g., when the Taylor series at a only equals f at a finite set of other points.

[12] Indeed, a constant function is defined by a single number, so a sequence of constant functions "is" a sequence of numbers. A reasonable definition of function convergence should generalize convergence for numbers.

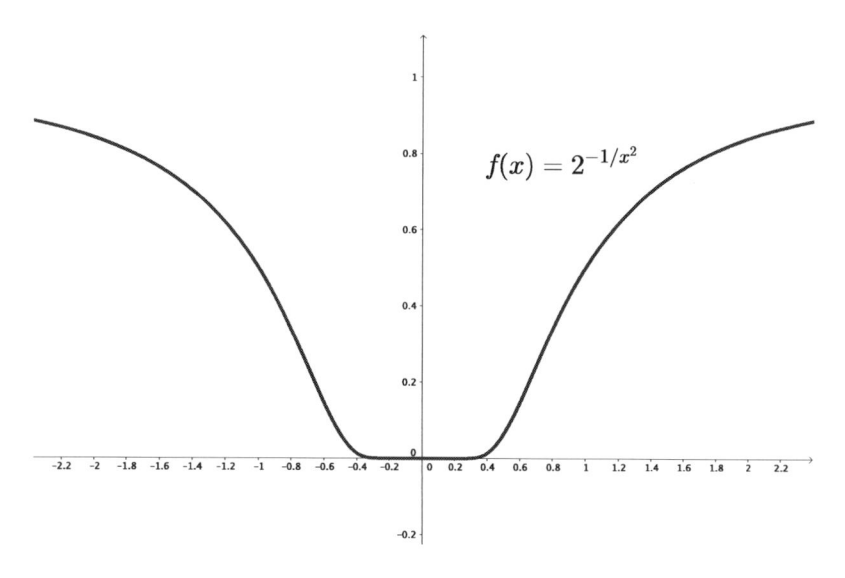

Figure 8.11: A function $f(x) = 2^{-1/x^2}$, all of whose derivatives are zero at $x = 0$.

of a function's "slope" (first derivative) and more informally its curvature (higher derivatives), provided you're willing to compute enough terms.

The Adolescent Taylor Theorem, however, doesn't allow us to quantify *how* good the approximation is. As we just saw, there are pesky functions whose Taylor polynomials at certain rotten points are all zero. They're so flat they tricked the poor polynomial!

As you might have guessed, there is an Adult Taylor Theorem—just called the Taylor Theorem—which gets one much closer to quantifying the error of the Taylor polynomial. Unfortunately, the proof of this theorem requires the Mean Value Theorem, which does not fit in this book, but we can state the Taylor theorem easily enough.

Theorem 8.14 (The Taylor Theorem). *Let $d \in \mathbb{N}$ and f be a $(d+1)$-times differentiable function. Let p_d be the degree d Taylor polynomial approximating f at a, and let x be an input to f. Then there exists some z between a and x for which*

$$f(x) = p_d(x) + \frac{f^{(d+1)}(z)}{(d+1)!}(x-a)^{d+1}$$

In words, the *exact* value of $f(x)$ can be computed from the Taylor polynomial $p_d(x)$ plus a remainder term involving a magical z plugged into the $(d+1)$-th derivative instead of x.

The dependence of the variables on each other are a bit confusing. Let's make it explicit with some pseudocode. In particular, the needed value of z depends on the specific input x.

```
def exact_value(f, d, a, x):
    '''Return the exact value of f at x.

    Arguments:
      f: the function to evaluate
      d: the degree for the taylor polynomial
      a: the input we can compute f at
      x: the input we'd like to compute f at
    '''
    p = taylor_polynomial(f, d, a)
    next_derivative = nth_derivative(f, n=d+1)

    z = find_magical_z_value(f, d, a, x)  # note z depends on all of these!

    remainder = (x-a)**(d+1) * next_derivative(z) / factorial(d+1)
    return p(x) + remainder
```

One important consequence of the remainder formula is that if $f^{(d+1)}$ is never large between a and x, then z is irrelevant. For the sake of concreteness, let's say that $f^3(z) < 100$ between a and x. Then $|f(x) - p_2(x)|$, the error in computing $f(x)$ from its Taylor polynomial at a, is bounded.

$$|f(x) - p_2(x)| < (100/6)(x-a)^3$$

In this case, if x is within 0.1 of a, then the error of the Taylor polynomial is only about 0.017. Often this coarse z-be-damned bound is enough. This is the viewpoint of Newton's method, this chapter's application.

8.6 Application: Finding Roots

Let's say you have a function $f(x)$ and you want to find its *zeros*,[13] that is, an input r producing $f(r) = 0$. Let's also say that you can compute both $f(x)$ and $f'(x)$ at any given input. An example of such a function is $x^5 - x - 1$. Try to algebraically solve for $f(x) = 0$, if you dare. On the other hand, $f'(x) = -1 + x^4$ is simple enough to compute.[14]

Figure 8.12 contains a plot of $f(x)$. The root is just under 1.2, but coming up with an algebraic formula for the root in terms of the coefficients is impossible in general (this is a deep theorem known as the Abel-Ruffini theorem).

One idea that should feel very natural by this point is to approximate the root of f by starting with some value close to the root (which we can guess), and progressively improving it. In theory, we want to find a sequence $x_1, x_2, \ldots$, such that $\lim_{n \to \infty} x_n = r$, where $f(r) = 0$.

---

[13] For polynomials, zeros are sometimes called roots, and I will use these terms interchangeably.
[14] Another good example is $f(x) = -1 + 2^x + 3^x$, but its derivative is more complicated: $f'(x) = 2^x \log(2) + 3^x \log(3)$

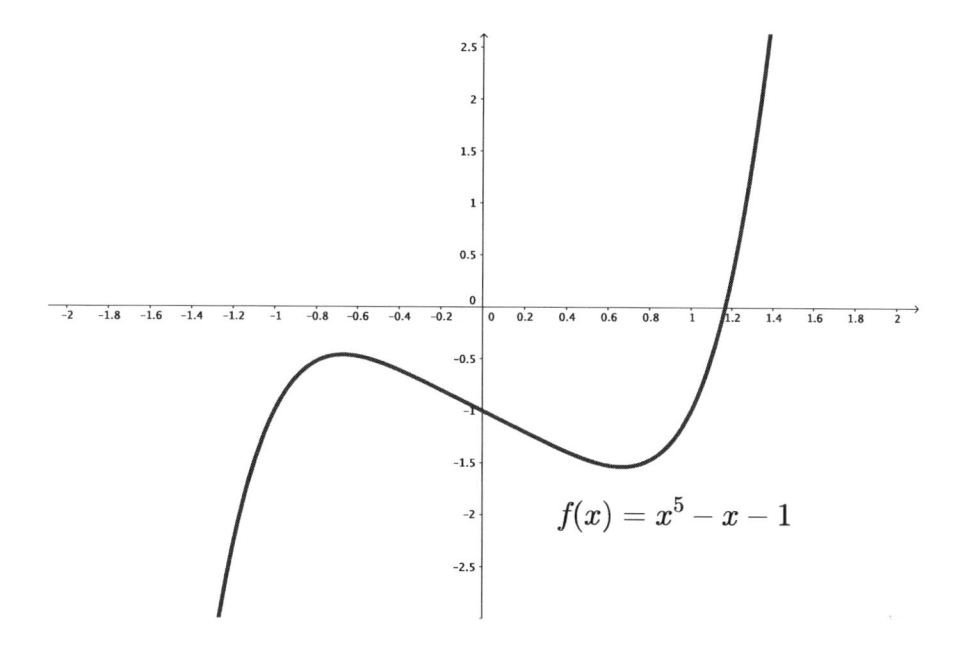

Figure 8.12: A function whose root does not have a nice formula.

One initial thought is obvious: perform a binary search. That is, pick two guesses c, d, where $f(c) < 0 < f(d)$, and then let your improved guess be the midpoint $(c+d)/2$, updating your upper and lower search bounds in the obvious way depending on whether $f((c+d)/2) > 0$.

Binary search does produce a sequence approaching a root of f, but it turns out to be much slower than the forthcoming Newton's method.[15] In Newton's method you choose your next guess x_{n+1} depending on the derivative of f at x_n. To convince you that this this could be faster than binary search, suppose you chose bad bounds for binary search as in Figure 8.13.

The tangent line at the point $(d, f(d))$ intersects the x-axis quite close to the root, whereas the midpoint between c and d is rather far away. A binary search would slowly approach the root from the left, whereas the tangent line guides us close to the root in the first step.

If this isn't convincing enough, we can provide something much better: a proof. But first, we have to make the algorithm explicit. Phrased geometrically, start from some intermediary x-value guess, calling it x_n for the n-th step in the algorithm. Draw the tangent line at x_n, which is $y = f(x_n) + f'(x_n)(x - x_n)$, and let x_{n+1} be the intersection of this line with the x-axis. This is illustrated in Figure 8.14. To find the intersection point, set $y = 0$ in the equation for the tangent line, and solve for x:

---

[15] To be precise, binary search requires k iterations to get k digits of precision, whereas Newton's method gets k^2 digits of precision in k steps, under the right starting conditions.

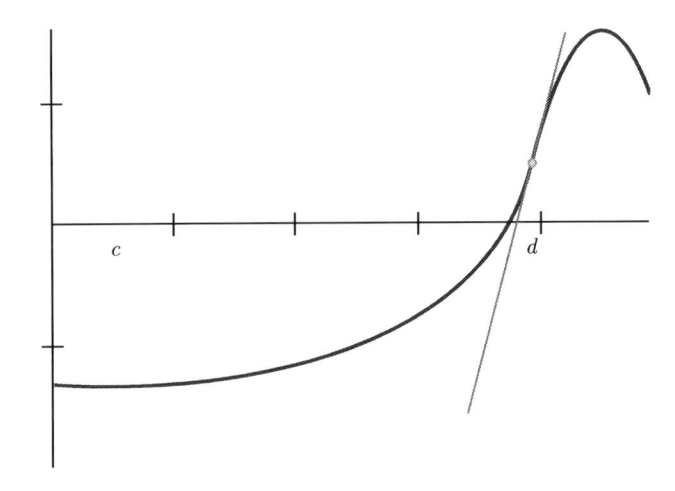

Figure 8.13: And example of Newton's method outperforming a binary search. The tangent line at d is better than the slow approach from c.

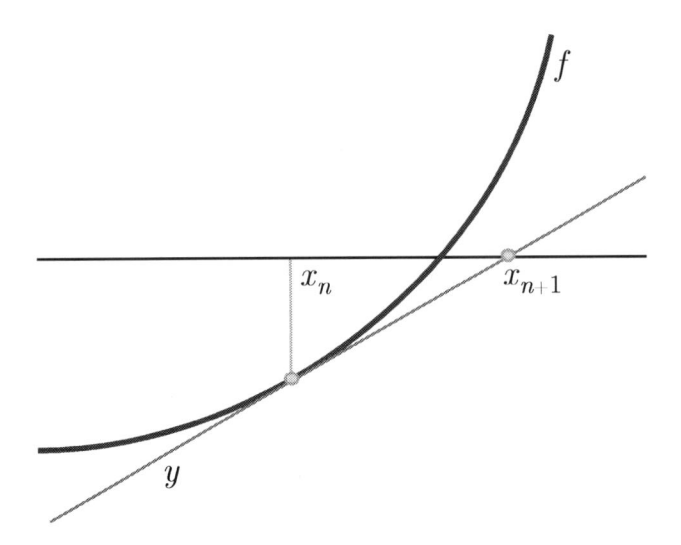

Figure 8.14: A generic illustration of Newton's method to get from x_n to x_{n+1}.

$$0 = f(x_n) + f'(x_n)(x - x_n)$$
$$0 = \frac{f(x_n)}{f'(x_n)} + (x - x_n)$$
$$x = x_n - \frac{f(x_n)}{f'(x_n)}$$

So set $x_{n+1} = x_n - \frac{f(x_n)}{f'(x_n)}$, and from a given starting x_1, use this formula to define a sequence $x_1, x_2, \ldots$. As a Python generator:

```
def newton_sequence(f, f_derivative, starting_x):
    x = starting_x
    while True:
        yield x
        x -= f(x) / f_derivative(x)
```

Obviously, if $f'(x_n) = 0$ then we're dividing by zero which is highly embarrassing. So let's assume $f'(x_n) \neq 0$, i.e., that the tangent line to f is never horizontal, and we'll make this formal in a moment.

When Taylor's theorem is your hammer, the world is full of nails. It takes no inspiration to come up with this algorithm. As we'll see in the proof below, literally all you do is rearrange the degree 1 Taylor polynomial and squint at the remainder. Still, without going through the proof it's not entirely clear that Newton's method should outperform binary search, other than the fuzzy reasoning that an algorithm that *somehow* uses the derivative should do better than one that does not.

Indeed, we'll wield a Taylor polynomial like a paring knife to prove Newton's method works. The theorem says that not only does x_n converge to a root r of f, but that if x_1 starts close enough, then in every step the number of correct digits roughly *doubles*. That is, the error in step $n + 1$, which is $|x_{n+1} - r|$, is roughly the square of the error in step n, i.e. $|x_n - r|^2$. Binary search, on the other hand, improves by only a constant number of digits in each step.

This theorem we'll treat like a cumulative review of proof reading. That is, we'll be more terse than usual and it's your job to read it slowly, parse the individual bits, and generate unit tests if you don't understand part of it.

Let $f : \mathbb{R} \to \mathbb{R}$ be a function which is "nice enough" (it has some properties we'll explain after the proof). Let $r \in \mathbb{R}$ be a root of f inside a known interval $c < r < d$, and pick a starting value x_1 in that interval. Define $x_2, x_3, \ldots$ using the formula $x_{n+1} = x_n - f(x_n)/f'(x_n)$. Call $e_k = |x_k - r|$ the error of x_k.

Theorem 8.15 (Convergence of Netwon's Method). *For every $k \in \mathbb{N}$, the error $e_{k+1} \leq Ce_k^2$, where C is a constant defined as*

$$C = \max_{c \leq z \leq d} \frac{|f''(z)|}{2|f'(z)|}$$

In other words, the error of Newton's method vanishes quadratically fast in the number of steps of the algorithm.

Proof. Fix step k. Compute the degree 1 Taylor polynomial for f at x_k. This is exactly the tangent line to f at x_k. Use that Taylor polynomial to approximate $f(r)$, the value of f at the unknown root r.

$$f(r) = f(x_k) + f'(x_k)(a - x_k) + R$$

Here R is the remainder from Theorem 8.14, and can be written as $R = \frac{1}{2}f''(z)(r - x_k)^2$ for some unknown z between r and x_k. Since r is a root, $f(r) = 0$ and we can rearrange.

$$0 = f(x_k) + f'(x_k)(r - x_k) + \frac{1}{2}f''(z)(r - x_k)^2$$

Recall we want to analyze the error of the approximation $e_{k+1} = |x_{k+1} - r|$, so at some point we must use use the formula for x_{k+1} in terms of x_k. The next three steps are purely algebraic rearrangements to enable this.

$$-f(x_k) - f'(x_k)(r - x_k) = \frac{1}{2}f''(z)(r - x_k)^2$$

$$\frac{f(x_k)}{f'(x_k)} + (r - x_k) = -\frac{f''(z)}{2f'(x_k)}(r - x_k)^2$$

$$\left[x_k - \frac{f(x_k)}{f'(x_k)}\right] - r = \frac{f''(z)}{2f'(x_k)}(r - x_k)^2$$

The bracketed term is x_{k+1}, and so we get

$$e_{k+1} = |x_{k+1} - r| = \left|\frac{f''(z)}{2f'(x_k)}(e_k)^2\right|$$

The fraction $\frac{f''(z)}{2f'(x_k)}$ is at most C, as defined in the statement of the theorem. □

Despite all the algebraic brouhaha in the proof above, all we did was take some value $x = x_k$ (though calling it x_k was only relevant in hindsight), write down the degree 1 Taylor polynomial that approximates f at x, and use that approximation to guess at the value of the unknown root r. We needed the notation and formalism to ensure that we weren't being tricked by our intuition, and to clearly outline the guarantees, and where those guarantees break down.

Speak of the devil! The proof allows us to identify the requirements of a "nice enough" function:

- $f'(x)$ can never be zero between c and d, except possibly at the root r itself. Otherwise we risk dividing by zero, or worse, getting stuck in a loop (as we'll see in the example below).

- f has to have first and second derivatives everywhere between c and d. Otherwise the claims in the proof that use those values are false.

- Realistically, $f'(x)$ should never be very close to zero, and $f''(x)$ should never be very far from zero, or else C will be impractically large.

Using our newtons_sequence generator from before, we can implement Newton's method for $f(x) = x^5 - x - 1$.

```
THRESHOLD = 1e-16

def newton_sequence(f, f_derivative, starting_x, threshold=THRESHOLD):
    x = starting_x
    function_at_x = f(x)
    while abs(function_at_x - x) > THRESHOLD:
        yield x
        x -= function_at_x / f_derivative(x)
        function_at_x = f(x)

def f(x):
    return x**5 - x - 1

def f_derivative(x):
    return 5 * x**4 - 1

starting_x = 1
approximation = []
i = 0
for x in newton_sequence(f, f_derivative, starting_x):
    print((x, f(x)))
    i += 1
    if i == 100:
        break
```

After only six iterations we have reached the limit of the display precision.

```
(1, -1)
(1.25, 0.8017578125)
(1.1784593935169048, 0.09440284131467558)
(1.16753738939611, 0.001934298548380342)
(1.1673040828230083, 8.661229708994966e-07)
(1.1673039782614396, 1.7341683644644945e-13)
(1.1673039782614187, 6.661338147750939e-16)
(1.1673039782614187, 6.661338147750939e-16)
(1.1673039782614187, 6.661338147750939e-16)
```

Let's see the same experiment with the starting_x changed to 0 instead of 1. This is an input which, as you can see from Figure 8.15, drives Newton's method in the *wrong* direction! By the end of a hundred iterations, Newton's method cycles between three points:

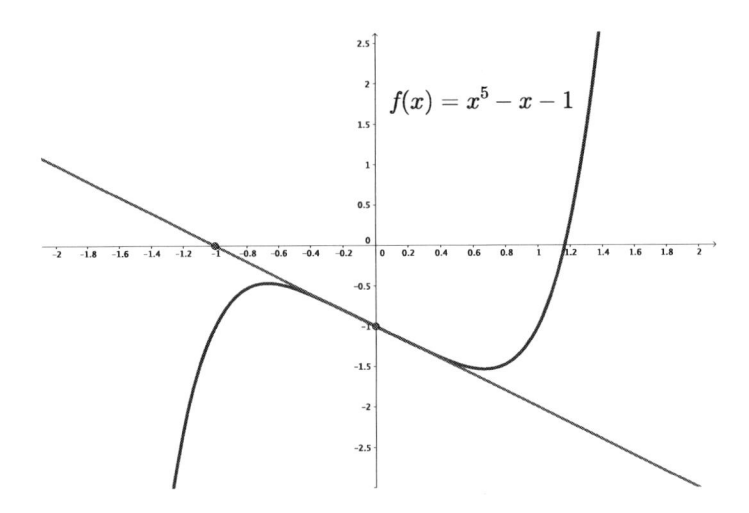

Figure 8.15: An example where the starting point of Newton's method fails to converge due to an unexpected loop.

```
...
(0.08335709970125815, -1.083353075191566)
(-1.0002575619492795, -1.001030911349579)
(-0.7503218281592572, -0.4874924386834848)
...
```

This behavior is allowed by Theorem 8.15, because in between the starting point and the true root, the derivative $f'(x)$ is zero, making the error bound C from Theorem 8.15 undefined (and indeed, unboundedly large for x values close to where $f'(x)$ is zero). Newton's method is very powerful, but take care to choose a wise starting point.

Newton's method stirs up a mathematical hankering: why stop at the degree 1 Taylor polynomial? Why not degree 2 or higher? All we did to "derive" Newton's method was take a random point, write down the degree 1 Taylor polynomial $p(x)$, and solve $p(x) = 0$. By rearranging to isolate the error terms, we got the formula for x_{k+1} for free. For degree 2, why not simply use the degree 2 Taylor polynomial instead?

$$0 = f(x_k) + f'(x_k)(x - x_k) + \frac{1}{2!}f''(x_k)(x - x_k)^2$$

There are two obstacles: (a) this polynomial might not even hit the x axis; it's trickier to nail down for quadratics than lines, and (b) even if it does, it might be hard to *find* the intersection, since finding roots is the problem we started with!

Admittedly, finding the root of a degree 2 polynomial isn't so hard (there's a formula with a sing-a-long mnemonic), but if you take this idea up to degree 3, 4, or higher, the formula approach eventually breaks down. For degree 5, the polynomial we want to approximate a root for *is* the Taylor polynomial, and we don't know how to find its roots.

Nevertheless, there is a technique called *Householder's method* that generalizes Newton's method to higher degree Taylor polynomials. Higher degrees unlock order-of-magnitude better convergence. The tradeoff, as expected, is that it takes progressively more work to compute each step in the update (and existence and good behavior of higher derivatives). Moreover, there are additional requirements at each step on the suitability of a starting point to guarantee convergence. The derivation and analysis of these methods is beyond the scope of this book, because it involves a more nuanced understanding of Taylor series.

8.7 Cultural Review

- Good definitions are designed to match a visual intuition while withstanding (or excluding) pathological counterexamples.

- Much of the murkiness of calculus comes from the fact that it must support a long history of manual calculations and pathological counterexamples. The "normal" case is usually easier to understand.

- A definition is well-defined if it doesn't depend on arbitrary choices used to show the definition holds. E.g., the limit of a function as the input approaches a point must not depend on which sequence you choose to approach that point.

- The Taylor polynomial is a mathematical hammer, and math is full of nails.

8.8 Exercises

8.1 Write down examples for the following definitions.

1. A sequence $x_1, x_2, \ldots$ is said to *diverge at a*, written $\lim_{x \to a} x_n = \pm\infty$, if for every $M > 0$, there is a $k \in \mathbb{N}$ so that if $n > k$, then $|x_n| > M$. Note that ∞ is not being used as a number, but rather notation for the concept, "x_n grows without bound." This unifies it with the usual limit definition.

2. A function $f : \mathbb{R} \to \mathbb{R}$ is called *concave up* at a if the second derivative $f''(x)$ is positive at $x = a$. Likewise, if $f''(a) < 0$, f is called *concave down* at a. How does the numerical property of being concave up/down relate to the geometric shape of a curve?

3. A function $f : \mathbb{R} \to \mathbb{R}$ is called *continuous at a* if for every $\varepsilon > 0$, there is a $\delta > 0$ such that whenever $|x - a| < \delta$, then $|f(x) - f(a)| < \varepsilon$. A function is called *continuous* if it it continuous at all inputs. Most functions in this book are continuous. Find an example function defined in this chapter which is not continuous according to this definition.

8.2 Prove the following basic facts using the definitions from Exercise 8.1.

1. Prove Theorem 8.9 that the map $f \mapsto f'$ is linear.

2. Using the definition of the limit of a function, prove that

$$\lim_{x \to a}[f(x)g(x)] = \left(\lim_{x \to a} f(x)\right)\left(\lim_{x \to a} g(x)\right),$$

provided both of the limits on the right hand side exist.

3. Prove that $a_n = \frac{2\sqrt{n}}{n^{10}}$ diverges.

4. Prove that a function $f(x)$ which is differentiable at a is also continuous at a.

5. Let x_n be a sequence of real numbers. Suppose that for every $\varepsilon > 0$, there is an $N \in \mathbb{N}$ (depending on ε), such that for every $n, m > N$ it holds that $|x_n - x_m| < \varepsilon$. Using this and the formal definition of a limit, prove that x_n converges. Such a sequence is called a *Cauchy sequence*.

8.3 Compute the Taylor series for $f(x) = 1/x$ at $x = 1$.

8.4 Compute the Taylor series for $f(x) = e^{-2x}$, and compare this to the procdure of plugging in $z = -2x$ into the Taylor series for e^x. Find an explanation of why this works.

8.5 Compute the Taylor series for $f(x) = \sqrt{1+x^2}$ at $x = 0$. We will use this in Chapter 12 to simplify a model for a physical system.

8.6 There are some functions which are challenging to compute limits for, but they aren't considered "pathological." One particularly famous function is

$$f(x) = x \sin(1/x).$$

Compute the limit for this function as $x \to 0$. The difficulty is that $\sin(1/x)$ is not defined at $x = 0$, and algebra doesn't provide a way to simplify $\sin(1/x)$. Instead, you have to use "common sense" reasoning about the sine function. This common-sense reasoning is made rigorous by the so-called Squeeze Theorem. Look it up after trying this problem, and note that this function is what best motivates the invention of the Squeeze Theorem. A plot will also help you understand how to prove this.

8.7 Find a differentiable function $f : \mathbb{R} \to \mathbb{R}$ with the property that $\lim_{x \to \infty} f(x) = 0$, but $\lim_{x \to \infty} f'(x)$ does not exist.

8.8 Let $f(x)$ be defined as

$$f(x) = \begin{cases} 2^{-1/x^2} & \text{if } x \neq 0 \\ 0 & \text{if } x = 0 \end{cases}$$

This function has derivatives of all orders at $x = 0$, and despite the fact that $f(x)$ is not flat, all of its derivatives are zero at $x = 0$. Prove this or look up a proof, as the computation is quite involved. These functions are sometimes called *flat* functions, since they're literally so flat that they avoid detection of any curvature by derivatives. Plot the function to see how flat it means to be flat. Taylor series provide no use at these points.

8.9 There are two definitions of the number e. One is the number used as an exponent base e^x, for which the derivative of e^x is e^x. The other is $e = \lim_{n \to \infty} \left(1 + \frac{1}{n}\right)^n$. First, prove the somewhat surprising fact that this limit is not equal to 1. Second, understand why these two definitions result in the same quantity.

8.10 Find the maximum of $f(x) = x^{1/x}$ for $x \geq 0$. One method: use an approximation given by the early terms of the Taylor series of e^x. Another: maximize the logarithm of f, which has the same maximizing input.

8.11 Look up a proof of the chain rule on the internet, and try to understand it. Note that there are many proofs, so if you can't understand one try to find another. Come up with a good geometric interpretation.

8.12 Write a program that implements the binary search root-finding algorithm and compare its empirical convergence to Newton's method. Find an example input for which (gasp!) they have the same convergence rate, and analyze the statement of Theorem 8.15 to determine why this is possible.

8.13 Look up a proof of the Taylor theorem, which may depend on other theorems in single-variable calculus like Rolle's theorem or the Intermediate Value Theorem.

8.14 Look up an exposition of the degree-2 Householder method for finding roots of differentiable functions, and implement it in code.

Chapter 9
On Types and Tail Calls

By relieving the brain of all unnecessary work, a good notation sets it free to concentrate on more advanced problems, and in effect increases the mental power of the race.

– Alfred Whitehead

So far we've studied functions with a single input and a single output. This was sufficient to whet our appetites for mathematics, introducing sets, graphs, and basic calculus, and exploring a few interesting algorithms. However, the overwhelming majority of applications of mathematics rely on linear algebra and multivariable optimization. Most of advanced mathematics reaches far beyond the confines of a single variable. We require both the construction of complicated types—often abstract spaces endowed with geometric structure—and structured functions mapping between these types. The remainder of this book will explore a variety of these settings.

In Chapter 8 we worked entirely with functions whose type signature was $\mathbb{R} \to \mathbb{R}$. Although we only implicitly understood the formal notion of 'continuity'—the fact that the graphs of these functions formed contiguous curves when plotted—we concentrated intently on the interplay between the algebra (computing limits, derivatives, and using Taylor series) and geometry (the intrinsic qualitative shapes of curves). There is much more to be said for single-variable calculus. One of the most common uses of calculus is to tune parameters. For example, a car manufacturer tunes how many of each car model to manufacture based on their costs and sales figures. Another example is tuning an algorithm that fails with some measurable probability depending on a parameter you can optimize.

The recipe for optimizing parameters is quite simple, bordering on monotonous. The key insight is that it reduces the optimal parameter choice from a continuum of options to a discrete set to check by hand.

- Define $f : \mathbb{R} \to \mathbb{R}$ whose input x is the parameter of interest, and whose output you'd like to minimize (maximizing is analogous). Select a range of interest[1] $a \leq x \leq b$.

---

[1] If you don't want to restrict to a range, you have to worry about the limiting behavior of f as the input tends to $\pm\infty$. When f blows up to ∞ or $-\infty$, these are sort of "trivial" optima, as well as being unattainable by

- Compute the values $a \leq x \leq b$ for which $f'(x) = 0$ or $f'(x)$ is undefined. These are called *critical points*.

- The optimal parameter x is the minimum value of $f(x)$ where x is among the critical points, or $x = a$ or $x = b$.

The analysis of an algorithm using the above recipe is so routine that authors seldom remark on it. In research papers they often skip the entire argument assuming the reader will recognize it! Life is similar for the poor Taylor polynomial, so ubiquitous it is almost forgotten. Such brevity can seem like malicious obfuscation, but it makes sense as a cognitive "tail call optimization" for proofs.[2]

The core of the proof is the primary focus, and requires all your working memory. Optimizing a parameter using standard tools is easy once you've done it enough times. So leave it to the end and compartmentalize the two jobs: big picture comprehension versus reproducing a rote computation. Indeed, the ability to maximize an elementary function rarely depends on memory of how you created that function, so why not shed a few stack frames full of mathematical baggage while you do the real work?

This is also a justification for why one might write the statement of a theorem like we did in the last chapter.

Theorem (Convergence of Netwon's Method). *For every $k \in \mathbb{N}$, the error $e_{k+1} \leq Ce_k^2$, where C is a constant defined as*

$$C = \max_{c \leq z \leq d} \frac{|f''(z)|}{2|f'(z)|}$$

The value of C, while it needs to be defined somewhere, is not crucially important to the first-glance understanding of the statement of the theorem. The big picture is that the error vanishes quadratically as opposed to linearly. The coefficient itself can be defined afterwards to emphasize the separation of concerns between the quadratic error rate and the exact data guiding the error.

In Chapter 5, we emphasized how overloading notation with context can help reduce cognitive overload. Here it's the organizational structure of a formula or proof that contributes. It guide's the reader's focus and keeps them awake. When contrasted with the humdrum of rote optimizations and detailed constants (which can be interesting but are often of secondary concern), we desperately want to return to the profound relationship between algebra and geometry. This is the life-blood of mathematical inspiration. We'll spend the next two chapters highlighting that relationship in the context of linear algebra and multivariable calculus.

---

a fixed input. But if, for example, you can compute that both infinite limits are $-\infty$, then that leaves open the possibility of a finite global maximum.

[2] For unfamiliar readers, tail call optimization is a feature of certain programming languages whereby a function whose last operation is a recursive call can actually shed its stack frame. It doesn't need it because there is no work left after the recursive call but to return. In this way, functions written in tail-call style will never cause a stack overflow.

The step between where we've been and where we want to go is graduating to functions with more complicated inputs and outputs. In the remainder of this interlude we'll introduce two techniques to make more complicated types (realized as sets): products and quotients.

We touched on products briefly when we introduced sets in Chapter 4. We defined the direct product of sets, $A \times B$, which is the most common mathematical way to make a compound data type. It's just the set of pairs of objects (a, b), where $a \in A$ and $b \in B$. To reiterate from Chapter 4, if we repeat this operation, we tend to ignore the grouping, so that $A \times B \times C$ isn't a pair-of-pairs, but rather a tuple of length 3. We're skipping the whole "linked" part of a linked list. Likewise, given a set A, most often $\mathbb{R}$ for calculus and linear algebra, we denote by A^n the tuples of length n for some fixed $n \in \mathbb{N}$. This is just $A \times A \times \cdots \times A$ with the product occurring n times. This may seem sloppy, but there is a way to make it rigorous using the concept of a *quotient*.

In order to define quotients, we need the notion of an *equivalence relation*.[3] Given a set A, an equivalence relation is a function $f : A \times A \to \{0, 1\}$ (where $\{0, 1\}$ are thought of as booleans) with the following three properties:

1. Reflexive: $f(a, a) = 1$ for all $a \in A$.

2. Symmetric: $f(a, b) = f(b, a)$ for all $(a, b) \in A \times A$.

3. Transitive: for all $a, b, c \in A \times A \times A$, if $f(a, b) = 1$ and $f(b, c) = 1$, then $f(a, c) = 1$.

In your mind you can replace $f(a, b) = 1$ with "a and b are equivalent." A more common notation for this is a squiggle $\sim$, so that $a \sim b$ if and only if $f(a, b) = 1$, with $a \not\sim b$ if $f(a, b) = 0$. The squiggle is supposed to remind you of the equal sign without asserting that it's an equivalence relation before that fact is established.

To define an equivalence relation is to say, "Here are the terms by which I want to think of different things as the same." We are literally overloading equality with a specific implementation. As long as the equivalence relation satisfies these three properties, you rest assured it has the most important properties of the equality operator.

Let's do a simple example with $\mathbb{R}$. Let $a \sim b$ if $a - b \in \mathbb{Z}$, and $a \not\sim b$ otherwise. Check that this indeed satisfies the three properties of an equivalence relation. This equivalence relation declares that $-1/2, 1/2, 3/2, 5/2$ are all equivalent, as are $-2, -1, 0, 1, 2$. But $1/2$ is not equivalent to 1. We call the set of all things equivalent to one object an *equivalence class*. So in this case $\mathbb{Z}$ is an equivalence class, as is the set of half-fractions $\{\ldots, -3/2, -1/2, 1/2, 3/2, \ldots\}$. An exercise to the reader: show that given a set X and an equivalence relation $\sim$, the equivalence classes partition X into disjoint subsets—i.e., every $x \in X$ is in exactly one equivalence class. No two classes may overlap.

---

[3] Most math books introduce the generic notion of a *relation*, and then use relations to define functions. We'll instead use functions as the primitive type and jump straight to an equivalence relation without defining relations at all.

An equivalence relation allows us to do math in a world (on a set) in which an equivalence relation is enforced as equality. This world is the quotient.

Definition 9.1. Let X be a set and $\sim$ an equivalence relation on X. The *quotient* of X by $\sim$, denoted $X/\sim$ is the set of equivalence classes of $\sim$ in X.

Back to our example with $\mathbb{R}$, the quotient $\mathbb{R}/\sim$ has a simpler representation. Since equivalence classes partition $\mathbb{R}$, and every real number shows up in some equivalence class, we can simply identify each equivalence class in $\mathbb{R}/\sim$ with our favorite "representative" from that class.

Concretely, let's choose the representative from each class in $\mathbb{R}/\sim$ that's between 0 and 1. For the equivalence class $\{\ldots, -2/3, 1/3, 4/3, 7/3, \ldots\}$, we choose $1/3$ as the representative. Some authors like to abbreviate the equivalence class represented by a particular element (say, $1/3$) using the notation $[1/3]$, so that $[1/3] = [-2/3] = [7/3]$ are all the same equivalence class. If we also recognize that $[0] = [1]$, then we can summarize:

$$\mathbb{R}/\sim = \{[x] : 0 \leq x < 1\}.$$

Curious plants spring from fertile soil. In this world $[1+1] = [0]$, and sequence which diverges in $\mathbb{R}$ converges: $x_n = \left[\frac{n+1}{2} + \frac{1}{n}\right]$.

$\mathbb{R}/\sim$ inherits operations from $\mathbb{R}$, as if $\mathbb{R}/\sim$ were a wrapper class or a subclass for $\mathbb{R}$. Define $[x] + [y]$ to be $[x+y]$ for any representatives x, y. We must prove this definition is well-defined, i.e., that any chosen representatives result in the same operation. We need to show that if $x \sim x'$ and $y \sim y'$, then $x + y \sim x' + y'$. Indeed, $(x + y) - (x' + y')$ is an integer because $(x - x')$ and $(y - y')$ both are. Note you cannot say the same of multiplication; find a counterexample!

We can also think of $\mathbb{R}/\sim$ geometrically. Imagine standing at 0 on $\mathbb{R}$ and walking in the positive direction, say, following a sequence $x_n = 0.001n$. On $\mathbb{R}$ you increase unboundedly. When we pass to the quotient, you cycle every thousand steps. This is an animated way to see that $\mathbb{R}/\sim$ is geometrically a circle. In fact, we can design a nice bijection that makes this formal. Call $C = \{(\cos(\theta), \sin(\theta)) : 0 \leq \theta < 2\pi\}$. Define $f : \mathbb{R}/\sim \to C$ by $f(t) = (\cos(2\pi t), \sin(2\pi t))$. Observe that f is a bijection.

This example generalizes nicely. Given a function $f : X \to Y$, define $\sim_f$ so that $a \sim_f b$ if and only if $f(a) = f(b)$. Show that this is always an equivalence relation, and notice that you get a new function $f : X/\sim_f \to Y$ defined by $f([x]) = f(x)$ that is guaranteed to be a bijection. Describing an equivalence relation in terms of a function has an advantage: the structure of the function f can be used to "move" properties between one space and the other. In the case of $\mathbb{R}/\sim$ and the circle, since f is differentiable,[4] it becomes obvious that functions defined on the circle can be converted to functions on $\mathbb{R}/\sim$ with most properties intact. This is how we can ultimately say that $\mathbb{R}/\sim$ has the "same geometry" as C, though to do this in general—connect two generic spaces in which

---

[4] We'll see more about what it means for a function with multiple inputs and outputs to have a derivative in Chapter 14, but in this case it just means each component of the output is differentiable as a single-variable function of the input.

one can make geometric statements—requires extensive groundwork beyond the scope of this book. You'll know you're treading in these waters if you hear the term "manifold" or "topology."

Nevertheless, equivalence relations will be meaningful even in less technical settings, such as vector spaces (Chapter 10) and groups (Chapter 16). There the structure of the function defining the relevant equivalence relations are *algebraic* in nature. This is all to explain the primary tool mathematicians use to assert that they want to consider two different things to be the same in a principled manner. You override equality, show it meets standards of decency, and then introduce it to your friends.

We can now trivially return to set products. Define the sets

1. $L = (A \times B) \times C = \{((a,b),c) : a \in A, b \in B, c \in C\}$ (left grouping)

2. $R = A \times (B \times C) = \{(a,(b,c)) : a \in A, b \in B, c \in C\}$ (right grouping)

3. $Z = \{(a,b,c) : a \in A, b \in B, c \in C\}$ (no grouping)

Now define an equivalence relation on $L \cup R$ so that $(a,(b,c)) \sim ((a,b),c)$ for any a, b, c. The resulting quotient $(L \cup R)/\sim$ is in bijective correspondence with Z.

Another useful example is when working with modular arithmetic. Working in $\mathbb{Z}$, define $a \sim_n b$ if, to use programming syntax, `a % n == b % n`. Equivalently, $a \sim_n b$ if and only if $a - b$ is a multiple of n. The quotient space for this equivalence relation is called $\mathbb{Z}/n\mathbb{Z}$ (where $n\mathbb{Z}$ is a shorthand for multiples of n; we'll revisit this in Chapter 16). The equivalence relation for modular arithmetic is usually denoted with an operator paired with "mod n," as in

$$a \equiv b \mod n$$

Arithmetic modulo n shares most properties with normal arithmetic on integers, which makes it extremely convenient. For example, a complex expression like 8^{3000} is extremely simple mod 9. From $8 \equiv -1 \mod 9$, you get $8^{3000} \equiv (-1)^{3000} \equiv ((-1)^2)^{1500} \equiv 1$ mod 9. This tells you that 8^{3000} is one plus a multiple of 9. Similar tricks with conveniently chosen moduli can extract useful information about 8^{3000} without computing it exactly.

Beyond allowing the study of new structures or enabling convenient computations, equivalence relations and quotients reduce the mental burden of overriding equality. You establish once that there's an equivalence relation, and you pick which new operations you want to define and prove they're well-defined in terms of the equivalence classes. Once that's done, you can safely continue your mathematical enterprise suppressing the type difference between $[x]$ and x (in fact, after defining a quotient and proving its well-definition, mathematicians immediately drop the brackets). As we saw with modular arithmetic, you can also freely choose the most advantageous equivalence class representative for your task, often eliminating costly computation. It's similar to the programmer's adage: work hard now to allow yourself to be lazy later. We set up equivalence relations that focus a mental laser on the aspects we care about. Eliminate annoying and irrelevant computations, or turn them into tail calls!

Chapter 10
Linear Algebra

There is hardly any theory which is more elementary [than linear algebra], in spite of the fact that generations of professors and textbook writers have obscured its simplicity by preposterous calculations with matrices.

– Jean Dieudonné

For a long time mathematicians focused on studying interesting sets, like numbers and solutions to various equations. In Chapter 6 we saw graphs, which you can think of as an interesting kind of set. In Chapter 8 we saw sets of numbers (sequences) and sets of pairs of numbers (functions $\mathbb{R} \to \mathbb{R}$).

One could spend a lifetime studying interesting graphs or interesting sets of numbers. However, more recent trends in mathematics have shifted the main focus from studying sets with interesting structure to studying *functions* with interesting structure.[1]

To ease into it, let's first consider the familiar concept of a compiler. A compiler is a function mapping the set of programs in a source language to the set of programs in a target language. Often the target is assembly. Too often it is Javascript. These functions preserve structure! Namely, a compiler preserves the semantic behavior of a valid input program in the target language when you run it.

Moreover, a computer program written in a compiled language like C is truly only defined by the behavior of the compiler. This is never more visible than when dealing with language forms that have "undefined behavior," on which different compilers produce programs that behave in myriad unanticipated ways. Languages like C, in which behavior can vary depending on the arbitrary contents of uninitialized memory, widen such pitfalls. A single program, compiled only once for the same target machine, depends on the behavior of other programs running on the same machine beyond its own code.

Of course, this isn't how we *want* to work with programs. We want to eliminate this tenuous disconnect as much as possible. Techniques addressing the problem, such as virtual machines,[2] are impressive feats of engineering. The ideal result is that we may

---
[1] In Chapter 8 we did study functions with interesting structure, i.e., differentiable functions, but we didn't describe them as structure preserving transformations.

[2] I admit, I am not an expert at low-level architecture. The one good example of this I have is the LLVM

peacefully ponder programs in their most natural environment: the semantics defined by a language's documentation.

Just as the burden of manually managing registers encumbered programmers of decades past, these issues of disparate compilers and uninitialized memory are digressions. It's nuance in the program-text-to-runtime transformation that we don't want to bother with. We judge programs by their semantics: two programs are logically the same if they behave the same on every input. This understanding shouldn't depend on idiosyncrasies of compilers or varying environments, and we feel unhygienic when it does.

In mathematics, when complexity and notational grime builds up we use essentially the same tool: abstraction. We add a layer of indirection that allows us to write arguments that say, "these two things are the same" in the context that matters for the task at hand, and we exhibit bijections and equivalence relations to formalize the connection. This allows us to identify and isolate structure in new settings, and mentally disregard impertinent information.

The *vector space* is a foundational example of such structure. It's the basic object of study in linear algebra, a subfield best studied from this function-centric perspective. The main tool that we use to relate two vector spaces is the *linear map*. As we will see, linear maps have a useful computational representation called *matrices* (singular, *matrix*). Matrices are "compiled" representations of a linear map in a particular environment (looking ahead, the particular choice of a basis for the vector space). The magic appears when we deeply understand how the operations on matrices translate back and forth to operations on linear maps, and how it all relates to geometry.

Better yet, because linear algebra is relatively elementary, one can appreciate it without years of prior study. The only machinery we need is the working terminology of sets and functions. And finally, linear algebra is obscenely practical. The application we'll see in this chapter, singular value decomposition, is a staple of data science and machine learning.

The second, more practical goal of this chapter is to prepare us for multivariable calculus and optimization. These subjects use vectors, vector spaces, and linear maps as primitive types.

So let's jump in.

10.1 Linear Maps and Vector Spaces

The definition of a linear map requires a bit of groundwork to nail down precisely, but the crucial underlying intuition is simple. A function $f : A \to B$ is called *linear* if the following identity[3] is always true, no matter what $x, y \in A$ are:

assembly language, which is an independent representation of assembly code that compilers may target, and then other compilers finish the job using platform-specific optimizations.

[3] We also need preservation of scalar multiples, but we are in inspiration mode. The formal definition is in Section 10.2

$$f(x+y) = f(x) + f(y)$$

Simple, right? There's something missing from this, so take a moment to identify what that is.

The problem is that we don't know what "+" means in this context. Because I used the + symbol you may have guessed that A and B are sets of numbers, but this need not be the case. Instead, we'll generalize the properties of addition that we care about, and the result will be called an abstract *vector space*. Any set might conceivably be a vector space, and we call the elements of a vector space *vectors*. Let's see a mostly complete definition that establishes the basic rules of a vector space. If you want to prove a set with a chosen + is a vector space, you need to establish that all these properties hold.

Definition 10.1. A set V is called a *vector space over* $\mathbb{R}$ if it has two operations + and $\cdot$ with the following properties:

1. $+ : V \times V \to V$ is a function on pairs of vectors[4] and $\cdot : \mathbb{R} \times V \to V$ is a function mapping a real number and a vector to a vector. Often the values in $\mathbb{R}$ are called *scalars*, and using the operation $\cdot$ is often called *scaling*. We mean "scaling" in the sense that this "stretches" or "shrinks" the vector by the amount represented by the scalar. Rather than denoting the operation by the strange prefix notation $+(x, y)$ and $\cdot(a, v)$, we'll use the usual infix notation $x + y$ and $a \cdot v$.

2. + obeys all the identities you expect it to, for example, that $v + w = w + v$ and $(v + w) + x = v + (w + x)$.

3. + and $\cdot$ distribute and "commute" with each other, i.e., identities you expect to be true from arithmetic, like $c \cdot (v + w) = c \cdot v + c \cdot w$, are required.

4. There is a special vector denoted 0 in V, which acts like zero should with respect to addition. In particular, $0 + v = v$ for every v.

5. Every $v \in V$ has an *additive inverse*, i.e., a vector w for which $v + w = 0$. This special vector is denoted $-v$, and is used in conjunction with + to perform subtraction: $u - v = u + (-v)$.

6. Finally, if I take 0 the *scalar*, and multiply it by any vector v, I must get the zero *vector* as a result. If I make the zero vector bold,[5] this is the same as requiring $0 \cdot v = \mathbf{0}$ for every $v \in V$. Likewise, $1 \cdot v = v$ for all $v \in V$.

---

[4] Another word commonly used here is that V is *closed* under this operation: applying + to vectors in V stays in V. We ensure this by stating the codomain of + is V, but it is a more stringent requirement if the vector space is built from a subset of some well-known set.

[5] Some authors write all vectors bold, but I will only do it when disambiguation is needed. More often than not the choice of letters suffices, u, v, w, x, y, z for vectors and a, b, c or Greek letters for scalars.

This is a monumental definition, and it's not even the most general definition (see the Chapter Notes for the most rigorous definition). There are a few things I want to remark about why the definition is what it is, but before we do let's write down some examples.

The simplest vector space is $\mathbb{R}$, with $\mathbb{R}$ also being the scalars. In this case vectors are just numbers, $+$ is addition of real numbers, and $\cdot$ is multiplication of real numbers. The number zero is the identity and the zero vector. Nothing about this should be surprising.

A more interesting example is one we're familiar with from Chapter 2, polynomials. Call V the set of all polynomials of a single variable. If t is our variable then $1 + t \in V$ as well as 7 (a degree-zero polynomial) and $\pi t + 700 t^{99}$. The operation $+$ is defined by adding coefficients term-wise, and $c \cdot p(t)$ by scaling each coefficient of p by c. The zero polynomial, $p(t) = 0$ for every t, is the zero vector.[6] As an aside, the secret sharing application from Chapter 2 can also be understood and proved by appealing to polynomials as a vector space; the evaluation-at-a-point function $\text{eval}_a(p)$ defined by $p \mapsto p(a)$ is a linear map. See the exercises for an exploration of this.

Even more general is the vector space of *all* functions $f : X \to \mathbb{R}$ for any set X. As an exercise to the reader: go through the conditions from Definition 10.1 and figure out what $+$ and $\cdot$ could mean. There should only be one natural and simplest option. As a sepcific example, the space of all *differentiable* functions $f : \mathbb{R} \to \mathbb{R}$ is a vector space, and the derivative operation $f \mapsto f'$ is a distinguished linear map for that space.

The final example is $\mathbb{R}^n = \mathbb{R} \times \mathbb{R} \times \cdots \times \mathbb{R}$, the set of all tuples of length n of real numbers. The elements of $\mathbb{R}^n$ are *vectors* in the sense the reader is probably used to. A vector is just a list of numbers, or in many programming languages a list whose elements have the same type. But for us vectors need these extra operations $+$ and $\cdot$, so let's define them now. The operation $+$ on tuples is entry-wise addition. This means

$$(a_1, a_2, \ldots, a_n) + (b_1, b_2, \ldots, b_n) = (a_1 + b_1, \ldots, a_n + b_n).$$

Similarly, $c \cdot (a_1, \ldots, a_n) = (ca_1, \ldots, ca_n)$, where on the right hand side the multiplication happening in each coordinate is the usual product of real numbers. The zero vector is $(0, 0, \ldots, 0)$, and the inverse of $(a_1, \ldots, a_n)$ is $-1 \cdot (a_1, \ldots, a_n) = (-a_1, \ldots, -a_n)$. All of the operations behave nicely because they're applied independently to each entry, and each entry is just arithmetic in $\mathbb{R}$, which has all of the desired properties.

As with all of the examples of vector spaces, the definition of the specific vector space is entirely contained in the implementation of the operations $+$ and $\cdot$. The miniature proofs that $+, \cdot$ have the needed properties constitute a proof that the chosen implementation is a vector space. This proof is rarely a challenge, as $\mathbb{R}^n$ is our main workspace for applications, so we'll brush over most of those details.

---

[6] Readers who know a bit about abstract algebra and number theory will protest: this is only true when the set of coefficients of the polynomial has certain properties. One could, for instance, define a family of polynomials where the coefficients are binary and addition behaves like binary XOR on each term. In this case there are polynomials with nonzero coefficients that are zero on all (binary) inputs. No such stumbling block exists for real numbers.

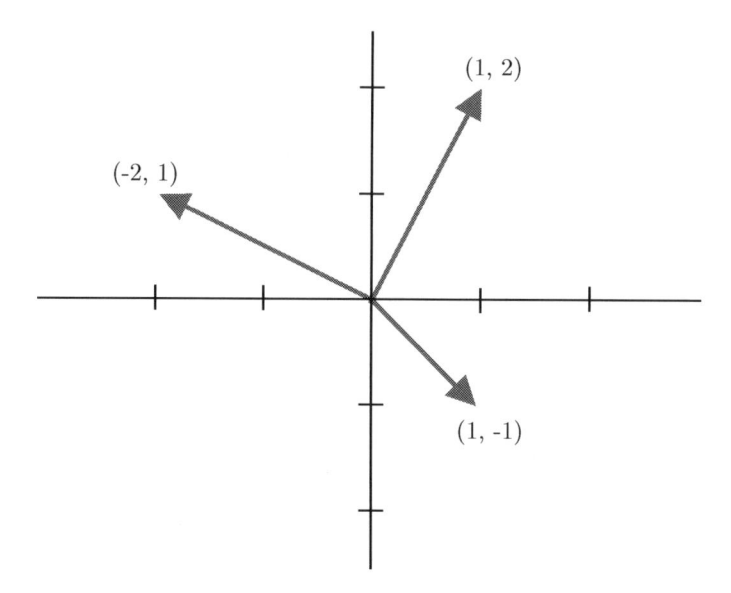

Figure 10.1: Examples of vectors.

With a few examples of vector spaces in our minds, let's return to why Definition 10.1 is the way it is. The reason is that it's the simplest way to define what addition means in a context that is useful for geometry (defining an "algebra" for geometric objects). The first thing a geometry needs is a set of points in space. Note that I'm using the words "points" and "space" informally, to appeal to your intuition that a grain of dirt "occupies a point" in the real world. A vector space is meant to appeal to that intuition, but with "point" replaced by "vector," and "space" just meaning the set of all vectors we allow.

Returning to our vector space, points are indeed simply vectors in $\mathbb{R}^n$. In Figure 10.1, we draw some vectors in $\mathbb{R}^2$ for the ease of visualization. For a reason we'll explain shortly, we also draw these points as arrows from the zero vector (the zero vector is called the "origin," in graphical parlance).

The "position" of a point specified by such an arrow is at the non-origin end of the drawn line segment. This choice of drawing from the origin also implies that every vector has a direction, which further implies there will be distances, angles, etc. More immediately, we can add two vectors by adding their coordinates. Geometrically this involves moving the tail of one arrow to the head of the other and drawing an arrow from the origin to the end of the resulting path. In Figure 10.2, we can add the two solid vectors to get the dashed vector. The transparent dotted vector shows this geometric motion of "moving the tail to the head."

In geometry we like lines, and scaling vectors allows us to have them here too. A *line* is the set of all ways to scale a single nonzero vector. In symbols, a line through the origin and v is the set $L_v = \{c \cdot v : c \in \mathbb{R}\}$. For example, drawn in Figure 10.3 you can scale $v = (1, 2)$ by a factor of 2 to get $(2, 4)$, shrink it down to $(0.5, 1)$, or scale it negatively to $(-2, -4)$. The set of all possible ways to do this gives you all the points on the line

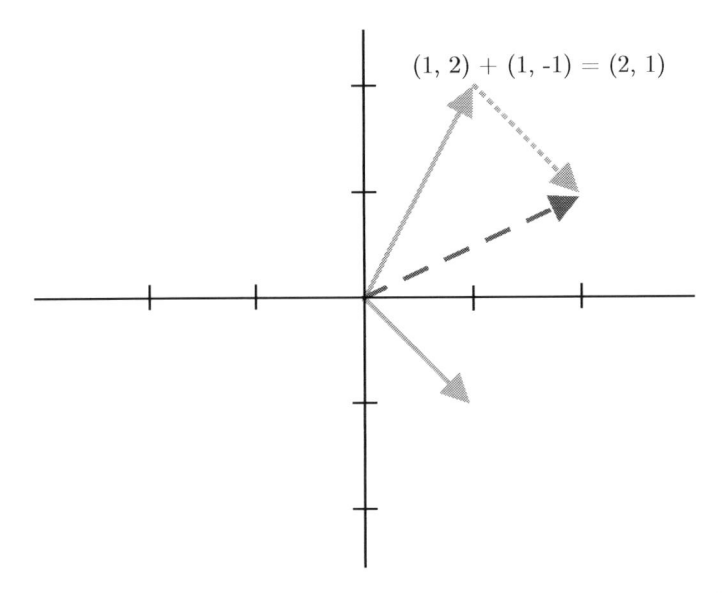

Figure 10.2: An example of vector addition. The dashed vector is the sum of the two solid vectors.

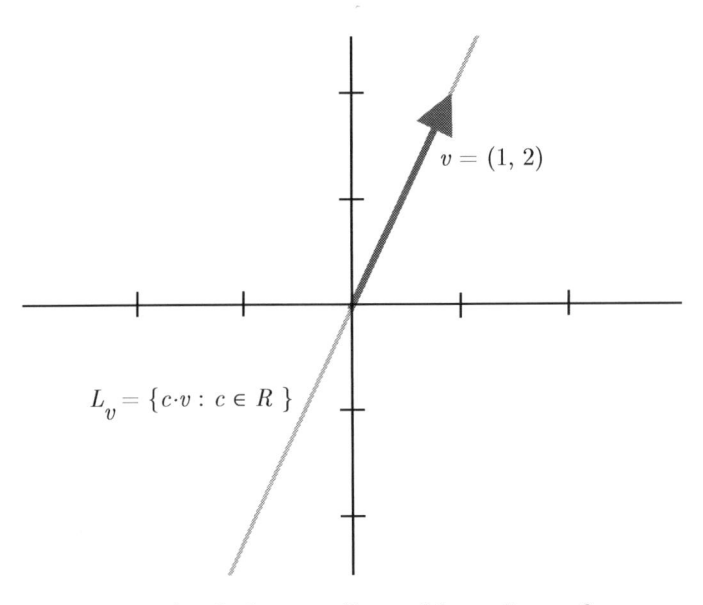

Figure 10.3: An example of a line as all possible scalings of a nonzero vector.

through $(1, 2)$.

You can further get a line *not* passing through the origin by taking some other vector w and adding it to every point on the line, i.e. $\{w + c \cdot v : c \in \mathbb{R}\}$. This is the line through the point w parallel to L_v, shown in Figure 10.4.

All this said, a plain vector space isn't *quite* enough to get all of geometry. For example, we can't compute distances or angles without more structure in the vector space. We will

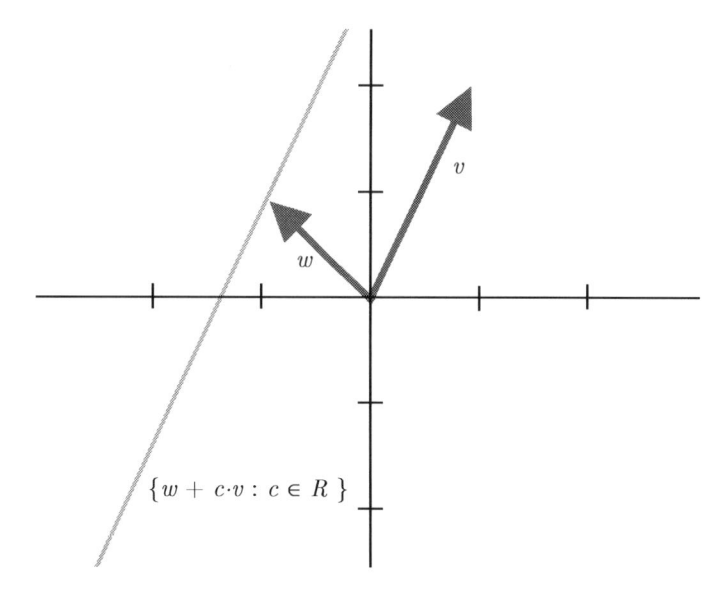

Figure 10.4: An example of a line as the span of a vector, shifted away from the origin by a second vector.

complete the geometric picture by the end of the chapter, but for now we see that connections between vectors and geometry make sense. We'll keep this geometric foundation in mind while dealing with linear maps more abstractly (which, to be frank, is the hard part of linear algebra). Our task for now is to study where Definition 10.1 takes us.

10.2 Linear Maps, Formally This Time

The simplest observation is that once we have a vector space, the definition of a linear map can apply to any vector space. It's just an iota more complicated than at the beginning of the chapter.

Definition 10.2. Let X, Y be vector spaces with $+_X, \cdot_X$ being the operations in X and $+_Y, \cdot_Y$ in Y. A function $f : X \to Y$ is called a *linear map* if the following two identities hold for every $v, w \in X$ and every scalar $c \in \mathbb{R}$:

1. $f(v +_X w) = f(v) +_Y f(w)$

2. $f(c \cdot_X v) = c \cdot_Y f(v)$

This notation $+_X, \cdot_X$ burns my eyes, so we'll drop it and understand that when I say $f(v + w) = f(v) + f(w)$, I mean that the $+$ on the left hand side is happening in X and the $+$ on the right hand side is happening in Y. Likewise for scaling, $f(cv) = cf(v)$. Any other interpretation would be a fatal type error. Moreover, as we go on I'll begin to drop the $\cdot$ in favor of "juxtaposition", so that if a is a scalar and v is a vector, it's understood that $av = a \cdot v$. I will use the dot only when disambiguation is needed.

Here's a simple example of a linear map. Let X be the vector space of polynomials, and $Y = \mathbb{R}$. Define the *evaluation at 7* function, which I'll denote by $\text{eval}_7 : X \to \mathbb{R}$, as $\text{eval}_7(p) = p(7)$. Let's check the two conditions hold. If p, q are two polynomials, then

$$\text{eval}_7(p) + \text{eval}_7(q) = p(7) + q(7) = (p+q)(7)$$

In just a little bit more detail at the expense of a big ugly formula, if $p = a_0 + a_1 x + \cdots + a_k x^k$ and $q = b_0 + b_1 x + \cdots + b_m x^m$, then $p+q$ is the polynomial formed by adding the coefficients together. If we suppose that m is the larger of the two degrees, then

$$(p+q)(7) = (a_0 + b_0) + (a_1 + b_1)7 + \cdots + (a_k + b_k)7^k + \\ + b_{k+1} 7^{k+1} + \cdots + b_m 7^m.$$

And we can distribute and rearrange all these terms to get exactly $p(7)+q(7)$. Likewise, $\text{eval}_7(c \cdot p) = c \cdot p(7)$. Since the number 7 was arbitrary, the same logic shows that eval_a for any scalar $a \in \mathbb{R}$ is a linear map.

A second and completely arbitrary example is the map $f : \mathbb{R}^3 \to \mathbb{R}^2$ defined by $(a, b, c) \mapsto (-2a + 3b, c)$. Verify as an exercise that this is a linear map.

For the rest of the chapter, linear maps are the only kind of function we care about for vector spaces. The reason, which we'll spend the rest of the chapter trying to understand, is that linear maps are the maps which preserve the structure of a vector space. Indeed, we *defined* them to preserve the two operations that define a vector space! But as we'll see this covers all the bases. For example, the following fact is true for any vector space: linear maps preserve the zero vector.

Proposition 10.3. *If X, Y are vector spaces and $f : X \to Y$ is a linear map, then $f(0) = 0$.*

As I did with $+$ and $\cdot$, I'm using the same symbol 0 for the additive identity in both vector spaces. In light of this fact it's not so surprising: if there's a "zero" in every vector space, and every linear map (the only maps we care about) preserve the "zero," then we can really call it "the" zero.

The proof of this fact is direct. That is, we'll directly use the definition of a linear map and a vector space, and the proof will just "fall out" from the definitions. In fact, I'll give two proofs. To distinguish 0 the vector from 0 the scalar, I'll make the vector bold, like $\mathbf{0}$.

Proof 1. Let's use the fact that $\cdot$ is preserved by a linear map. First, $f(\mathbf{0})$ is the same as $f(0 \cdot \mathbf{0})$. Since f is linear, this is the same as $0 \cdot f(\mathbf{0})$. But $0 \cdot v = \mathbf{0}$ no matter what v is. Putting these two together,

$$f(\mathbf{0}) = f(0 \cdot \mathbf{0}) = 0 \cdot f(\mathbf{0}) = \mathbf{0},$$

which is what we wanted to prove.

Proof 2. This proof is similar, but uses the fact that $\mathbf{0} + \mathbf{0} = \mathbf{0}$. Indeed,

$$f(\mathbf{0}) = f(\mathbf{0} + \mathbf{0}) = f(\mathbf{0}) + f(\mathbf{0}),$$

i.e., $f(\mathbf{0}) = f(\mathbf{0}) + f(\mathbf{0})$, and subtracting $f(\mathbf{0})$ from both sides gives $\mathbf{0} = f(\mathbf{0})$.

So linear maps preserve zero. Now it's your turn: go do the similar proofs in Exercises 10.1-10.4 which claim basic facts about linear maps.

10.3 The Basis and Linear Combinations

Though we defined a vector space as a set with two operations, in reality you can't do much with that mental model. We need more concrete and computational tools to work with a vector space. The first tool is called a *basis*. In short, a basis for a vector space V is a minimal set of vectors $B = \{v_1, \ldots, v_n\}$ from which you can get all vectors in V by adding and scaling vectors in B.

The simplest example of this is for $V = \mathbb{R}^2$. Let $e_1 = (1, 0)$ and $e_2 = (0, 1)$. Then any vector (a, b) can be written as $a \cdot (1, 0) + b \cdot (0, 1)$. More generally, $\mathbb{R}^n$ has a basis of the n vectors which have a 1 in a single coordinate and zeroes elsewhere. E.g., $e_2 = (0, 1, 0, \ldots, 0)$. This is often called the *standard basis* of $\mathbb{R}^n$ and denoted with e's as $\{e_1, \ldots, e_n\}$.

Two things to note about the $\mathbb{R}^2$ example. First, this is far from the only basis. Almost any two vectors you can think of form a basis. Say, $\{(3, 4), (-1, -5)\}$, and one way to show this is a basis is to write a known basis like $(1, 0)$ and $(0, 1)$ in terms of these two vectors:

$$(1, 0) = \frac{5}{11}(3, 4) + \frac{4}{11}(-1, -5)$$

From the above, one can write $(0, 1)$ as $\frac{1}{4}((3, 4) - 3 \cdot (1, 0))$. Once $(1, 0)$ and $(0, 1)$ are expressed in terms of your basis, you can get any vector by using $(c, d) = c(1, 0) + d(0, 1)$. Convince yourself of this by expressing $(2, -1)$ in terms of our example basis. By the way, I calculated the fractions $5/11$ and $4/11$, by writing down the equation

$$a(3, 4) + b(-1, -5) = (1, 0),$$

which is really a set of two equations, one for each coordinate:

$$3a - b = 1$$
$$4a - 5b = 0$$

Solving for a and b gives $a = 5/11$ and $b = 4/11$. The fact that this works for most pairs of vectors you can think of is no coincidence, but we'll return to that later in the chapter. The point for now is that there are many possible bases ("BAY-sees", the plural of basis) of a vector space, and each basis allows you to write any vector in the vector space by summing and scaling the vectors in the basis.

The second note is that a basis can be thought of as an alternative coordinate system for a vector space. In $\mathbb{R}^2$ we usually think of coordinates for points by specifying their x- and

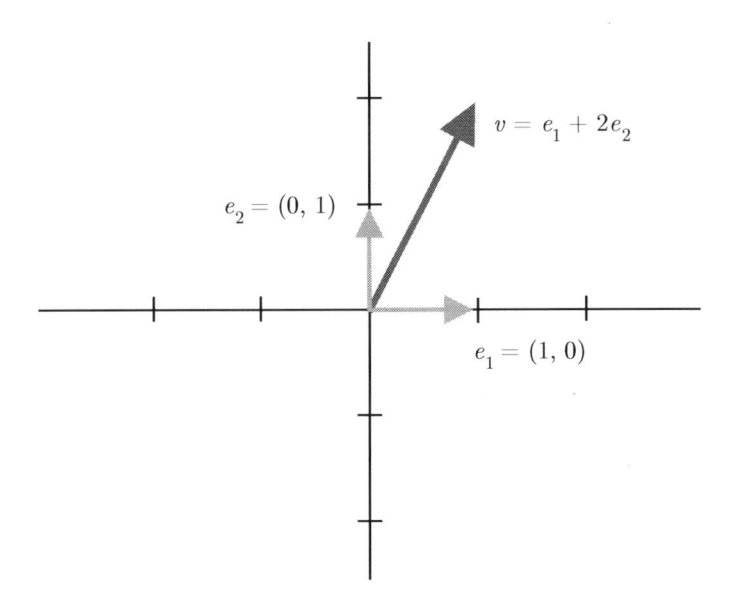

[Figure: vector diagram showing $e_2 = (0,1)$, $e_1 = (1,0)$, and $v = e_1 + 2e_2$]

Figure 10.5: Assembling a point $(1, 2)$ as the linear combination of basis vectors representing x and y coordinates.

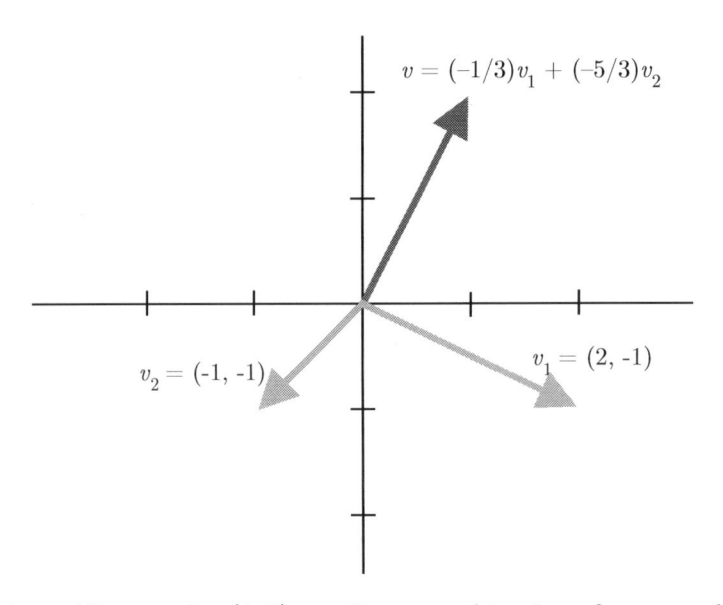

[Figure: vector diagram showing $v_2 = (-1,-1)$, $v_1 = (2,-1)$, and $v = (-1/3)v_1 + (-5/3)v_2$]

Figure 10.6: Assembling a point $(1, 2)$ as a linear combination of two new basis vectors.

y-coordinates (i.e., using the standard basis, e_1, e_2). However, once we're fluid with linear algebra we realize that saying "the x- and y-coordinate" is an arbitrary choice, and one could just as easily have chosen $v_1 = (2, -1)$, $v_2 = (-1, 1)$ as a basis and expressed the same points by their v_1-coordinate and v_2-coordinate, the coefficients needed to write a point using sums-and-scales of v_1, v_2. In this case, the vector in the diagram in Figure 10.6 is represented as $(-\frac{1}{3}, -\frac{5}{3})$.

This process of expressing a vector's coordinates with respect to a different basis is analogous to the process of writing integers in a different number base, such as binary or hexadecimal. You choose a base that's useful to you. And just like with numbers, if you find a basis with useful properties, you study it in depth and learn its computational secrets.

The brief and formal way to say a vector v "can be written using sums and scales of other vectors" is the following definition.

Definition 10.4. Let $v_1, v_2, \ldots, v_n$ be a set of vectors in a vector space V, and let x be a vector in V. We say x is a *linear combination* of $v_1, \ldots, v_n$ if there are scalars $a_1, \ldots, a_n \in \mathbb{R}$ with

$$x = a_1 v_1 + \cdots + a_n v_n = \sum_{i=1}^{n} a_i v_i$$

In particular, any way one could "add and scale" vectors reduces to this form, provided one is willing to distribute scalar multiplication over addition, expand, and group all the terms. This is the standardized way to express the existential claim that x can be "built" up from the v_i, like how a polynomial has a regularized form, even though polynomials generically encode all ways to add and multiply a number.

A bit of common terminology is the *span* of a set B of vectors, which is the set of all linear combinations of those vectors. That is,

$$\text{span}(v_1, \ldots, v_k) = \{a_1 v_1 + \cdots + a_k v_k : a_i \in \mathbb{R}\}$$

When we said informally that a basis is a set of vectors from which you can "get all vectors in V," we really meant that a basis is a set of vectors whose span is V. This is almost complete, but we need minimality.

Definition 10.5. Let V be a vector space. A set $\{v_1, \ldots, v_n\} \subset V$ is called a *basis* of V if its span is V and if it is *minimal* in the property of spanning V. That is, if you remove any vector from a basis $\{v_1, \ldots, v_n\}$, the resulting set does not span V.

This definition makes it clear why we don't say things like "$\{(1,0), (2,0), (3,0), (0,1)\}$ is a basis for $\mathbb{R}^2$." Because while it does span $\mathbb{R}^2$, it includes superfluous information. We want our definitions to capture a notion as efficiently as possible.

We will have a lot more to say about vector space bases. Many insights and applications of linear algebra revolve around computing a clever basis of a vector space. But first we need a few more tools. One of the most important definitions in elementary linear algebra is related to the existence and uniqueness of linear combinations.

Definition 10.6. Let V be a vector space, and $v_1, \ldots, v_n \in V$ be nonzero vectors. The set $\{v_1, \ldots, v_n\}$ is said to be *linearly independent* if no v_i is in the span of the other vectors $\{v_j : j \neq i\}$. Informally we will also say the *list* $v_1, \ldots, v_n$ is linearly independent, though the ordering of the vectors has no consequence.

Another, equivalent definition of linear independence, and one that's easier to work with in proofs, is that the only way to write the zero vector as a linear combination of $v_1, \ldots, v_n$ is if all the coefficients a_i are zero. In other words, there is no *nontrivial* way to write zero as a linear combination.

$$0 = a_1 x_1 + \cdots + a_n x_n \Rightarrow a_i = 0 \text{ for all } i$$

Another equivalent (but seemingly more restrictive) way to express linear independence is to say that B is linearly independent if every vector in span(B) has a *unique* expression as a linear combination of vectors in B. Indeed, if some vector x could be written as both $\sum_{i=1}^n a_i v_i$ and $\sum_{i=1}^n b_i v_i$, then the difference $\sum_{i=1}^n (a_i - b_i) v_i$ would be a nontrivial way to write the zero vector! It's nontrivial because some a_i and b_i have to be different, by our assumption that x has two different representations.

For example, in $\mathbb{R}^2$ the set $\{(1,0), (0,1)\}$ is linearly independent, as is the set $\{(3,4), (-1,-5)\}$. However, $\{(1,0), (3,4), (-1,-5)\}$ is not linearly independent (we call it linearly *dependent* to avoid the double-negative) because, as we saw, $(1,0)$ is a linear combination of the other two vectors.

Linear independence provides a different perspective on the concept of a basis, which will lead us to Theorem 10.8 and allow us to have a coherent definition of a vector space's dimension.

Theorem 10.7. *Let V be a vector space. Let $B = \{v_1, \ldots, v_n\}$ be a set of linearly independent vectors in V, and suppose it's maximal in the sense that if you add any new vector to B, then the resulting set is linearly dependent. Then B is a basis for V.*

Proof. Suppose $B = \{v_1, \ldots, v_n\}$ is maximally linearly independent. Our task is to prove that B is a basis of V. By definition, this means we need to show both that span(B) = V and that one cannot remove any vectors from B and still span V.

For the first, let $x \in V$ be a vector, and our task is to write x as a linear combination of the vectors in B. First, we form the set $C = B \cup \{x\}$ by adding x to B. Since B is maximally independent, C is a linearly dependent set. That means there are some $a_i \in \mathbb{R}$ that allow us to write

$$\mathbf{0} = a_0 x + a_1 v_1 + \cdots + a_n v_n,$$

and not all the a_i are zero. Note a_0 is the cofficient of x, the newly added vector. Moreover, $a_0 \neq 0$ since, if it were, that would provide a nontrivial linear combination of $\mathbf{0}$ using only the vectors in B, which contradicts the assumption that B is linearly independent.

We can then safely rearrange to solve for x:

$$x = -\frac{1}{a_0}(a_1 v_1 + \cdots + a_n v_n)$$

This proves that $x \in \text{span}(B)$. Beacuse x was chosen arbitrarily from V, this proves that $V \subset \text{span}(B)$. Since $\text{span}(B) \subset V$ by definition of a vector space,[7] we've shown $\text{span}(B) = V$ (cf. Definition 4.2 for a reminder on using subsets to prove set equality).

Second, we need to show that B is minimal with respect to spanning V. Indeed, you cannot write v_1 as a linear combination of $v_2, \ldots, v_n$, because $v_1, \ldots, v_n$ form a linearly independent set! Hence, removing v_1 from B would make the resulting set not span V; ($v_1 \notin \text{span}\{v_2, \ldots, v_n\}$). The same goes for removing any v_i.

□

The above proof makes it clear that for any $x \notin B$, the statements "$x \in \text{span}(B)$" and "$B \cup \{x\}$ is a linearly dependent set" are logically equivalent. This theorem also provides a simple algorithm to construct a basis (though it's not quite concrete enough to implement). Start with $B = \{\}$. While there exists some vector not in $\text{span}(B)$, find such a vector and add it to B. When this loop terminates, B is a basis.

With linear independence, spanning, and bases in hand, we can define dimension and finally the matrix.

10.4 Dimension

While the concept of a basis seems relatively underwhelming at first, it unlocks a world of use. The first thing it allows us to do is measure the size of a vector space. We can do this because of the following fact:

Theorem 10.8. *Let V be a vector space. Then every basis of V has the same size.*

Proof. This proof hinges on the claim that if $U = \{u_1, \ldots, u_n\}$ is a list of n linearly independent vectors in V (perhaps not maximal), and $W = \{w_1, \ldots, w_m\}$ is a list of m vectors which span V (perhaps not minimally), then $n \leq m$. The theorem follows because if U and W are both bases, then they are both independent and spanning, meaning both $n \leq m$ and $m \leq n$, so $n = m$. To prove the claim, we use an iterative algorithm that transforms W into U as much as possible.[8] This will work by replacing each item from W by one from U until we run out of vectors from U. Connecting to the fancy and useful terminology from Section 4.1, we're building an injection $U \to W$ one element at a time, and the existence of an injection $U \to W$ implies $|U| \leq |W|$.

Start by taking u_1, removing it from U, and adding it to W. By the fact that W spans V, we can write u_1 as a linear combination of the w_i in which some coefficient, say a_1 for w_1, is nonzero.[9]

---

[7] $B \subset V$ is a set of vectors, and the closure properties of a vector space ensure they stay in V.
[8] The only other proof of this theorem I'm aware of uses all kinds of needless machinery regarding homogeneous systems of linear equations. Algorithms save the day!
[9] This is another example of the mathematical sleight of hand called "without loss of generality." What we really mean is: take whichever w_i has a nonzero coefficient, and use that going forward. However, since we're planning to do this step iteratively, if we wanted to be precise we'd have to keep track of which indices were selected, and writing that down is painful (with a sub-index like $w_{i_1}, w_{i_2}, \ldots$). Instead we say, "let's

$$u_1 = a_1 w_1 + a_2 w_2 + \cdots + a_m w_m$$

This means we can rearrange the above to solve for w_1 in terms of $u_1, w_2, w_3, \ldots, w_m$, and hence we can remove w_1 from $W \cup \{u_1\}$ without changing the fact that what remains spans V. Call this resulting set $W_1 = \{u_1, w_2, w_3, \ldots, w_m\}$, and call $U_1 = V - \{u_1\}$. Repeat this process with u_2, forming W_2, U_2, and keep doing it until you get to $U_n = \{\}$, and W_n. In each step we can always remove a new w_i—that is, we can find a w_i with a nonzero coefficient—because all of the u's that we're adding are linearly independent, while W_i is still spanning. So the algorithm will reach the n-th step, at which point either all of W is replaced by all of U (i.e. $n = m$), or there are some w_i left over ($n < m$).

□

Definition 10.9. The *dimension* of a vector space V is the size of a basis. Denote the dimension of V by $\dim(V)$.

Theorem 10.8 hence provides well-definition for the notion of the dimension of a vector space. Dimension is an invariant, because it does not depend on which basis you choose. This reinforces our intuitive understanding of what dimension should be for $\mathbb{R}^n$, i.e., how many coordinates are needed to uniquely specify a point. So $\mathbb{R}$ is one-dimensional, the plane $\mathbb{R}^2$ is two-dimensional, physical space at a fixed instant in time is 3-dimensional, etc. The dimension of the space doesn't (and shouldn't) depend on the perspective, and for linear algebra the perspective is the choice of a basis.

We end this section with the notion of a subspace.

Definition 10.10. Let V be a vector space, and let $W \subset V$ be a subset. We call W a *subspace* if the same operations from V also make W a vector space.

In particular, to be a subspace all operations involving only vectors in W must evaluate to vectors in W, and W must have the same zero vector as V.

The simplest nontrivial example of a subspace is in $V = \mathbb{R}^2$. A subspace here is a line through $(0, 0)$, equivalently the span of a single nonzero vector $v \in V$. Likewise, the span of two linearly independent vectors $v, w \in \mathbb{R}^3$ forms a two-dimensional subspace. Geometrically the subspace is the plane containing $(0, 0, 0)$ and v and w. In general, any set of $k \leq n$ linearly independent vectors in $\mathbb{R}^n$ spans a k-dimensional subspace of $\mathbb{R}^n$, which corresponds to a k-dimensional plane. Such things are impossible to visualize, but we understand them simply as a set, the span of the chosen vectors.

As these two examples suggest, subspaces can be formed easily by taking a basis B of V, and picking any subset of B to form a basis of $W \subset V$. The converse also works: if you start with a set of vectors $A = \{v_1, \ldots, v_k\}$ spanning a k-dimensional subspace of an

---

just relabel the vectors post-hoc so that w_1 is one of the vectors with a nonzero coefficient." You often need a mental spot-check to convince yourself this doesn't break the argument; in this case, the order of the w_i is irrelevant. If we had to program this, we might be forced to keep track, perhaps for efficiency gains (relabeling would require a full loop through the w_i). But in mathematical discourse we can flexibly and usefully change the data to avoid crusty notation and get to the heart of the proof.

n-dimensional vector space V, you can iteratively add vectors not in the span of A until the resulting set spans all of V. This process, though not well-defined algorithmically, is existentially possible, and it's called *extending* A to a basis of V. In Chapter 12 we'll see a concrete algorithm for it called the *Gram-Schmidt process,* which produces extra useful properties of the resulting basis.

10.5 Matrices

Now we can finally get to the heart of linear algebra.

Linear maps seem relatively complicated at first glance, but in fact they have a rigid structure uniquely determined once you fix a basis in the domain and codomain. Let's draw this out and discover what that structure is. In this section English letters v, w, x, and y will always be vectors, while Greek letters α, β, and γ will be scalars.

Start with a linear map $f : V \to W$, maybe given by some formula. We want to compute f on an input x. You choose a basis $\{v_1, \ldots, v_n\}$ and a basis $\{w_1, \ldots, w_m\}$ for V and W, respectively.[10] Now fix $x \in V$. Since the v_i form a basis, there is some way to write x as a linear combination of the v_i, say

$$x = \alpha_1 v_1 + \alpha_2 v_2 + \cdots + \alpha_n v_n$$

Crucially, f is a linear map, so we can break $f(x)$ up across the input.

$$f(x) = \alpha_1 f(v_1) + \cdots + \alpha_n f(v_n)$$

If we know what f does to the basis vectors, the above formula tells us how f behaves on x, or any arbitrary vector. In other words, a linear map is completely determined by how it acts on a basis. This is such an important revelation that I want to shout it from the mountaintops! Chisel it on the forearm of the Statue of Liberty! Put a fuchsia HTML marquee on the front page of Google!

Theorem 10.11. *A linear map is completely determined by its behavior on a basis!*

This implies the data representation of any linear map $f : V \to W$ can be reduced to a fixed number $\dim(V)$ of vectors in W: the output of f for each input basis vector.

Now let's say we know that $f(v_1) = y_1$, $f(v_2) = y_2$, etc., the vectors y_i now being in W. We can do the same decomposition of each y_i in terms of the chosen basis for W.

$$f(v_1) = y_1 = \beta[1,1]w_1 + \cdots + \beta[1,m]w_m$$
$$f(v_2) = y_2 = \beta[2,1]w_1 + \cdots + \beta[2,m]w_m$$
$$\vdots$$
$$f(v_n) = y_n = \beta[n,1]w_1 + \cdots + \beta[n,m]w_m$$

---

[10] I just want to point out how, even though I'm casually defining this basis here, you will remember that the lower-case v's are the basis of V while the w's are the basis of W. This is the kind of notational mnemonic mentioned earlier that mathematicians use everywhere.

I'm using familiar array-index notation to hint at where we're going. The structure of the matrix will fall out from our analysis. The point of the notation is that the first index, the i in $\beta[i, j]$, tells you which basis vector v_i of V you're mapping through f to get y_i, and the second index j identifies the coefficient of the basis of W in the output (that of w_j).

To write $f(x)$ in terms of the basis for W, we substitute the expansion of each $f(v_i)$ into the formula $f(x) = \sum_i \alpha_i f(v_i)$.

$$f(x) = \alpha_1(\beta[1,1]w_1 + \cdots + \beta[1,m]w_m)$$
$$+ \alpha_2(\beta[2,1]w_1 + \cdots + \beta[2,m]w_m)$$
$$+ \cdots$$
$$+ \alpha_n(\beta[n,1]w_1 + \cdots + \beta[n,m]w_m)$$

If you expand and regroup the terms so that the w_j's are on the outside (so you can read off their coefficients), you get

$$f(x) = (\alpha_1\beta[1,1] + \alpha_2\beta[2,1] + \cdots + \alpha_n\beta[n,1])w_1$$
$$+ (\alpha_1\beta[1,2] + \alpha_2\beta[2,2] + \cdots + \alpha_n\beta[n,2])w_2$$
$$+ \cdots$$
$$+ (\alpha_1\beta[1,m] + \alpha_2\beta[2,m] + \cdots + \alpha_n\beta[n,m])w_m$$

Using summation notation, the coefficient of w_j is $\sum_{i=1}^n \alpha_i \beta[i,j]$.

This is a mouthful of notation, but it's completely generic. The α_i's let you specify an arbitrary input vector $x \in V$, and the n-by-m array $\beta[i,j]$ contains all the data we need to specify the linear map f. We've reduced this initially enigmatic operation f to a simple table of numbers. Provided we've fixed a basis, that is.

We've only cracked the tip of the iceberg. The problem with the notational mess above is it adds too much cognitive load. It's hard to keep track of so many indices! You could make it more succinct by writing it in summation notation, but we can do better. What we really need is a well-chosen abstraction.

The abstraction we're about to see (the matrix) has two virtues. First, it eases the cognitive burden of doing a calculation by representing the operations visually. Second, it provides a rung on the ladder of abstraction which you can climb up when you want to consider the relationship between matrices, linear maps, and the basis you've chosen more abstractly. It does this by defining a new *algebra* for manipulating linear maps.

Both the visual representation and the algebra merge seamlessly with the functional description of linear maps. As we'll see, composition of functions corresponds to matrix multiplication. Natural operations on linear maps correspond to operations on the corresponding matrices, and conversely operations on matrices correspond to new, useful operations on functions. We will explore this in even more detail in Chapter 12.

So here's the abstraction that works for any linear map $f : V \to W$. Again, we fix a basis $\{v_i\}$ for V and $\{w_j\}$ for W. Write the numbers from β describing the linear map $f : V \to W$ in a table according to the following rule. The columns of the table correspond to the basis of V, and the rows correspond to basis vectors of W. We call this construction $M(f)$, and the mapping $f \mapsto M(f)$ will be a bijection from the set of linear maps (all using the same fixed basis) to the set of matrices. The underscores denote the part of the construction I haven't specified yet.

$$M(f) = \begin{array}{c} \\ w_1 \\ w_2 \\ \vdots \\ w_m \end{array} \begin{pmatrix} \overset{v_1}{-} & \overset{v_2}{-} & \cdots & \overset{v_n}{-} \\ - & - & \cdots & - \\ \vdots & \vdots & \ddots & \vdots \\ - & - & \cdots & - \end{pmatrix}$$

The entries of a column i are defined as the expansion of $f(v_i)$ in terms of the w_j. That is, take the basis vector v_i for that column, and expand $f(v_i)$ in terms of the w_j, getting $f(v_i) = \beta[i,1]w_1 + \cdots + \beta[i,m]w_m$. The numbers $\beta[i,j]$ (where j ranges from 1 to m) form the i-th column of $M(f)$.

$$M(f) = \begin{array}{c} \\ w_1 \\ w_2 \\ \vdots \\ w_m \end{array} \begin{pmatrix} \overset{v_1}{\beta[1,1]} & \overset{v_2}{\beta[2,1]} & \cdots & \overset{v_n}{\beta[n,1]} \\ \beta[1,2] & \beta[2,2] & \cdots & \beta[n,2] \\ \vdots & \vdots & \ddots & \vdots \\ \beta[1,m] & \beta[2,m] & \cdots & \beta[n,m] \end{pmatrix}$$

You will have noticed that we've flipped the indices $\beta[i,j]$ from their normal orientation so that i is the column instead of the row. This is an occupational hazard, but we trust the competent programmer can handle index wizardry. One clever way to express the construction of $M(f)$ with fewer indices is like this:

$$M(f) = \begin{array}{c} \\ w_1 \\ \vdots \\ w_m \end{array} \begin{pmatrix} \overset{v_1}{|} & \cdots & \overset{v_n}{|} \\ f(v_1) & \cdots & f(v_n) \\ | & & | \end{pmatrix}$$

The vertical lines signal that $f(v_i)$ is "spread out" over column i by its expansion in terms of $\{w_j\}$.

The computational process of mapping an input vector x to $f(x)$ is called a *matrix-vector product*, and it works as follows. First, write x in terms of the basis for V as before, $x = \alpha_1 v_1 + \cdots + \alpha_n v_n$, this time writing the coefficients in a column:

$$x = \begin{pmatrix} \alpha_1 \\ \alpha_2 \\ \vdots \\ \alpha_n \end{pmatrix}$$

Sometimes people call this a "column vector" to distinguish it from the obvious analogue of writing the entries in a row. Let's just call it a vector. Now to compute $f(x)$ using $M = M(f)$, you write M and x side by side (as if the operation were multiplication of integers).

$$Mx = \begin{array}{c} \\ w_1 \\ w_2 \\ \vdots \\ w_m \end{array} \begin{array}{cccc} v_1 & v_2 & \cdots & v_n \end{array} \\ \left(\begin{array}{cccc} \beta[1,1] & \beta[2,1] & \cdots & \beta[n,1] \\ \beta[1,2] & \beta[2,2] & \cdots & \beta[n,2] \\ \vdots & \vdots & \ddots & \vdots \\ \beta[1,m] & \beta[2,m] & \cdots & \beta[n,m] \end{array}\right) \begin{pmatrix} \alpha_1 \\ \alpha_2 \\ \vdots \\ \alpha_n \end{pmatrix}$$

Recall, the output is a vector $f(x) = z \in W$, which, if written in the same column style as x, would have m entries. We'll denote these entries by the Greek gamma $(\gamma_1, \ldots, \gamma_m) = z$.

$$Mx = \begin{array}{c} \\ w_1 \\ w_2 \\ \vdots \\ w_m \end{array} \begin{array}{cccc} v_1 & v_2 & \cdots & v_n \end{array} \\ \left(\begin{array}{cccc} \beta[1,1] & \beta[2,1] & \cdots & \beta[n,1] \\ \beta[1,2] & \beta[2,2] & \cdots & \beta[n,2] \\ \vdots & \vdots & \ddots & \vdots \\ \beta[1,m] & \beta[2,m] & \cdots & \beta[n,m] \end{array}\right) \begin{pmatrix} \alpha_1 \\ \alpha_2 \\ \vdots \\ \alpha_n \end{pmatrix} = \begin{pmatrix} \gamma_1 \\ \gamma_2 \\ \vdots \\ \gamma_m \end{pmatrix} = z$$

The computation to get from the left-hand side of this equation to the right is the same as how we grouped terms to get the coefficient of w_i earlier. Take the row of M corresponding to w_i, compute an entrywise product with x, and sum the result.[11]

$$\gamma_i = \beta[1,i]\alpha_1 + \beta[2,i]\alpha_2 + \cdots + \beta[n,i]\alpha_n$$

Visually it has always helped me to imagine picking up the first row and rotating it 90 degrees clockwise; that motion lines up the β entry with the α entry that it should be multiplied by. Then the sum gives you the first entry γ_1, and you continue down the rows of M. Here's an example with a 2×3 matrix.

$$\begin{pmatrix} 9 & 2 & 1 \\ 7 & -2 & 0 \end{pmatrix} \begin{pmatrix} 3 \\ -1 \\ 4 \end{pmatrix} = \begin{pmatrix} a \\ b \end{pmatrix}$$

The first step:

$$\begin{pmatrix} 9 & 2 & 1 \end{pmatrix} \begin{pmatrix} 3 \\ -1 \\ 4 \end{pmatrix} \longrightarrow \begin{pmatrix} 9 \\ 2 \\ 1 \end{pmatrix} \begin{pmatrix} 3 \\ -1 \\ 4 \end{pmatrix}$$

$$\longrightarrow a = 9 \cdot 3 + 2 \cdot (-1) + 1 \cdot 4 = 29$$

[11] As we'll see later in this chapter, this "entrywise product with sum" is called the inner product.

The second:

$$\begin{pmatrix} 7 & -2 & 0 \end{pmatrix} \begin{pmatrix} 3 \\ -1 \\ 4 \end{pmatrix} \longrightarrow \begin{pmatrix} 7 \\ -2 \\ 0 \end{pmatrix} \begin{pmatrix} 3 \\ -1 \\ 4 \end{pmatrix}$$

$$\longrightarrow b = 7 \cdot 3 + (-2) \cdot (-1) + 0 \cdot 4 = 23$$

It's easy to get lost in the notation and miss the bigger picture. We've defined a mechanical algebraic process for computing the output $f(x) \in W$ from the input $x \in V$, provided we have chosen a basis for V and W and provided we can express vectors in terms of a given basis. This is a new type of "multiplication" operator that has very nice properties. For example:

Definition 10.12. Let A, B be two $n \times m$ matrices and let $c \in \mathbb{R}$ be a scalar.

1. Define by cA the matrix A with all its entries multiplied by c.

2. Define by $A + B$ the matrix whose i, j entry is $A[i, j] + B[i, j]$.

Theorem 10.13. *Let V, W be vector spaces and $f, g : V \to W$ two linear maps. The mapping $f \mapsto M(f)$ is linear. That is, if $f + g$ is the function $x \mapsto f(x) + g(x)$, then $M(f + g) = M(f) + M(g)$, and likewise $M(cf) = cM(f)$ for every scalar c.*

Proof. The proof is left as an exercise to the reader.[12]

□

Beyond being linear, the mapping $f \mapsto M(f)$ is a bijection (again, for a fixed choice of a basis). Injectivity: every f maps to a different $M(f)$, since f is completely determined by how it acts on the basis, and two matrices $M(f)$ and $M(g)$ with the same entries act the same on a basis. If that's not convincing enough, consider $M(f - g) = M(f) + (-1)M(g)$. If that's the matrix of all zeroes, then, because linear maps preserve zero, $f - g$ must be the zero map. Surjectivity: if you specify a matrix A, the f mapping to A is the one with $f(v_i)$ equal to the linear combination defined by the i-th column of A.

This bijection allows us to say that linear maps and matrices are "the same thing" without angry mathematicians throwing chalkboard erasers at us.[13] The matrix representation of a linear map is unique, so we can freely switch back and forth between a linear map and its matrix, provided the basis does not change.

---

[12] This generally means the proof is not complicated, but it may contain a mess of notation required to write it out properly and doesn't make for good reading. In any event, the statement of the theorem is the enlightening part, while the proof is purely mechanical.

[13] This actually happened to a friend of mine, and there's an apocryphal tale of the irascible wunderkind Évariste Galois, who, during an admittance exam to a prestigious French universty, was so frustrated by the examiner's inability to recognize his genius that he threw a chalkboard eraser at him. Needless to say, Galois was not admitted.

Matrix-vector multiplication continues to surprise: given two matrices A and B, one can define the *product* of the two matrices by applying the matrix vector product of A to each column of B separately.

$$B = \begin{pmatrix} | & & | \\ \mathbf{b}_1 & \cdots & \mathbf{b}_m \\ | & & | \end{pmatrix} \qquad AB = \begin{pmatrix} | & & | \\ A\mathbf{b}_1 & \cdots & A\mathbf{b}_m \\ | & & | \end{pmatrix}$$

Then we have the following astounding theorem.

Theorem 10.14. *Let U, V, W be three vector spaces. Let $f : U \to V$ and $g : V \to W$ be linear maps. Then*

$$M(g \circ f) = M(g)M(f),$$

where $g \circ f$ denotes the function composition $x \mapsto g(f(x))$, and $M(g)M(f)$ denotes matrix multiplication.

So the matrix representation of a linear map allows us to compute the composition of functions. If you reflect on this fact (before attempting a rigorous and index-intensive proof), it could not be any other way: the matrix-vector product using $M(g)$ details how to take a basis vector $v_i \in V$ and express $g(v_i)$ in terms of the basis of W, while the columns of $M(f)$ express how to do the same with f from U to V.

This whole process we've undertaken, going from an abstractly defined theory of vector spaces and linear maps to the concrete world of matrices, is analogous to the process of building a computational model for a real-world phenomenon. It's like we're taking light, something which we observe obeys certain behaviors such as reflecting on various surfaces, and casting it to a type where we can quantitatively answer how *much* it reflects. We can say, without observation, what its different components are in our model, and how two types of light we've never observed interacting would interact. All of these things are possible because of the computational model.

In some more concrete and advanced terminology, we've defined an *algebra* for linear maps. We showed how to add and "multiply" (compose) linear maps, and these operations hold true to standard algebraic identities (distributive and associative properties). We then did the same for matrices—after fixing a basis—where adding and multiplying are matrix addition and multiplication. The map $f \mapsto M(f)$ provides a way to say these two perspectives behave identically. A linear map f and $M(f)$ are the "same" object, represented two different ways.[14]

The task of finding a route from a conceptually intuitive land (linear maps) to a computationally friendly world (matrices) is one of the chief goals in much of mathematics. This is the same goal of calculus—it's namesake is "calculate"—to convert computations on curves with an infinite nature to a domain where one can do mechanical calculations.

---

[14] The map M provides an *isomorphism of algebras*, but rather than introduce this term now, we will discuss it at length in Section 10.7, and again in later chapters.

And we aren't yet done doing this with linear algebra! Because while we have said how to compute once you have chosen a basis, we haven't discussed the means of actually finding such bases. Many applications of linear algebra are based on computing a useful basis, and that will be the subject of both this chapter's application and the next. As such, it behooves us to dive deeper.

10.6 Conjugations and Computations

One assumption I've been leaning on so far is that, given a basis $\{v_1, \ldots, v_n\}$ for V and a vector $x = (\alpha_1, \ldots, \alpha_n) \in V$, one can find the unique expression of x in terms of the basis. In fact, the way we defined a basis ensures existence, but the only example I gave so far to compute this decomposition was, for $V = \mathbb{R}^2$, to set up a system of two linear equations with two variables, and solve them.

$$3a - b = 1$$
$$4a - 5b = 0$$

Here $v_1 = (3, 4)$ and $v_2 = (-1, -5)$ were the two vectors acting as our basis, and we wanted to express the vector $x = (1, 0)$ in terms of them. The variables a, b are the unknown coefficients of v_1, v_2 we solved for.

One important thing to point out: even though we want to write $x = (1, 0)$ in terms of v_1, v_2, we actually *had* a representation of x in terms of a basis already! To even write x down in this coordinate-form, we implicitly used the standard basis for $\mathbb{R}^2$, $e_1 = (1, 0), e_2 = (0, 1)$. In the example above $x = 1e_1 + 0e_2$. In order to express x in terms of a given basis, you have to have already expressed it in terms of some (maybe easy) basis.

This strategy generalizes. Let's say we have an n-dimensional vector space V with two bases:

$$E = \{e_1, e_2, \ldots, e_n\}$$
$$B = \{v_1, v_2, \ldots, v_n\}$$

Say E is the "easy" basis, often the standard basis in $\mathbb{R}^n$, and B is the target basis we wish to express some vector $x = (\alpha_1, \ldots, \alpha_n)$ in. Write down a system of n equations with n unknowns, as follows. I'm going to use the notation (e.g.) $v_{2,4}$ to denote the 4th entry of v_2, which is the standard way to do double-indexing in mathematics. Note that all symbols here represent numbers in $\mathbb{R}$.

$$\beta_1 v_{1,1} + \cdots + \beta_n v_{n,1} = \alpha_1$$
$$\beta_1 v_{1,2} + \cdots + \beta_n v_{n,2} = \alpha_2$$
$$\vdots$$
$$\beta_1 v_{1,n} + \cdots + \beta_n v_{n,n} = \alpha_n$$

We can rewrite the system of equations as a single matrix equation.

$$\begin{pmatrix} v_{1,1} & \cdots & v_{n,1} \\ v_{1,2} & \cdots & v_{n,2} \\ \vdots & \ddots & \vdots \\ v_{1,n} & \cdots & v_{n,n} \end{pmatrix} \begin{pmatrix} \beta_1 \\ \beta_2 \\ \vdots \\ \beta_n \end{pmatrix} = \begin{pmatrix} \alpha_1 \\ \alpha_2 \\ \vdots \\ \alpha_n \end{pmatrix}$$

This makes it clear that expressing a vector in terms of a basis, while originally posed as solving a system of equations, really is computing the unknown input of a linear map, $y = (\beta_1, \ldots, \beta_n)$, given a specified output $x = (\alpha_1, \ldots, \alpha_n)$. It's worthwhile to break this down a bit further.

The matrix $A = (v_{i,j})$ defined above converts a vector from the domain basis to the codomain basis. The domain basis—which indexes the columns of A—is the target basis. It's the one we want to express x in terms of. The codomain basis—indexing the rows—is the "easy" basis E, the basis used to write $x = (\alpha_1, \ldots, \alpha_n)$. Finally, y is the vector of coefficients $(\beta_1, \ldots, \beta_n)$ that expresses x in terms of $v_1, \ldots, v_n$, which is what we want.

This entire matrix-vector equation $Ay = x$ expresses the conversion of a vector in the hard basis to a vector in the easy basis. This is mildly strange, since if we think of A as the matrix of a linear map, that linear map is $x \mapsto x$, a no-op! Much like a change of a number basis from binary to decimal or hexadecimal, the semantic meaning of the input is unchanged by the operation, just its data representation and interpretation. Linear maps are semantic, matrices are data interpretations. Nevertheless, these so-called *change of basis* matrices are crucial to every computational endeavor. In particular, to write an expression for x expressed in the basis $(v_1, \ldots, v_n)$, we simply form the change of basis matrix P whose columns are the v_i, and write $y = P^{-1}x$.

As an aside, it should be intuitively clear that P has an inverse as a function: every vector has exactly one representation in terms of a basis. Even if we didn't know how the conversion works computationally, it must be a bijection. More usefully, and not that we have a matrix multiplication operation, the inverse of a matrix A is defined in terms of an identity. The *identity matrix*, denoted I_n or 1_n, is the square $n \times n$ matrix defined by having 1's on the diagonal and zeros elsewhere.

$$I_n = \begin{pmatrix} 1 & 0 & 0 & \cdots & 0 \\ 0 & 1 & 0 & \cdots & 0 \\ 0 & 0 & 1 & \cdots & 0 \\ \vdots & \vdots & \vdots & \ddots & \vdots \\ 0 & 0 & 0 & \cdots & 1 \end{pmatrix}$$

The matrix multiplication operation ensures that $I_n A = A I_n = A$ for any matrix A. Then the inverse A^{-1}, if it exists, is defined as the matrix B for which $AB = BA = I_n$. As an exercise, prove that if a linear map is a bijection, then its inverse is also a linear map, and the linear-map-to-matrix correspondence preserves inverses.

More generally, a pattern used everywhere in mathematics is to change basis for a limited-scope operation. In other words, given a change of basis matrix P which changes from basis B to basis E, and some linear map A expressed in terms of E, you can apply A to a vector w expressed in B-coordinates as

$$P^{-1}APw$$

This expression works in sequence: express w in basis E, apply A, and convert the result back to B. The matrix $P^{-1}AP$ is exactly the linear map for A expressed in terms of the B basis. It's also true that any invertible map is a change of basis to some basis (the basis formed by the columns of the inverse).

This general pattern of doing $P^{-1}AP$ is called *conjugation* of A by P. If two matrices can be equated by conjugation, they are often called *similar*. I personally hate the term "similar" because we're really saying they're identical. If you look at a laptop on your desk and then pick it up and hold it sideways above your head, it's not "similar" to the laptop on your desk, it's the same thing from two different perspectives! That's exactly what happens when you conjugate a matrix; it may not be ergonomic to type that way, but it's the same machine. Taking a cue from Chapter 9, matrix similarity is an equivalence relation, and the equivalence classes correspond to linear maps.

Now to actually compute $P^{-1}x$ is a different pickle. From the perspective of a system of n equations, the standard principle of solving the matrix-vector equation $Ab = x$ by isolating a single variable, substituting, and solving works, but it's extremely tedious. To help with the tedium, mathematicians came up an algorithm called *Gaussian elimination* that formalizes the tedium and uses the matrix-form above to help organize. Gaussian elimination is important, but it's both inefficient[15] and it computes a lot of extra information.

Gaussian elimination is a general-purpose algorithm that works no matter what your basis is. A shrewder approach, which many applications of linear algebra utilize, is to think hard about the best basis for your intended application, and convert to that basis once at the beginning of a computation. See the exercises for further references and pointers to industry-standard techniques for changing bases, and Chapter 12 for an extended parable on the value of a good basis.

10.7 One Vector Space to Rule Them All

Now we turn to a classification theorem, that $\mathbb{R}^n$ is the "only" vector space of finite dimension. We make this formal by showing that all n-dimensional vector spaces are isomorphic to each other.

Discussing vector spaces of infinite dimension is quizzical, given our insistence that matrices—inherently finite objects built for computation—are the geese that lay the golden eggs. Suffice it to note here that we have seen an example of such an exotic vector space: polynomials. Let V be the set of all polynomials in a single variable t. Then the following set is a basis:

---

[15] It's polynomial-time in $n = \dim(V)$, but in the worst case its runtime is more than n^3. Here's a more complete story: http://cstheory.stackexchange.com/questions/3921

$$B = \{1, t, t^2, t^3, \ldots\} = \{t^j : j \in \mathbb{Z} \text{ and } j \geq 0\}$$

Indeed, any polynomial can be uniquely written as a linear combination of polynomials in B by specifying their coefficients. The operations of adding two polynomials and scaling a polynomial are applied to each term by degree, as expected. There are other bases, to be sure (see the exercises), but questions about infinite dimensional vector spaces are much harder to answer without more advanced techniques.[16]

Let's restrict our attention back to finite-dimension. We'll argue why $\mathbb{R}^n$ is the only vector space by an illuminating example. Define by P_m the vector space of polynomials of degree at most m. Note that the obvious basis is $\{1, t, \ldots, t^m\}$, making $\dim P_m = m+1$. Recall from Chapter 2 the "data definition" of a polynomial as a list of coefficients. This perspective naturally inclines us to think that it's "the same" as a *usual* list of numbers, that is, a vector in $\mathbb{R}^{m+1}$.

In fact, we can make this formal by constructing an isomorphism between P_m and $\mathbb{R}^{m+1}$.

Definition 10.15. Let V and W be vector spaces. A linear map $f : V \to W$ is called an *isomorphism* if it is a bijection. If an isomorphism exists $V \to W$, then we say V and W are *isomorphic*, often denoted by $V \cong W$.

An isomorphism f preserves all structure in mapping elements from V to W. As far as linear-algebraic structure is concerned, V and W are identical, and the elements of W can be thought of as a "relabeling" of the elements of V. Whereas previously we described the linear-map-to-matrix function M as an isomorphism of algebras, this is an isomorphism of vector spaces. The concept of isomorphism is the same (preserving structure both forward and backward), but what is being preserved is different.

Note first that if a linear map f is a bijection, then the inverse f^{-1} is also a linear map. This is because if $f(v) = x + y$ and $f(x') = x$, $f(y') = y$, then by injectivity $v = x' + y'$, and so

$$f^{-1}(x+y) = f^{-1}(f(x') + f(y')) = f^{-1}(f(x' + y')) = x' + y'.$$

Proposition 10.16. *Let P_m be the vector spaces of polynomials in one variable with degree at most m. Then $\mathbb{R}^{m+1} \cong P_m$.*

Proof. Let $\{1, t, t^2, \ldots, t^m\}$ be the usual basis for P_m, and fix the standard basis of $\mathbb{R}^{m+1}$, i.e., $\{e_1, \ldots, e_{m+1}\}$. Define $f : P_m \to \mathbb{R}^{m+1}$ as

$$f(a_0 + a_1 t + \cdots + a_m t^m) = (a_0, a_1, \ldots, a_m)$$

---

[16] In particular, without using the Axiom of Choice, a somewhat unintuitive postulate, one cannot even conclude that all infinite dimensional vector spaces *have* bases! This fact led to an amusing—if somewhat off-color—t-shirt designed by my undergraduate math club, which emblazoned the slogan, "Pro Axiom of Choice: because every vector space deserves a basis."

First, f is a linear map: when you add polynomials you add their same-degree coefficients together, and scaling simply scales each coefficient. Second, f is a bijection: if two polynomials are different, then they have at least one differing coefficient (injection); if $(b_0, b_1, \ldots, b_m)$ is a vector in $\mathbb{R}^{m+1}$, then it is the image of $p(t) = \sum_{k=0}^{m} b_k t^k$ under f.

□

This theorem isn't meant to conclude that polynomials are the same as lists in every respect. Quite the opposite, a polynomial comes with all kinds of extra interesting structure (as we saw in Chapter 2). Rather, to phrase polynomials as a vector space is to ignore that additional structure. It says: if all you consider about polynomials is their linearity, then they have the same linear structure as lists of numbers. At times it can be extremely helpful to "ignore" certain unneeded aspects of a problem. As you'll see in an exercise, the polynomial interpolation problem relies only on the linear structure of polynoimals. As such, it can inspire other (perhaps more efficient) techniques for doing secret sharing.

This exploration suggests that *all* data representations of finite-dimensional vector spaces can be thought of as lists of numbers. Those numbers are the coefficients of the basis vectors.

Theorem 10.17. *Every n-dimensional vector space is isomorphic to $\mathbb{R}^n$.*

Proof. Let $\{v_1, \ldots, v_n\}$ be a basis for an n-dimensional vector space V, and let $\{e_1, \ldots, e_n\}$ be the standard basis for $\mathbb{R}^n$. Define $f : V \to \mathbb{R}^n$ as follows. Let $x \in V$ be the input, write $x = \alpha_1 v_1 + \cdots + \alpha_n v_n$, and let $f(x) = (\alpha_1, \ldots, \alpha_n)$.

An analogous argument as in Proposition 10.16 shows f is a linear bijection.

□

10.8 Geometry of Vector Spaces

In studying matrices, we saw the elegant relationship linear algebra provides between the functional and algebraic perspectives on a linear map. Geometry is the final ingredient. To that end, we need to be able to compute distances and angles. Because all finite-dimensional vector spaces are isomorphic to $\mathbb{R}^n$, it makes sense to define angles and distances for vectors in $\mathbb{R}^n$ with its standard basis. Subsequently, angles in a vector space V can be defined using the isomorphism between V and $\mathbb{R}^n$.

There is a small wrinkle in this plan. The primitive we're about to define, the inner product, defines angles in $\mathbb{R}^n$. However, the standard inner product might not be preserved by an isomorphism! As it turns out—and it's not hard to prove this—if V has a reasonable definition of angles (i.e., it has its own inner product) then there is an isomorphism that converts it to the standard inner product we're about to define.[17] So suffice it to say, the specificity in this section generalizes. We'll see this happen in Chapter 12.

---

[17] In formal terms: all finite-dimensional vector spaces with inner products are "isometric" to $\mathbb{R}^n$ with the standard inner product.

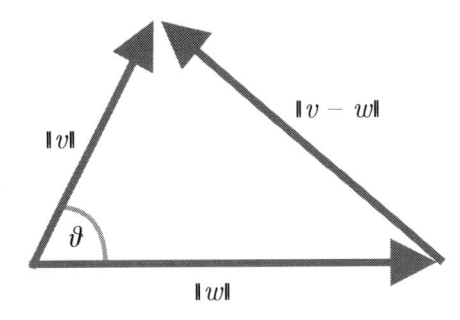

Figure 10.7: The lengths of the sides of the triangle satisfy the law of cosines.

Definition 10.18. Let v, w be vectors in $\mathbb{R}^n$, and let $\{e_1, \ldots, e_n\}$ be the standard basis for $\mathbb{R}^n$, so that $v = \sum_{i=1}^{n} \alpha_i e_i$ and $w = \sum_{i=1}^{n} \beta_i e_i$. The *standard inner product* of v and w, denoted $\langle v, w \rangle$, is a scalar given by the formula

$$\langle v, w \rangle = \alpha_1 \beta_1 + \cdots + \alpha_n \beta_n = \sum_{i=1}^{n} \alpha_i \beta_i.$$

This formula is special because it has a geometric interpretation. Indeed, it can even be defined geometrically without any appeal to the basis, which we'll do now. Note that to understand this proof requires some "elementary" geometry which we haven't covered in this book, namely the idea of a cosine and the law of cosines. If you're unfamiliar with these topics, look them up online.

First, a special case of the inner product: the *norm* of a vector v, denoted $\|v\|$, is defined as $\|v\| = \sqrt{\langle v, v \rangle}$. This quantity is the geometric length or magnitude of v. Its formula, $\|v\| = \sqrt{\alpha_1^2 + \cdots + \alpha_n^2}$, is the generalization of the Pythagorean theorem to n dimensions.

Theorem 10.19. *The inner product $\langle v, w \rangle$ is equal to $\|v\| \|w\| \cos(\theta)$, where θ is the angle between the two vectors.*[18]

Proof. If either v or w is zero, then both sides of the equation are zero and the theorem is trivial, so we may assume both are nonzero. Label a triangle with sides v, w and the third side $v - w$ as in Figure 10.7. The length of each side is $\|v\|, \|w\|$, and $\|v - w\|$, respectively. Assume for the moment that θ is not 0 or 180 degrees, so that this triangle has nonzero area.

The law of cosines states that

$$\|v - w\|^2 = \|v\|^2 + \|w\|^2 - 2\|v\|\|w\| \cos(\theta).$$

---
[18] This angle is computed in the 2-dimensional subspace spanned by v, w, viewed as a typical flat plane.

The left hand side is the inner product of $v-w$ with itself, i.e. $\|v-w\|^2 = \langle v-w, v-w\rangle$. We'll expand $\langle v-w, v-w\rangle$ using two facts. The first is trivial from the formula, that inner product is symmetric: $\langle v, w\rangle = \langle w, v\rangle$. Second is that the inner product is linear in each input. In particular for the first input: $\langle x+y, z\rangle = \langle x, z\rangle + \langle y, z\rangle$ and $\langle cx, z\rangle = c\langle x, z\rangle$. The same holds for the second input by symmetry of the two inputs.[19] Hence we can split up $\langle v-w, v-w\rangle$ as follows.

$$\begin{aligned}\langle v-w, v-w\rangle &= \langle v, v-w\rangle - \langle w, v-w\rangle \\ &= \langle v, v\rangle - \langle v, w\rangle - \langle w, v\rangle + \langle w, w\rangle \\ &= \|v\|^2 - 2\langle v, w\rangle + \|w\|^2\end{aligned}$$

Combining our two offset equations, subtract $\|v\|^2 + \|w\|^2$ from each side and get

$$-2\|v\|\|w\|\cos(\theta) = -2\langle v, w\rangle,$$

Which, after dividing by -2, proves the theorem if $\theta \notin \{0, 180\}$.

Now if $\theta = 0$ or 180 degrees, the vectors are parallel and $\cos(\theta) = \pm 1$. That means we can write $w = cv$ for some scalar c. In particular, $c < 0$ when $\theta = 180$ and $c > 0$ for $\theta = 0$, and $\|w\| = c\|v\|$ when $c > 0$ and $\|w\| = -c\|v\|$ when $c < 0$. So the inner product is

$$\langle v, cv\rangle = c\langle v, v\rangle = c\|v\|^2 = (c\|v\|)\|v\| = \pm\|w\|\|v\|,$$

where the sign matches up with $\cos(\theta) \in \{\pm 1\}$.

$\square$

The inner product is important because it allows us to describe perpendicularity of vectors in terms of algebra.

Theorem 10.20. *Two vectors $v, w \in \mathbb{R}^n$ are perpendicular if and only if $\langle v, w\rangle = 0$.*

When I say, "P is true *if and only if* Q is true," I am claiming that the two properties are logically equivalent. In other words, you cannot have one without the other, nor can you exclude one without excluding the other. Proving such an equivalence requires two sub-proofs, that P implies Q and that Q implies P. Because logical implication is often denoted using arrows—"P implies Q" being written $P \to Q$, and "Q implies P" being written $P \leftarrow Q$—these sub-proofs are informally called "directions." So one will prove an if-and-only-if by saying, "For the forward direction, assume P...and hence Q", and "For the reverse/other direction, assume Q...and hence P." Authors will also often mix in proof by contradiction to complete the sub-proofs. The combined if-and-only-if is often denoted with double-arrows: $P \leftrightarrow Q$, and when pressed for brevity, mathematicians abbreviate "if and only if" with "iff" using two f's. So "iff" is the mathematical cousin of a classic Unix command: 2-3 letters and a long man page to explain it.

Let's prove the if and only if for perpendicular vectors now.

---

[19] We will see in Chapter 12 how these properties become a definition.

Proof. For the forward direction, assume v and w are perpendicular. By definition the angle θ between them is 90 or 270 degrees, and $\cos(\theta) = 0$. Hence $\langle v, w \rangle = \|v\|\|w\|\cos(\theta) = 0$. For the reverse direction, if $\langle v, w \rangle = 0$ then so is $\|v\|\|w\|\cos(\theta)$, meaning one of $\|v\|, \|w\|$, or $\cos(\theta)$ must be zero. Perpendicularity is not defined if one of the two vectors is zero,[20] so both vectors must be nonzero and have a nonzero norm. This leaves $\cos(\theta) = 0$. The vectors are perpendicular.

□

As a side note, we'll need the fact that two nonzero perpendicular vectors are linearly independent. Suppose for contradiction that $\langle x, y \rangle = 0$ but $ax + by = 0$ for some scalars a, b. Suppose without loss of generality that $b \neq 0$ (i.e., $ax + by = 0$ is a nontrivial linear dependence). In this case, a is also nonzero, since $a = 0$ implies $by = 0$, which implies $y = 0$, and y was assumed to be nonzero. Then

$$0 = \langle x, y \rangle = \langle x, -(a/b)x \rangle = -(a/b)\|x\|^2,$$

meaning that $\|x\| = 0$, which implies x is the zero vector, a contradiction. A similar proof shows that if x is a vector perpendicular to the plane (or any subspace) spanned by two vectors y, z, then the set $\{x, y, z\}$ is a linearly independent set. So if you have a set of linearly independent vectors, and you add a vector that's perpendicular to their span, you increase the dimension of the spanned subspace by one.

Next we define the projection of one vector onto another.

Definition 10.21. Let v, w be vectors in $\mathbb{R}^n$. The *projection* of w onto v, denoted $\text{proj}_v(w)$, is defined as $\text{proj}_v(w) = cv$ where $c \in \mathbb{R}$ is a scalar defined as[21] $c = \frac{\langle v, w \rangle}{\|v\|}$.

Let me depict this formula geometrically. Say that v, the vector being projected onto, is special in that it has magnitude 1. Such a special vector is called a *unit vector*.[22] In this case the formula defined above for the projection is just $\langle v, w \rangle v$. Now (trivially) write

$$w = \text{proj}_v(w) + [w - \text{proj}_v(w)]$$

The terms above are labeled on the diagram in Figure 10.8, with v and w solid dark vectors, and the terms of the projection formula as dotted lighter vectors perpendicular to each other. To convince you that the inner product computes the pictured projection, I need to prove to you that the two terms $\text{proj}_v(w)$ and $w - \text{proj}_v(w)$ are geometrically perpendicular. Indeed, I need to show you that

---

[20] One can either say that perpendicularity as a concept only applies to nonzero vectors, or establish (by convention) that the zero vector is perpendicular to all vectors.

[21] Another example of tail-call optimization: I want to make it obvious, formula be damned, that projecting w onto v results in a vector on the line spanned by v.

[22] The words "unit" and "unity" refer to the multiplicative identity 1, and their etymology is the Latin word for one, *unus*. The word also shows up in complex numbers when we speak of "roots of unity," being those complex numbers which are n-th roots of 1. Someday they'll make a biopic about collaborating mathematicians called "Roots of unity," and Cauchy will roll over in his grave.

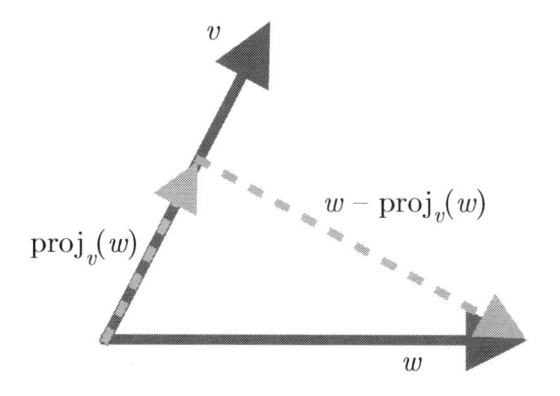

Figure 10.8: The orthogonal projection of w onto v.

$$\langle w - \text{proj}_v(w), \text{proj}_v(w) \rangle = 0$$

Indeed, since $\text{proj}_v(w) = \langle v, w \rangle v$, let's call $p = \langle v, w \rangle$ and expand:

$$\begin{aligned}
\langle w - \text{proj}_v(w), \text{proj}_v(w) \rangle &= \langle w - pv, pv \rangle \\
&= \langle w, pv \rangle - \langle pv, pv \rangle \\
&= p\langle w, v \rangle - p^2 \|v\|^2 \\
&= p^2 - p^2 = 0
\end{aligned}$$

The last step used the assumption that $\|v\| = 1$, and again that $p = \langle w, v \rangle = \langle v, w \rangle$. You can prove the same fact with the version of the projection formula that does not require unit vectors, if you keep track of the extra norms. The essence of the proof is the same.

Figure 10.8 is not a lie: the two vectors are actually perpendicular. The extra term in the formula for $\text{proj}_v(w)$ dividing by $\|v\|$ is just to make v a unit vector. Ideally you never project onto something which is not a unit vector, but if you must you can normalize it as part of the formula.

By virtue of being perpendicular to the projection, the vector $w - \text{proj}_v(w)$ can be thought of as measuring the distance of w from $\text{proj}_v(w)$. Or, more geometrically, the distance of the point represented by w from the line spanned by v. This is useful for obvious reasons in the kind of geometry used in computer graphics. But it's also useful for us because the data we compute from the projection allows us to measure a "best fit." Finding the line of best fit for a collection of points is the base case of the SVD algorithm, the application for this chapter. More generally, given a subspace $V \subset \mathbb{R}^n$ spanned by $\{v_1, \ldots, v_k\}$, the *distance* from w to the subspace can be thought of as the minimal distance from w to any vector in $\text{span}\{v_1, \ldots, v_k\}$. You can also define the projection of a vector w onto a subspace as the sum of projections onto each vector in the subspace basis:

$$\text{proj}_V(w) = \sum_{i=1}^{k} \text{proj}_{v_i}(w).$$

Then the distance from w to the subspace V is $w - \text{proj}_V(w)$, as expected.

10.9 Application: Singular Value Decomposition

A brief summary of this chapter would rephrase the relationship between a matrix and a linear map. A matrix is a natural representation of a linear map that is fixed after choosing a basis, and the algebraic properties of a matrix correspond to the functional properties of the map. That, and certain operations on vectors have nice geometric interpretations.

We save the juiciest properties for Chapter 12, where we will discuss eigenvalues and eigenvectors. Nevertheless, we have access to fantastic applications. The technique for this chapter, the singular value decomposition (SVD), is a ubiquitous data science tool. It was also a crucial part of the winning entry for the million dollar Netflix Prize. The Netflix Challenge, held from 2006-2009, was a competition to design a better movie recommendation algorithm. The winning entry improved on the accuracy of Netflix's algorithm by ten percent. The singular value decomposition was used to represent the data (movie ratings) as vectors in a vector space, and the "decomposition" part of SVD chooses a clever basis that models the data. After finding this useful representation, the Netflix Prize winners used the vector representation as input to a learning algorithm.[23]

Though true movie ratings require dealing with issues we will ignore (like missing data), we'll couch the derivation of the SVD in a discussion of movie ratings. The geometric punchline is: treat the movie ratings as points in a vector space, and find a low-dimensional subspace which all the points are close to. This low-dimensional subspace "approximates" the data in a way that makes subsequent operations like clustering and prediction easier.

A Linear Model for Rating Movies

Let's start with the idea of a movie rating database to understand the modeling assumptions of the SVD. We have a list of people, say Aisha, Bob, and Chandrika, who rate each movie with an integer 1-5. These intrepid movie lovers have watched and critiqued every single movie in the database. We write their ratings in a matrix A as in Figure 10.9.

Each person's ratings is an a priori complicated function, not entirely determined by the movies alone. Aisha likes Thor but not Skyfall, but the *reason* is not in the data. By writing the ratings in a matrix we are implicitly adding a "linear model" to the ratings. That is, we're saying the input is $\mathbb{R}^3$ and the basis vectors are people:

$$\{x_{\text{Aisha}}, x_{\text{Bob}}, x_{\text{Chandrika}}\}$$

---

[23] Ironically, most of the hard work beyond the standard SVD and subsequent learning algorithm was not ultimately used by Netflix, even after declaring the winner.

$$\begin{array}{c} \\ \text{Up} \\ \text{Skyfall} \\ \text{Thor} \\ \text{Amelie} \\ \text{Snatch} \\ \text{Casablanca} \\ \text{Bridesmaids} \\ \text{Grease} \end{array} \begin{array}{c} \text{Aisha} \quad \text{Bob} \quad \text{Chandrika} \\ \left(\begin{array}{ccc} 2 & 5 & 3 \\ 1 & 2 & 1 \\ 4 & 1 & 1 \\ 3 & 5 & 2 \\ 5 & 3 & 1 \\ 4 & 5 & 5 \\ 2 & 4 & 2 \\ 2 & 2 & 5 \end{array} \right) = A \end{array}$$

Figure 10.9: An example movie rating matrix for three people.

The codomain is $\mathbb{R}^8$ (if there are only 8 movies, as in this toy example), and the basis vectors are $y_{\text{Up}}, y_{\text{Skyfall}}$, etc. By representing the ratings this way, we're imposing the hypothesis that the *process* of rating movies is linear in nature. That is, the map A computes the decision making process from people to ratings. The coefficients of $A(x_{\text{Aisha}})$ written in terms of the basis of movies, forms the first column of the matrix in Figure 10.9. It is also assumed to be one *combined* function, as opposed to different for each person.

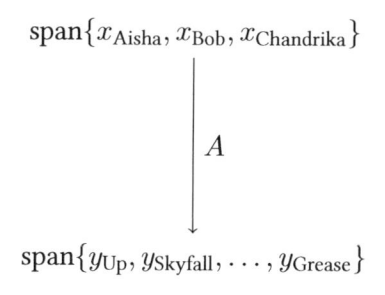

$$\text{span}\{x_{\text{Aisha}}, x_{\text{Bob}}, x_{\text{Chandrika}}\}$$
$$\bigg\downarrow A$$
$$\text{span}\{y_{\text{Up}}, y_{\text{Skyfall}}, \ldots, y_{\text{Grease}}\}$$

These assumptions should give us pause. Beyond the sociological assumptions made here, the linear model also grants us strange new mathematical abilities. We started with a dataset of ratings, which is included in the linear-algebraic world as $A(x_{\text{Aisha}})$, $A(x_{\text{Bob}})$, and $A(x_{\text{Chandrika}})$. But since we represent movies and people as vectors, we may form linear combinations. We may construct the movie $0.5y_{\text{Up}} + 0.5y_{\text{Snatch}}$, which we might think of as the abstract equivalent of a movie that is "half-way" between *Up* and *Snatch*. We may also ask for a "person" whose movie-rating preferences are half-way in between Aisha and Bob, and ask how this person would rate *Amelie*. Indeed, the fact that A is a linear map provides an immediate answer to this question: average the ratings of Aisha and Bob. The behavior of A on any vector is determined by its behavior on the basis.

We can also create nonsense when we *subtract* people, or scale them beyond reasonable interpretations. What would the movie $75y_{\text{Grease}} - 8y_{\text{Thor}}$ look like? You may conjure a cohesive explanation, but you'd be straining logic to fit the image of gibberish. Very off brand.

Of course, the goal of a rating system is to predict the ratings of people on movies they have not seen, based on how two people's ratings align. So a valid answer is, "we don't care about weird linear combinations." That said, more likely than not your chosen linear algebraic hammer *relies* on strange linear combinations. It's worthwhile to illustrate the necessary assumptions entailed by imposing linear algebra on a real world problem, and the curious luggage this stranger brings along.

The central point is that we can represent a movie (or a person) formally as a linear combination in some abstract vector space. But we don't represent a movie in the sense of its *content*, only those features of the movie that influence its rating. We don't know what those features are, but we can presumably access them indirectly through the data of how people rate movies. We don't have a legitimate mathematical way to understand that process, so the linear model is a proxy. What's amazing is how powerful a dumb linear proxy can be.

It's totally unclear what this means in terms of real life, except that you can hope (or hypothesize, or verify), that if the process of rating movies is "linear" in nature then this formal representation will accurately reflect the real world. It's like how physicists all secretly know that mathematics doesn't literally dictate the laws of nature, because humans made up math in their heads and if you poke nature too hard the math breaks down. But math as a language is so convenient to describe hypotheses (and so accurate in most cases!), that we can't help but use it to design airplanes. We haven't yet found a better tool than math.

Likewise, movie ratings aren't literally a linear map, but if we pretend they are we can make algorithms that accurately predict how people rate movies. So if you know that Skyfall gets ratings 1, 2, and 1 from Aisha, Bob, and Chandrika, respectively, then a new person would rate Skyfall based on a linear combination of how well they align with these three people on other ratings. In other words, up to a linear combination, in this example Aisha, Bob, and Chandrika epitomize the process of rating movies.

The idea in SVD is to use a better choice of people than Aisha, Bob, and Chandrika, and a better choice of movies, by isolating the "orthogonal" aspects of the process into separate vectors in the basis. Concretely this means the following:

1. Choose a basis $p_1, \ldots, p_n$ of the space of people. Every person in the database can be written as a linear combination of the p_i, and all the p_i are perpendicular. This is true of our starting basis, but (3) will clarify why this new basis is special.

2. Do the same for movies, to get $q_1, \ldots, q_m$.

3. Do (1) and (2) in such a way that the resulting representation of A only has entries on the diagonal.[24] I.e., $A(p_1) = c_1 q_1$ for some constant c_1, likewise for p_2, p_3, etc.

[24] Matrices with only nonzero entries on the diagonal are often called "diagonal" matrices, and if a matrix is diagonal with respect to some choice of basis, it's called "diagonalizable."

One might think of the p_i as "idealized critics" and the q_j as "idealized movies." If the world were unreasonably logical, then q_1 might correspond to the "ideal action movie" and p_1 to the "idealized action movie lover." The fact that A only has entries on the diagonal means that p_1 gives a nonzero rating to q_1 and *only* q_1. A movie is represented by how it decomposes (linearly) into "idealized" movies. To make up some arbitrary numbers, maybe Skyfall is $2/3$ action movie, $1/5$ dystopian sci-fi, and $-6/7$ comedic romance. A person would similarly be represented by how they decompose (via linear combination) into a action movie lover, rom-com lover, etc.

To be completely clear, the singular value decomposition does not find the ideal action movie. The "ideality" of the singular value decomposition is with respect to the inherent linear structure of the rating data. In particular, the "idealized genres" are related to how closely the data sits in relation to certain lines and planes. This is the crux of why the SVD algorithm works, so we'll explain it shortly. But nobody has a strong idea of how the movie itself relates to the geometric structure of this abstraction. It almost certainly depends on completely superficial aspects of the movie, such as how much it was advertised or whether it's a sequel. Indeed, one could add these features to a learning model! Nevertheless, much of the usefulness of the abstraction relies on not being domain-specific. The more a model encodes about movie-specific features, the less it applies to data of other kinds. One sign of a deep mathematical insight is domain-agnosticism.

The takeaway is that this mental model of an idealized genre movie and an idealized genre-lover grounds our understanding of the SVD. We want to find bases with special structure related to the data. We know the analogy is wrong, but it's a helpful analogy nonetheless.

Earlier I said that the SVD is about finding a low-dimensional subspace that approximates the data well. It won't be clear until we dive into the algorithm, but this is achieved by taking our special basis of idealized people, $p_1, \ldots, p_n$ (likewise for movies), and ordering them by how well they capture the data. There is a single best line, spanned by one of these p_i, that the points are collectively closest to. Once you've found that, there is a second best vector which, when combined with the first, forms the best-fitting plane (two-dimensional subspace), and so on.

The approximation aspect of the SVD is to stop at some step k, so that you have a k-dimensional subspace that fits the data well. The matrix P whose rows are the chosen $p_1, \ldots, p_k$ is the linear map that projects the input vector x to the closest point in the subspace spanned by $p_1, \ldots, p_k$. This is simply because the matrix-vector multiplication Px involves an inner product $\langle p_i, x \rangle$—the projection formula onto a unit vector p_i—between each row of P and x.

Hopefully, k is much less than m or n, but still captures the "essence" of the data.[25] Indeed, it turns out that if you *define* the special basis vectors in this way—spanning the best-fitting subspaces in increasing order of dimension—you get everything you want. And what's astounding is that you can build these best-fitting subspaces recursively. The

---

[25] One useful perspective is that the "truth" is a low-dimensional subspace, but the observations you see are jostled off that subspace by noise in a predictable fashion. This is a modeling assumption.

best-fitting 2-dimensional subspace is formed by taking the best line and finding the next best vector you could add. Likewise, the best 3-dimensional subspace is that best plane coupled with the next best vector. We're glomming on vectors greedily.

It should be shocking that this works. Why should the best 5-dimensional subspace be at all related to the best 3-dimensional subspace? For most problems, in math and in life, the greedy algorithm is far from optimal. When it happens, once in a blue moon, that the greedy algorithm is *the* best solution to a natural problem—and not obviously so—it's our intellectual duty to stop what we're doing, sit up straight, and really understand and appreciate it.

Minimizing and Maximizing

First we'll define what it means to be the "best-fitting" subspace to some data. Below, by the "distance from a vector x to a subspace W," I mean the minimal distance between x and any vector in W.

Definition 10.22. Let $X = \{w_1, \ldots, w_m\}$ be a set of m vectors in $\mathbb{R}^n$. The best approximating k-dimensional linear subspace of X is the k-dimensional linear subspace $W \subset \mathbb{R}^n$ which minimizes the sum of the squared distances from the vectors in X to W.

Next we study this definition to come up with a suitable quantity to optimize. Say I have a set of m vectors $w_1, \ldots, w_m$ in $\mathbb{R}^n$, and I want to find the best approximating 1-dimensional subspace. Given a candidate line spanned by a unit vector v, measure the quality of that line by adding the sum-of-squares distances from w_i to v. Using the projection function defined earlier,

$$\text{quality}(v) = \sum_{i=1}^{m} \|w_i - \text{proj}_v(w_i)\|^2$$

This formula, in a typical math writing fashion, exists only to help us understand what we're optimizing: squared distances of points from a line. To make it tractable, we convert it back to the inner product. I'll describe this process in a fine detail, with sidebars to explain some notational choices.

We want to find the unit vector v that minimizes the quality function. We'd write the goal of minimizing this expression as

$$\arg\min_{v} \sum_{i=1}^{m} \|w_i - \text{proj}_v(w_i)\|^2.$$

A sidebar on notation: when I write $\min_v$ EXPR I am defining an anonymous function whose input is v and whose output is EXPR (depending on v), and the total expression (with the min) evaluates to[26] the minimal output value considered over all possible inputs v. The domain of v is usually defined in the prose, but if it's helpful and fits, the conditions on v can be expressed in the subscript, such as

---

[26] I'm using programming-language parlance here. A mathematician would say "is."

$$\min_{\substack{v \in \mathbb{R}^n \\ \|v\|=1}} \text{EXPR},$$

which is the minimum value of EXPR considered over all possible unit vectors in $\mathbb{R}^n$. Just to drive the point home, this is existentially equivalent to the Python snippet:

```
min(EXPR for v in domain if norm(v) == 1)
```

The analogous expression which evaluates to the input vector v (instead of the expression being optimized) is called "arg min." The arg prefix generally means, get the "argument," or input, to the optimized expression. Note that there can be multiple minimizers of an expression, so we are implicitly saying we don't care which minimizer is chosen. It's a highly context-dependent bit of notation. If I replaced min with arg min in the offset equation above, it would correspond to the following Python snippet.

```
min(v for v in domain if norm(v) == 1, key=lambda v: EXPR)
```

I introduced the argmin because we actually want to find the minimizing vector. It's false to claim $\min_{x \geq 0}(x^2 + 1) = \min_{x \geq 0} x^2$, even though the argmins are unique and equal. So our line-of-best-fit problem is most rigorously written as:

$$\arg\min_{\substack{v \in \mathbb{R}^n \\ \|v\|=1}} \sum_{i=1}^{m} \|w_i - \text{proj}_v(w_i)\|^2$$

Now we continue to convert it to the inner product. Since $\text{proj}_v(w_i)$ and $w_i - \text{proj}_v(w)$ are perpendicular, we can apply the Pythagorean theorem, in this case that $\|\text{proj}_v(w)\|^2 + \|w_i - \text{proj}_v(w)\|^2 = \|w\|^2$, rearranging to replace each term in the sum:

$$\arg\min_{v} \sum_{i=1}^{m} \left(\|w_i\|^2 - \|\text{proj}_v(w_i)\|^2 \right)$$

Next, notice that the $\|w_i\|^2$ don't depend on the input v, meaning we can't optimize them and can remove them from the expression without changing the argument of the minimum (it does change the *value* of the min). The minimization problem is now

$$\arg\min_{v} \left(- \sum_{i=1}^{m} \|\text{proj}_v(w_i)\|^2 \right)$$

And because minimizing something is the same as maximizing its opposite, we can swap the optimization. Let's also put in the inner product formula instead of the squared-norm. We've reduced the best fitting line optimization to finding a unit vector v which maximizes

$$\arg\max_v \sum_{i=1}^{n} \langle w_i, v \rangle^2$$

If we place the vectors w_i as the rows of a matrix A, the matrix-vector multiplication formula gives us (almost) exactly these inner products! That is, Av as a vector has the values $\langle w_i, v \rangle$ as its entries, and taking a squared norm $\|Av\|^2$ gives the quantity we're trying to optimize. So our problem can be written as

$$\arg\max_v \|Av\|^2$$

Maximizing the square of a non-negative value is the same as maximizing the non-squared thing, so we can equivalently write: $\arg\max_v \|Av\|$.

To summarize, we started with a dataset of m vectors w_i which we interpreted as points in $\mathbb{R}^n$. These are the rows of the movie rating matrix, the vector of ratings per movie. We saw that the best approximating line for the vectors $\{w_i\}$ is spanned by the unit vector $v \in \mathbb{R}^n$ which maximizes $\|Av\|$, where A is a matrix whose rows are the w_i. This v will end up being one of our "idealized people," the so-called first singular vector of A.

There are many algorithms that solve this optimization problem. We'll use a particularly simple one, and defer implementing it until after we see how this problem can be used as a subroutine to compute the full singular value decomposition.

Singular Values and Vectors

Here is the main theorem that makes the SVD work:

Theorem 10.23 (The SVD Theorem). *Computing the best k-dimensional subspace fitting a dataset reduces to k applications of the one-dimensional optimization problem.*

This is so astounding and useful that the solutions to each one-dimensional problem are given names: the *singular vectors*. I will define them recursively. Let A be an $m \times n$ matrix (m rows for the movies, and n columns for the people) whose rows are the data points w_i. Let v_1 be the solution to the one-dimensional problem

$$v_1 = \arg\max_{\substack{v \in \mathbb{R}^n \\ \|v\|=1}} \|Av\|$$

Call v_1 the *first singular vector* of A. Call the *value* of the optimization problem, i.e. $\|Av_1\|$, the *first singular value* and denote it by $\sigma_1(A)$, or just σ_1 if A is understood from context.

Informally, $\sigma_1(A)$ is larger if we capture the data better by v_1. So as the points in A move toward the line spanned by v_1, $\sigma_1(A)$ increases. If all the data points lie on the line spanned by v_1, then $\sigma_1(A)$ is exactly the sum of squared-norms of the rows of A. Indeed, if $x \in \text{span}(v_1)$ and v_1 is a unit vector, then $v_1 = \pm x/\|x\|$ and $\text{proj}_{v_1}(x) = \langle x, v_1 \rangle v_1 = x$.

Now we can move up in dimension. To find the best 2-dimensional subspace, you first take the best line v_1, and you look for the next best line, ignoring all lines that are in the span of v_1. That optimization problem is written as (assuming henceforth that the domain is $\mathbb{R}^n$)

$$v_2 = \arg \max_{\substack{\|v\|=1 \\ \langle v, v_1 \rangle = 0}} \|Av\|$$

The solution v_2 is called the *second singular vector*, along with the second singular value $\sigma_2(A) = \|Av_2\|$.

Often writers will use the binary operator $\perp$ to denote perpendicularity of vectors instead of the inner product. So $v \perp v_1$ is the assertion that v and v_1 are perpendicular. The $\perp$ symbol has many silly names ("up tack" on Wikipedia). In my experience most people call it the "perp" symbol, since in mathematical typesetting it's denoted by \perp.

Continuing with the recursion, the k-th singular vector v_k is defined as the solution to the optimization problem $\|Av\|$ for unit vectors v perpendicular to every vector in span$\{v_1, \ldots, v_{k-1}\}$. The corresponding singular value is $\sigma_k(A) = \|Av_k\|$. You can keep going until either you reach $k = n$ and you have a full basis, or else some $\sigma_k(A) = 0$, in which case all the vectors in your data set lie in the span of $\{v_1, \ldots, v_{k-1}\}$.

As a side note, by the way we defined the singular values and vectors,

$$\sigma_1(A) \geq \sigma_2(A) \geq \cdots \geq \sigma_n(A) \geq 0.$$

This should be obvious, and if it's not take a moment to do a spot check and see why. Now we can prove the SVD Theorem.

Proof. Recall we're trying to prove that the first k singular vectors are actually the k-dimensional subspace of best fit for the vectors that are the rows of A. That is, they span a linear subspace W which maximizes the squared-sum of the projections of the data onto W. For $k = 1$ this is trivial, because we defined v_1 to be the solution to that optimization problem. The case of $k = 2$ contains all the important features of the general inductive step. Let W be any best-approximating 2-dimensional linear subspace for the rows of A. We'll show that the subspace spanned by the two singular vectors v_1, v_2 is at least as good (and hence equally good as W).

Let w_1, w_2 be a basis of unit vectors of W, and require $w_1 \perp w_2$. Note $\|Aw_1\|^2 + \|Aw_2\|^2$ is the quantity we need to maximize, and any unit-vector-basis of W maximizes this quantity by assumption. Moreover, we're going to pick w_2 so that it's perpendicular to the first singular vector v_1. Justify this by considering two cases: either by happenstance v_1 is already perpendicular to every vector in W, in which case any choice for w_1, w_2 will do, or else v_1 isn't perpendicular to W and you can choose w_1 to be the unit vector spanning $\text{proj}_W(v_1)$, with w_2 being any unit vector in W perpendicular to w_1. The resulting w_2 is perpendicular to v_1. (If it's hard to visualize that this can be done, draw a picture in 3 dimensions.)

By definition v_1 maximizes $\|Av\|$, implying $\|Av_1\|^2 \geq \|Aw_1\|^2$. Moreover, since we chose w_2 to be perpendicular to v_1 (and hence a possible candidate for the second singular vector), the second singular value v_2 satisfies $\|Av_2\|^2 \geq \|Aw_2\|^2$. Hence the objective by $\{v_1, v_2\}$ is at least as good as W:

$$\|Av_1\|^2 + \|Av_2\|^2 \geq \|Aw_1\|^2 + \|Aw_2\|^2.$$

The right hand side of this inequality is maximal by assumption, so they must actually be equal and both be maximizers.

For the general case of k, the inductive hypothesis tells us that the first k terms of the objective for $k+1$ singular vectors is maximized, and we just have to pick any vector w_{k+1} that is perpendicular to all $v_1, v_2, \ldots, v_k$, and the rest of the proof is just like the 2-dimensional case. We encourage the skeptical reader to fill in the details. $\square$

The singular vectors v_i are elements of the domain. In the context of the movie rating example, the domain was people, and so the singular vectors in that case are "idealized people." As we said earlier, we also want the same thing for the codomain, the "idealized movies," in such a way that A is diagonal when represented with respect to these two bases.

Say the singular vectors are $v_1, \ldots, v_n$, and the singular values are $\sigma_1, \ldots, \sigma_n$. That gives us two pieces of the puzzle: the diagonal representation Σ (the Greek capital letter sigma, since its entries are the lower case singular values σ_i) defined as follows:

$$\Sigma = \begin{pmatrix} \sigma_1 & 0 & \cdots & 0 & 0 \\ 0 & \sigma_2 & \cdots & 0 & 0 \\ \vdots & \vdots & \ddots & \vdots & \vdots \\ 0 & 0 & \cdots & \sigma_{n-1} & 0 \\ 0 & 0 & \cdots & 0 & \sigma_n \\ 0 & 0 & \cdots & 0 & 0 \\ \vdots & \vdots & \ddots & \vdots & \vdots \\ 0 & 0 & \cdots & 0 & 0 \end{pmatrix}$$

And the domain basis: a matrix V whose columns are the v_i, or equivalently V^T whose rows are the v_i.[27] If we want to write A in this diagonal way, we just have to fill in a change of basis matrix U for the codomain.

$$A = U\Sigma V^T$$

Indeed, there's one obvious guess (which we'll later scale to unit vectors): define $u_i = Av_i$. Let's verify the u_i form a basis. Note they form a basis of the *image* of A (the set $\{Av : v \in \mathbb{R}^n\}$), since it can happen that $m > n$. To get a full basis, just extend the

---

[27] Here the superscript T denotes the *transpose* of V; that is, V^T has as its i, j entry the j, i entry of V. It swaps rows and columns but we'll have much more to say in Chapter 12. For now, it's enough to note (and easy to verify) that if V has perpendicular unit vectors as columns, then $V^T = V^{-1}$, so we can use V^T as a change of basis from the standard basis to the basis defined by V.

partial basis of u_i's in any legal way to get a full basis. To show the u_i form a basis, take any vector w in the image of A, write it as $w = Ax$, and write x as a linear combination of the v_i:

$$\begin{aligned} w &= A(c_1 v_1 + \cdots + c_n v_n) \\ &= c_1 A v_1 + \cdots + c_n A v_n \\ &= c_1 u_1 + \cdots + c_n u_n \end{aligned}$$

It can be proved that the u_i are perpendicular, but the only proof I have seen is somewhat technical and for brevity's sake I will skip it. But taking this on faith, the u_i form a basis and one can express $A = U\Sigma V^T$, as desired. The fact that $A = U\Sigma V^T$ is why SVD is called a "decomposition." The U, Σ, V are the components that A is broken into, and each are particularly simple.

The One-dimensional Problem

Now that we've seen that the SVD can be computed by greedily solving a one-dimensional optimization problem, we can turn our attention to solving it. We'll use what's called the *power method for computing the top eigenvector*. The next chapter will be all about eigenvectors, but we don't need to know anything about eigenvectors to see this algorithm. In lieu of knowledge about eigenvectors, the algorithm will just appear to use a clever trick.

The idea is to take A, the original input data matrix, and instead work with $A^T A$. Why is this helpful? Using our decomposition from the previous section, we can write $A = U\Sigma V^T$, where U, V are change of basis matrices (whose columns are perpendicular unit vectors!) and V actually contains as its columns the vectors we want to compute. So we can do a little bit of matrix algebra to get

$$A^T A = (U\Sigma V^T)^T (U\Sigma V^T) = V\Sigma U^T U \Sigma V^T = V\Sigma^2 V^T$$

We're using Σ^2 to denote $\Sigma^T \Sigma$, which is a square matrix whose diagonals are the squares of the singular values $\sigma_i(A)^2$. Also note that because the columns of U are perpendicular unit vectors, the product $U^T U$ is a matrix with 1's on the diagonal and zeros elsewhere; i.e., the identity matrix.

Using $A^T A$ isolates the V part of the decomposition. Now for the algorithm:

Theorem 10.24 (The Power Method). *Let x be a unit vector that has a nonzero component of v_1 (a random unit vector has this property with high probability). Let $B = A^T A = V\Sigma^2 V^T$. Define $x_k = B^k x$, the result of k applications of B to x. Then as long as $\sigma_1(A) > \sigma_2(A)$, the limit $\lim_{k \to \infty} \frac{x_k}{\|x_k\|} = v_1$.*

Proof. I will use σ_i as a shorthand for $\sigma_i(A)$. First expand x in terms of the singular vectors $x = \sum_{i=1}^n c_i v_i$. Applying B gives $Bx = \sum_{i=1}^n c_i \sigma_i^2 v_i$. Applying it repeatedly gives

$$x_k = B^k x = \sum_{i=1}^{n} c_i \sigma_i^{2k} v_i$$

Notice that, since σ_1 is larger than σ_2 (and hence all other singular values), the coefficient for σ_1 grows faster than the others. Normalizing x_k causes the coefficient of σ_1 tends to 1 while the rest tend to 0.

□

The intuition to glean from this proof is that $B = A^T A$, when applied to a vector, "pulls" that vector a little bit toward the top singular vector. If you normalize after each step, then the magnitude of the vector doesn't change, but the direction does.

The relevant quantity tracking the growth is the ratio between the two biggest singular values, $(\sigma_1/\sigma_2)^{2n}$. Even if σ_1 is only marginally bigger, say $\sigma_1 = (1+\varepsilon)\sigma_2$, the resulting growth rate is exponential in the number of iterations. The growth rates will be terrible, convergence will be swift. Most importantly, this lets us compute! Solving the 1-dimensional optimization problem is now as simple as computing a matrix-vector product and normalizing at each step.

Code It Up

Here's the python code that solves the one-dimensional problem, using the numpy library for matrix algebra. Note that numpy uses the dot method for all types of matrix-matrix and matrix-vector and inner product operations.[28] Also note the .T property returns the transpose of a matrix or vector.

First, some setup and defining a function that produces a random unit vector.

```
from math import sqrt
from random import normalvariate

def random_unit_vector(n):
    unnormalized = [normalvariate(0, 1) for _ in range(n)]
    the_norm = sqrt(sum(x * x for x in unnormalized))
    return [x / the_norm for x in unnormalized]
```

And now the core subroutine for solving the one-dimensional problem.

---

[28] They, along with most applied linear algebraists, view vectors as matrices with one column.

```
def svd_1d(A, epsilon=1e-10):
    n, m = A.shape
    x = random_unit_vector(min(n, m))
    last_v = None
    current_v = x

    if n > m:
        B = np.dot(A.T, A)
    else:
        B = np.dot(A, A.T) # spot check: why is this okay?

    iterations = 0
    while True:
        iterations += 1
        last_v = current_v
        current_v = np.dot(B, last_v)
        current_v = current_v / norm(current_v)

        if abs(np.dot(current_v, last_v)) > 1 - epsilon:
            return current_v
```

Since, as we saw in Chapter 8, the sequence will never *quite* achieve its limit, we stop after x_n changes its angle (as computed using the inner product) by less than some threshold.

Now we can use the one-dimensional subroutine to compute the entire SVD. The helper function we need for this is how to exclude vectors in the span of the singular vectors you've already computed. Unfortunately, to solve this question opens up questions about a new topic, namely the rank of a matrix, which I've found hard to fit into this already very long chapter. As much as it hurts me to do so, we will save it for an exercise, and present the formula here.[29]

The idea is this: to exclude vectors in the span of the first singular vector v_1 with corresponding u_1, subtract from the original input matrix A the rank 1 matrix B_1 defined by $b_{i,j} = u_{1,i} v_{1,j}$ (the product of the i-th and j-th entries of u_1, v_1, respectively). The name for this matrix is the "outer product" of u_1 and v_1, and it's closely related to a concept called the *tensor product*. Likewise, you can define B_i for each of the singular vectors v_i. To exclude all the vectors in the span of $\{v_1, \ldots, v_k\}$, you replace A with $A - \sum_{i=1}^{k} B_i$.

In the following code snippet, we do this iteratively when we loop over svd_so_far and subtract. The following assumes the case of $n > m$, with the other case handled similarly in the complete program.[30] The parameter k stores the number of singular values to compute before stopping.

---

[29] And, again, I would like to stress that this book is far too small to provide a complete linear algebra education. The fantastic text "Linear Algebra Done Right" is an excellent such book for the aspiring mathematician. In that I mean, they exhaustively prove every fact about linear algebra from the ground up.

[30] See pimbook.org

```
def svd(A, k=None, epsilon=1e-10):
    A = np.array(A, dtype=float)
    n, m = A.shape
    svd_so_far = []
    if k is None:
        k = min(n, m)

    for i in range(k):
        matrix_for_1d = A.copy()

        for singular_value, u, v in svd_so_far[:i]:
            matrix_for_1d -= singular_value * np.outer(u, v)

        v = svd_1d(matrix_for_1d, epsilon=epsilon) # next singular vector
        u_unnormalized = np.dot(A, v)
        sigma = norm(u_unnormalized) # next singular value
        u = u_unnormalized / sigma
        svd_so_far.append((sigma, u, v))

    singular_values, us, vs = [np.array(x) for x in zip(*svd_so_far)]
    return singular_values, us.T, vs
```

Let's run this on some data. Specifically, we'll analyze a corpus of news stories and use SVD to find a small set of "category" vectors for the stories. These can be used, for example, to suggest category labels for a new story not present in our data set. We'll sweep a lot of the data-munging details under the rug (see the Github repository for full details), but here's a summary:

1. Scrape a set of 1000 CNN stories, and a text file one-grams.txt containing a list of the most common hundred-thousand English words. These files are in the data directory of the Github repository.

2. Using the natural language processing library nltk, convert each CNN story into a list of (possibly repeated) words, excluding all stop words and words that aren't in one-grams.txt. The output is the file all-stories.json.

3. Convert the set of all stories into a *document-term* matrix A, with m rows (one for each word) and n columns (one for each document), where the $a_{i,j}$ entry is the count of occurrences of word i in document j.

Then we run SVD on A to get a low-dimensional subspace of the vector space of words. Indeed, if the above recipe is factored out into functions, then the entire routine is:

```
data = load(filename)
matrix, (index_to_word, index_to_document) = make_document_term_matrix(data)
matrix = normalize(matrix)
sigma, U, V = svd(matrix, k=10)
```

Here U is the basis for the subspace of documents, V for the words. However, these basis vectors are very difficult to understand! If we go back to our interpretation of such a word vector as an "idealized" word, then it's a "word" that best describes some large set of documents in our linear model. It's represented as a linear combination of a hundred thousand words!

To clarify, we can project the existing words onto the subspace, and then we can cluster those vectors into groups and look at the groups. Here we use a black-box clustering algorithm called `kmeans2`, provided by the `scipy` library.

```
projectedDocuments = np.dot(matrix.T, U)
projectedWords = np.dot(matrix, V.T)

documentCenters, documentClustering = kmeans2(projectedDocuments)
wordCenters, wordClustering = kmeans2(projectedWords)
```

Once we've clustered, we can look at the output clusters and see what words are grouped together. As it turns out, such clusters often form topics. For example, after one run the clusters have size:

```
>>> Counter(wordClustering)
Counter({1: 9689, 2: 1051, 8: 680, 5: 557, 3: 321,
         7: 225, 4: 174, 6: 124, 9: 123})
```

The first cluster, as it turns out, contains all the words that don't fit neatly in other clusters—such as "skunk," "pope," and "vegan"—which explains why it's so big.[31] The other clusters have more reasonable interpretations. For example, after one run the second largest cluster contained primarily words related to crime:

```
>> print(wordClusters[1])
['accuse', 'act', 'affiliate', 'allegation', 'allege', 'altercation',
 ... 'dead', 'deadly', 'death', 'defense', 'department', 'describe',
 ... 'investigator', 'involve', 'judge', 'jury', 'justice', 'kid', 'killing', ...]
```

This is just as we'd expect, because crime is one of the largest news beats. Other clusters include business, politics, and entertainment. We encourage the reader to run the code themselves and inspect the output.

A natural question to ask is why not just cluster to begin with? Efficiency! In this model, each word is a vector of length 1000 (one entry for each story), and each document has length 100,000! Clustering on such large vectors is slow. But after we compute the SVD and project, we get clusters of length $k = 10$. We trade off accuracy for efficiency, and the SVD guarantees us that it's extracting the most important (linear) features of the data. Because of this, SVD is often called a "dimensionality reduction" algorithm: it reduces the dimension of the data from their natural dimension to a small dimension, without losing too much information.

[31] It could also occur like this because we chose too few clusters: we have to pick ahead of time how many clusters we want `kmeans2` to attempt to find, which I omitted from the simplified code above.

But there's more to the story. Recall our modeling assumption, that word meanings "have the structure of" a low-dimensional vector space, but the values we see are perturbed by some noise. A crime story might use the word "baseball" for idiosyncratic reasons, but most crime stories do not. The low-dimensional subspace captures the "essence" of the data, ignoring noise, and the projection of the input word vectors onto the SVD subspace provide a "smoothed" representation of the data. This new representation has some strikingly useful properties, which are a direct consequence of the linear model doing its job well in representing the most influential aspects of the English language.

Before I explain what that means, I need a caveat. What I'm about to describe doesn't strictly work for the code presented in this chapter. Since I wrote this code with the goal to group news articles by topic, I counted frequency of terms occurring in documents (and the dataset I used is quite small!). If you want to reproduce the behavior below, you need a larger dataset and a different preprocessing technique, which is basically to count how often word pairs co-occur in a document. Check out Chris Moody's lda2vec,[32] which does this.

Now the fun stuff. The vector representation of words produced by the SVD has a *semantic* linear structure. For example, if you take the vector for the word "king," subtract the vector for "man" and add the vector for "woman," the result approximates the vector for "queen." Indeed, the SVD representation has reproduced the gender aspect of language. This occurs for all kinds of other properties of words that fit into typical word-association style tests like "Paris is to France as Berlin is to..."

This is surprising, and it tells us that some aspect of this SVD representation of words is much better than the original input of raw word counts. It's surprising because we think of language as a highly quirky, strange, perhaps nonlinear thing. But when it comes to the relationships between words, or the semantic meaning of document topics, these linear methods work well. One might argue that the core insight behind this is that for language, *context is linear* in nature. And then it's immediately clear why this works: if you see a document with "child" and "she" in it, and those words occur close together, you intuitively know, that you're more likely to be talking about a daughter than a son. Replace the "she" with a "he" and you expect to see the word son instead. The SVD captures this.

This fascinates me philosophically. Because while I certainly unconsciously understood that semantic meaning is roughly additive, I never consciously knew it until I saw these linear models and asked why they work. Math imitates life, but it can also teach us about life as it drives us to explore, refine, and build. In fact, I was confused for a long time because the original "additive word vector" ideas came from neural network research, which typically involves models that are highly nonlinear. It wasn't until I talked with some experts in natural language processing that the additive roots of the model became apparent.

---

[32] https://github.com/cemoody/lda2vec, forked at https://github.com/pim-book/lda2vec just in case the original is removed. Also note that these techniques can also be produced by neural networks, the application of Chapter 14.

10.10 Cultural Review

1. The heart of linear algebra is a very concrete connection between linear maps and matrices. The former is intuitive, useful for thinking about linear algebra geometrically. The latter is computationally tractable, allowing us to discover and apply useful algorithms. Operations on linear maps, such as function composition, correspond pleasingly to operations on matrices, such as matrix multiplication.

2. Coordinate systems are arbitrary, and linear algebra gives you the power to change coordinate systems—change the basis of the vector space—at will. A useful basis is a treasure.

3. The matrix representation hides the difficult notation of working with linear maps, reducing the cognitive burden of the mathematician.

4. The linear model is a powerful abstraction for working with real-world data, and understanding linear algebra allows us to pinpoint the assumptions of this model, and in particular where those assumptions might break down or limit the applicability of the model.

10.11 Exercises

10.1 Prove the **0** (the zero vector) is unique; that is, if there are two vectors v, w both having the properties of the zero vector, then they are equal.

10.2 Prove that the composition of two linear maps is linear. I.e., the map $x \mapsto g(f(x))$ is linear if g and f are linear.

10.3 Prove that the image of a linear map $f : V \to W$ is a subspace of the codomain of W. Prove that the subset $\{v \in V : f(v) = 0\}$ is a subspace of V.

10.4 Let V, W be two vector spaces. Show that the direct product $V \times W$ is also a vector space by defining the two operations $+$ and $\cdot$. How does the dimension of $V \times W$ compare to the dimensions of V and W?

10.5 In $\mathbb{R}^2$ we have colorful names for special classes of linear maps that correspond to geometric transformations. Look up defintions and pictures to understand matrices that perform *rotation*, *shearing*, and *reflection* through a line.

10.6 Research definitions and write down examples for the following concepts:

1. The *column space* and *row space* of a matrix.
2. The *rank* of a matrix.
3. The *rank-nullity* theorem.
4. The *outer product* of two vectors.

5. The *direct sum* of two subspaces of a vector space.

10.7 Prove that the standard inner product on $\mathbb{R}^n$ (Definition 10.18) is linear in the first input. I.e., if you fix $y \in \mathbb{R}^n$, then $\langle x, y \rangle : \mathbb{R}^n \to \mathbb{R}$ is a linear map. Argue by symmetry that the same is true of the second coordinate.

10.8 Prove that for two matrices A, B, we have $(AB)^T = B^T A^T$.

10.9 Given two (possibly negative) integers $a, b \in \mathbb{Z}$, the *Fibonacci-type sequence* is a sequence $f_{a,b}(n)$ defined by

$$f_{a,b}(0) = a$$
$$f_{a,b}(1) = b$$
$$f_{a,b}(n) = f_{a,b}(n-1) + f_{a,b}(n-2) \quad \text{for } n > 1$$

Prove that the set of all Fibonacci-type sequences form a vector space (under what operations?). Find a basis, and thus compute its dimension.

10.10 In Chapter 2 we defined and derived an algorithm for polynomial interpolation. Reminder: given a set of $n+1$ points $(x_0, y_0), \ldots, (x_n, y_n)$, with no two x_i the same, there is a unique degree-n polynomial passing through those points. Rephrase this problem as solving a matrix-vector multiplication problem $Ay = x$ for y. Hint: A should be an $(n+1) \times (n+1)$ matrix.

10.11 The *Bernstein basis* is a basis of the vector space of polynomials of degree at most n. In an exercise from Chapter 2, you explored this basis in terms of Bezier curves. Like Taylor polynomials, Bernstein polynomials can be used to approximate functions $\mathbb{R} \to \mathbb{R}$ to arbitrary accuracy. Look up the definition of the Bernstein basis, and read a theorem that proves they can be used to approximate functions arbitrarily well.

10.12 Look up the process of Gaussian Elimination, and specifically pay attention to the so-called elementary row operations. Each of these operations corresponds to a change of basis, and is hence a matrix. Write down what these matrices are for $\mathbb{R}^3$, and realize that every change of basis matrix is a product of some number of these elementary matrices.

10.13 The *LU decomposition* is a technique related to Gaussian Elimination which is much faster when doing batch processing. For example, suppose you want to compute the basis representation for a change of basis matrix A and vectors $y_1, \ldots, y_m$. One can compute the LU decomposition of A once (computationally intensive) and use the output to solve $Ax = y_i$ many times quickly. Look up the LU decomposition, what it computes, read a proof that it works, and then implement it in code.

10.14 Look up the definition of an inner product space (a vector space equipped with an inner product), and the definition of an isometry between two inner product spaces.

Find, or discover yourself, the aforementioned proof that all n-dimensional inner product spaces are isometric.

10.15 Linear independence has applications and generalizations all over mathematics. One fruitful area is the concept of a *matroid*. Matroids have a special place in computer science, because they are the setting in which one studies greedy algorithms in general. That is, every problem that can be solved optimally with a greedy algorithm corresponds to some matroid, and every matroid can be optimized using the greedy algorithm. Look up an exposition on matroids and understand this correspondence. Apply this to the problem of finding a minimum spanning tree in a weighted graph. See Chapter 6, Exercise 6.11 for an introduction to weighted graphs.

10.16 The *k-means* clustering algorithm is an algorithm for splitting a set of n vectors $\{x_1, \ldots, x_n\} \subset \mathbb{R}^d$ into $k < n$ sets. The algorithm works as follows: choose k random input vectors that are considered as "centers" of their clusters. Then repeat the following: label each vector x_i with its closest center ("assign" the vector to that cluster). Then compute a new center for each cluster as the center of all the vectors in the cluster (add up all the vectors and divide by the number of vectors added). Repeat this until there is a round in which the centers don't change, or you exceed a predetermined number of rounds. Look up this algorithm and read about what goal it's trying to achieve, and how it can fail.

10.17 The singular value decomposition code in this chapter has at least one undesirable property: numerical instability. In general, *numerical instability* is when an algorithm is highly sensitive to small perturbations in the input. The SVD of a matrix which is not full rank (Cf. Exercise 10.6) contains values that are zero. The algorithm in this chapter does not output these properly, and instead produces non-deterministic mumbo-jumbo. Audit the algorithm to verify this undesirable behavior occurs, and research a fix.

10.12 Chapter Notes

Vector Spaces, Rigorously

The rigorous definition of a vector space first requires a rigorous definition of the scalar type, which goes by the name of *field*.

Definition 10.25. A *field* is a set K with addition $+ : K \times K \to K$ and multiplication $\cdot : K \times K \to K$ (or just juxtaposition) operations having the following properties.

- Both operations are commutative and associative.

- Addition and multiplication have identity elements which are distinct. Call them zero and one, respectively.

- Addition and multiplication both have inverses, and every element is invertible, with the exception that zero may have no multiplicative inverse.

- Multiplication distributes over addition, i.e. $x \cdot (y + z) = (x \cdot y) + (x \cdot z)$ for all $x, y, z \in K$.

The field is the triple $(K, +, \cdot)$, or just K if the operations are clear from context.

By convention, multiplication has higher operator precedence than addition, regardless of the definition of the operations. The letter K is stands for *Körper*, the German term for this mathematical object (which literally translates to "body"). Obviously, $\mathbb{R}$ is a field, but there are many others. For example, the set of fractions of integers (rational numbers) forms a field denoted $\mathbb{Q}$ with the normal addition and multiplication. Another example is the binary field $\{0, 1\}$ with the logical AND and OR operations.

Now a *vector space* can be defined so that its scalars come from some field K in the same way we used scalars from $\mathbb{R}$. We say that V is a *vector space over K* to mean that the scalars come from K. As long as the operations in K have the properties outlined above, you can do all the same linear algebra we've done in this chapter. To be particularly clear, a linear combination of vectors in V requires coefficients coming from F, and so they're called F-linear combinations. Also note that F-linear combinations must be *finite* sums.

Linear algebra can have more nuance for some special fields, but to understand when and how they are different you need to study a bit of field theory. If you're interested, look up the notion of *field characteristic* and in particular what happens when fields have characteristic 2.

To leave you with one example of an interesting vector space over a field that's not $\mathbb{R}$, consider $V = \mathbb{R}$ as a vector space over $K = \mathbb{Q}$. This might not seem interesting at first until you ask what a basis might be. Take the set $C = \{1, 2, 3, 4, 5\}$, for example. Is it possible to write π (an element of V) as a $\mathbb{Q}$-linear combination of the vectors in C? You could only do so if π itself was rational, which it's not. So how, then, might one find a basis so that π (and every other irrational number) can be written as a finite $\mathbb{Q}$-linear combination of the elements in the basis? A curious thought indeed.

Bias in Word Embeddings

The process of turning English language words into vectors in such a way that arithmetic on vectors corresponds to semantic transformations of words ("king" - "man" + "woman" = "queen") is called *semantic word embedding*. This approach has roots in linguistics and information retrieval, and was popularized in computer science in the early 2000's by Yoshua Bengio and others. In 2013, Google released an open source tool called "word2vec" that constructs embeddings using neural networks, and there are many other tools (such as GloVe) that have become popular since then.

Semantic word embeddings are an interesting case study into the shortcomings of linear models. In a 2016 paper, "Man is to Computer Programmer as Woman is to Homemaker?" a team of researchers at Microsoft Research studied how human bias expressed itself through word embeddings. Here a corpus of documents is used to train a linear model, in which pairs of words like "woman" and "receptionist" show up more often than, say, "woman" and "architect." These associations (implicit or not) will manifest themselves in the resulting embedding. As a consequence, any system based on these word embeddings is likely to associate women with receptionists more than architects. This outcome is not surprising, considering the adage, "a word is characterized by the company it keeps."

Whether one is willing to accept this outcome depends on the goal of the application, but awareness is crucial. Mathematical assumptions baked into algorithms and models—even simple ones line linearity—can dupe the unwitting. Take care when applying them to situations that involve people's lives or livelihoods.

Chapter 11
Live and Learn Linear Algebra (Again)

Good mathematicians see analogies between theorems or theories. The very best ones see analogies between analogies.

– Stephen Banach

During my PhD studies, my thesis advisor Lev and I would occasionally talk about teaching. Among others, he taught algorithms and I taught calculus and intro Python. One algorithms topic he covered was the Fast Fourier Transform.

For those who don't know (and apropos to an essay between two linear algebra chapters) the Fourier Transform is a linear map that takes an input function $f : \mathbb{R} \to \mathbb{R}$ and outputs the coefficients for a representation of f with respect to a special basis of sine and cosine functions.[1] The input functions are often thought of as "signals," such as sound waves, and the output representation is thought of as tonal frequencies. The Fast Fourier Transform, or FFT for short, is a particularly efficient algorithm for writing (finite approximations of) signals in this special basis. It's fast because it takes advantage of the symmetries in sines and cosines. The discovery of this algorithm has been described as the beginning of the information age.

Lev was well familiar with the FFT, as the insights from the algorithm relate to deep and important advances in theoretical computer science, his field of expertise. FFT is a cornerstone of electrical engineering, but the technique is much deeper than simply interpreting electrical signals. For example, FFT can be used to multiply large integers much faster than the usual algorithm. He was frustrated by students who didn't understand the basic FFT, and who didn't *care* that they didn't get it. It's boring to teach people who don't care. I can sympathize.

But then he excitedly explained a new insight! It was something he learned about the FFT while preparing his lecture notes. The details are irrelevant, but my advisor also attempted to explain this new insight to his students. This was probably not helpful for them. Instead of focusing on basic syntax and properties of the Fourier Transform, Lev tried to convey insights he had learned over his career. This would have been great for a graduate seminar, but unfortunately it was levels above his students ability to comprehend. They were still missing the foundational tools needed to express these thoughts.

[1] This is nontrivial because the vector spaces involved are infinite dimensional.

Lev was tapping the beat of a song that played clearly in his head, but which his students had never heard before.

Pedagogical critiques aside,[2] after that conversation I synthesized what felt like an obvious truth in hindsight, about math, programming, and surely all endeavors worth pursuing. Understanding comes in levels of insight. And as you learn—but more importantly as you *re-learn*—you gain meta insights. Insights about insights. You learn what parts of a thing to appreciate and what parts are cruft.

Most experienced programmers understand these levels well. You start with the basic syntax and semantics of a given programming language. You move up to the basic tenets of designing and maintaining software, such as how to extract and organize functions for reuse, proper testing and documentation, and the role of various protocols interfacing with your system. From there it grows to insights about a particular area of specialization, such as how the choice of database affects the performance of a web application, how to manage an ecosystem of interdependent services, or the tradeoffs between development speed, maintainability, and extensibility.

When you switch to a new language, syntactic scaffolding can initially mask the core idea of a program as you become acquainted with the basic paradigms. This can be complex type declarations, or a strange new package management style, or the orthodoxy of a particular pattern (promises, streams, coroutines, etc.), which are foundationally important, but mostly orthogonal to the core logic of a program. Over time—and with experience, an improved mental model, and useful tooling—the cruft becomes invisible. You see a program for its core logic while still taking advantage of the features of the language.

In software, once an engineer is experienced in the lower levels of the hierarchy, for the most part they're not encouraged to relearn them. There are exceptions to this, for example, when one learns a new programming language or is submitted to code review by senior engineers with too much time on their hands. But usually one doesn't spend a lot of time revisiting the foundations of programming language design to pick up Go, nor dive deep into the design of a database when deciding what to use for a new app. You learn SQL once, and don't revisit the technicalities of relational algebra unless absolutely necessary.

In mathematics, relearning one's field is routine. The prevalence of teaching in the research mathematician's profession has a large impact on this. Mathematicians spend an unusual amount of time learning and relearning the basics of their field because they prepare lectures for undergraduates, run seminars and reading groups, and induct clueless graduate students into the world of their research. It's an entrenched part of the culture, and perhaps it explains why so many advanced math books have "Introduction" in their title.

---

[2] Collegiate education at research institutions is a snake's nest of competing incentives and demands on one's time. Having been on the academic job market and seen what constitutes success in research, I can understand the need to conduct teaching as Lev did even if I want the world to be better.

Terry Tao summarizes it well in his essay[3] "There's more to mathematics than rigour and proofs."

The point of rigour is not to destroy all intuition; instead, it should be used to destroy bad intuition while clarifying and elevating good intuition. It is only with a combination of both rigorous formalism and good intuition that one can tackle complex mathematical problems; one needs the former to correctly deal with the fine details, and the latter to correctly deal with the big picture. Without one or the other, you will spend a lot of time blundering around in the dark (which can be instructive, but is highly inefficient). So once you are fully comfortable with rigorous mathematical thinking, you should revisit your intuitions on the subject and use your new thinking skills to test and refine these intuitions rather than discard them. One way to do this is to ask yourself dumb questions; another is to relearn your field.

This is a worthwhile endeavor for anyone who wants to understand mathematics more deeply than copying a formula from a book or paper. One aspect of this is that it's difficult to fully appreciate a definition or theorem the first time around. Veterans of college calculus will appreciate our discussion of the motivation for the "right" definition of a limit in Chapter 8, because typical calculus courses are more about the mechanics—the syntax and basic semantics—of limits and derivatives. A deep understanding of the elegance and necessity of the "supporting" definitions, and how they generalize to ideas all across mathematics, is nowhere to be found. To do so requires equal parts elementary proofs and sufficient time to discuss counterexamples, neither of which are present for college freshmen in computer science and engineering.

Another aspect is that mathematical definitions and theorems create a complex web of generalization, specialization, and adaptation that is too vast to keep in your head at once. As one traverses a career, and studies some topics in more detail, reevaluating the same ideas can produce new inspiration. While gnawing on a tough problem, returning to teach basic calculus and thinking about limits might spur you to frame the problem in the light of successively better approximations, providing a new avenue for progress. While many researchers may find this more grueling than it's worth—dealing with the added distractions of grading, course design, and cheating students—in theory it has benefits beyond the education of the pupils. My advisor's foray into Fourier Analysis is another example. He may not have found that insight were he not required to prepare a lecture on the topic.

Linear algebra, even the basic stuff, is a perfect example of the web of variation and generalization. One can take the idea of linear independence of vectors, and generalize it to the theory of *matroids*, which turns out is a cozy place to study greedy algorithms (Cf. Chatper 10, Exercise 10.15). Or, if one is interested in number theory, you have the idea of *transcendental numbers*, those numbers like e and π which can't be represented as the root of a polynomial with rational coefficients. Independence plays an analogous

---

[3] https://terrytao.wordpress.com/career-advice/

role via the idea of a *transcendence basis*, since $\mathbb{R}$ is a vector space over $\mathbb{Q}$ (cf. Chapter 2, Exercise 2.4). In fields like algebraic geometry or dynamical systems, a central tool is to take a complicated object and "linearize" it, via a transformation that, say, adds new variables and equations, so that techniques from linear algebra can be applied. The form and function of the applications and generalizations shapes one's understanding of the underlying theory.

Linear algebra has higher levels of abstraction as well. We spent time, and will continue to spend time, discussing how to cleverly choose a basis. But there is a whole other side of linear algebra that builds up the entire theory basis-free. As we discussed about the definition of the limit, the "right" definition of a concept shouldn't depend on arbitrary choices. But almost *everything* we've seen about linear algebra depends on the choice of a basis! Recreating linear algebra without a basis requires more complicated and nuanced definitions, but often results in more enlightening proofs that generalize well to harder problems. As the mathematician Emil Artin once said, "Proofs involving matrices can be shortened by 50% if one throws the matrices out." Though we don't have the bandwidth in this book to cover this perspective, it's clearly a higher rung on the ladder.

One might expect such an elegant theory could completely replace linear algebra with their messy basis choices and matrix algebra. It could hardly be further from the truth. There is a famous quote of Irving Kaplansky, an influential 20th century mathematician who worked in abstract algebra (among other topics), discussing how he and his colleagues approach problems that use linear algebra.

We share a philosophy about linear algebra: we think basis-free, we write basis-free, but when the chips are down we close the office door and compute with matrices like fury.

That humorous scene is a microcosm of mathematical attitudes toward the various levels of abstraction. When it comes down to it, mathematicians will pick the most effective tool for the job, despite any additional mess or a high-horse preference for elegance. Or, as my father-in-law likes to say, "Sometimes you gotta stick your hand in the toilet." Kaplansky understands the depth and limitations of "thinking basis-free," and part of the meta-insight is to know which situations call for which tools, and why. One nice feature of matrices (and most computationally-friendly representations) is you can let the syntax bear the weight of most of the cognition. Fluency with notation and mechanics lets you write a thing down (be certain it was correct when you wrote it) and forget about it until you need it again.

In that respect, "cumbersome" syntax is like the manuals, READMEs, and automated scripts that you write for yourself and refer to every time you forget how to configure your web server. Writing things down in a precise, computational syntax also has the benefit of isolating and clarifying the nuance and essential characteristics of difficult examples. It's much easier to focus on the bigger picture, to look at a mess and point to the interesting core—as one would with a large program—once one can freely create and manipulate the atomic units. It's the same reason I say (fully aware of the irony) that the primary goal of a calculus class is to learn algebra.

You don't learn calculus until you do differential equations. And then you don't learn calculus until you study smooth manifolds. And then you don't learn calculus until you write programs that do calculus. And then you don't learn calculus until you teach calculus. You basically never learn calculus, and every time you use it in a new setting you get new insights about it. I learned calculus while writing this book! As you mature, those insights become more nuanced, and your continued appreciation for that nuance is what keeps mathematics fresh and enjoyable. This isn't a unique feature to mathematics (appreciation for nuance is as important over a long career in politics or tennis as it is in mathematics), but the layman's attitude toward mathematics is that of stark facts. In reality, theories evolve and take on new colors over time.

Learning and re-learning is continuous in mathematics. When you return to an old subject, you must repeat the useful mechanism I've been touting throughout this book: to write down characteristic examples that serve as your mental model for a general pattern. Keeping examples in mind—picturesque examples with enough detail that you can descend the ladder of abstraction to compute if necessary—is what fortifies an idea and fertilizes the orchard from which you can pick ripe analogies.

The final aspect is that relearning one's field allows one to revisit the proofs of the central theorems of that subject. The maturity afforded by not spending most of one's effort trying to understand the proof allows one to then *judge* the proof on its merits. It's like reading the code for a system you designed, long after you've implemented and maintained it. You have a much better understanding of the *real* requirements and failures of the system. Such considerations often result in alternative proofs, which generalize and adapt in new and novel ways. Or one can gain a deeper understanding of the benefits and limitations of a proof technique, and how they apply (or don't) to a problem in the back of one's head.

Back down to earth, this book is roughly a second or third level of insight. The first level would be functional fluency with symbol manipulation. Though it sounds like it's quite basic, most of college mathematics education for engineers does not tread far off this path. This includes even differential equations, statistics, and linear algebra, often considered the terminal math courses for future software engineers.

The second level is largely about proof. Can you logically prove that the symbolic manipulations in the first level are correct? It's a meta level of insight, but in another sense it's still a kind of basic fluency. For many undergraduate mathematics majors, becoming fluent in the language of proof is the central goal of their studies. This is why almost all advanced math courses are proof-based courses, and why we've spent so much time in this book proving and discussing methods for proof.

The next level of insight, usually which comes after being able to prove the basic facts about an object, are the insights about why the existence and prevalence of that object makes sense. This occurs often through proof, but also through a non-rigorous hodge-podge of examples, discussion, connections to other objects, and the consideration of alternatives by which one becomes accommodated with a thing.

Further tiers revolve around new research. Understanding what questions are interesting, sketching why a theorem should be true before a proof is found, generalizing families of proofs into a theory that makes all those proofs trivial. And all the while one traverses the ladder of abstraction as needed, sometimes diving into the muddy waters to crack a tough integral, other times honing in on the importance of one particular property of an object.

It sounds negligent to speak about math in such an imprecise manner, and mathematicians like to poke fun at themselves. John von Neumann (of computer architecture fame) once told a physicist colleague, "In mathematics you don't understand things. You just get used to them." How deliciously blasphemous! More seriously, my interpretation is that this quote continues, "...until you find that next level of insight." It's true, at least, in my experience, that one must gain sufficient comfort in mechanics before one can attempt proof, and one must gain some level of comfort with proof before the next-level insights about definitions can be appreciated.

It's not just professional mathematicians who experience this. This happens at every level of the hierarchy. My wife is a math professor at a community college, and despite having spent years of her undergraduate career doing proofs by induction, it was not until she taught it a few times that the deeper understanding of why it worked dawned on her. She had a similar experience re-learning algebraic topology for a qualifying exam, and I distinctly recall her gleeful yelp when she realized that the intimately understood what she was doing and why it worked.

The cognitive scientist Douglas Hofstadter asserts that analogies are the core mechanism of human cognition. Part of his evidence is the wealth of analogies that surround us in every day life: the commonplace concept of an airport "hub" relies on analogies between the spokes of a bicycle wheel and notions of centrality in a network, each of which rely on lower-level analogies of position and motion. These ideas are paired with ideas about corporations, a brand, and not to mention the myriad web of analogies that go into human conceptions of airplane flight. This is all summarized by the single word "hub."

The quote at the beginning of this interlude suggests that mathematics is no different. Mathematical cognition is also largely built on analogies between analogies built on top of lower level analogies. And just like humans understand the concepts of motion or a wheel long before we're able to understand the concept of an airport hub, we're able to understand the lower levels of mathematical abstraction (and must become comfortable with them) before we can draw the analogies necessary to make use of the more complex and nuanced abstractions. And then, much later, we can look back at the bicycle wheel with a new appreciation for its purpose and use. Mathematical intuition in particular is the graduation from purely analytical and mechanical analysis to a more visceral feeling of why a thing should behave the way it does.

No matter where you currently stand, there are insights to be found and analogies to draw. Don't underestimate their value, even if they lie among "simple" things that you think you should have mastered years ago.

Chapter 12
Eigenvectors and Eigenvalues

The notion of eigenvalue is one of the most important in linear algebra, if not in algebra, if not in mathematics, if not in the whole of science.

– Paolo Aluffi

If you polled mathematicians on what the "most interesting" topic in linear algebra was, they'd probably agree on eigenvalues. The definition of an eigenvalue is so simple that I can state it now without further ado.

Definition 12.1. Let V be a vector space and let $f : V \to V$ be a linear map. A scalar λ is called an *eigenvalue* for f if there is a nonzero vector $v \in V$ such that $f(v) = \lambda v$. The associated vector v is called an *eigenvector* of f with the corresponding eigenvalue λ.

A more concise, less precise rephrasing is to find a "nontrivial" solution[1] to $f(v) = \lambda v$. Note that $\lambda = 0$ is a valid choice of $f(v) = 0$, so long as v is nonzero. As you would infer from our discussion in Chapter 10, the same definition holds for a matrix A, where the condition is equivalently written $Av = \lambda v$.

The question of why eigenvalues are so central to linear algebra and its applications is a deep one, and there is no easy answer. In a vague sense, the eigenvectors and eigenvalues of a linear map encode the most important data about that map in a natural, efficient way. More concretely, in the scope of this chapter eigenvectors provide the "right" basis in which to study a linear map $V \to V$. They transform our perspective so that the important features of a map can be studied in isolation. If you accept that premise, it's no surprise that eigenvalues are useful for computation. But to say anything more concrete than that, to explain the universality of eigenvalues, is difficult.

The application for this chapter is a deep dive into how eigenvectors and eigenvalues explain the dynamics of a particular physical system describing one-dimensional waves. In no uncertain terms, eigenvalues *are* the scientific theory that reveals the inner nature

---

[1] "Trivial" gets new meaning in this context that is partially subjective. To conjure "the nontrivial solutions" means to ignore the obvious counterexamples. For eigenvalues and eigenvectors, if 0 denotes the zero vector, it's clear that $f(0) = \lambda \cdot 0$ for *every* λ. It would make the definition useless if we included these "trivial" solutions. In this book we will state explicitly what the "trivial" solutions are, but elsewhere you may have to infer.

of the system. As a bonus, the clarification provided by eigenvectors gives naturally efficient algorithms to determine the state of the dynamical system at any future time. In Chapter 14 we'll see how eigenvalues encode information about smooth surfaces in a way that enables optimization. And the singular values we saw in Chapter 10 are closely related to eigenvectors and eigenvalues in a way we didn't have the language to explain in that chapter (see the exercises for more on that).

I could spend all day giving *examples* of how eigenvectors are used in practice. But to get to the heart of what makes them useful is another task entirely. The word eigenvalue itself doesn't have any intrinsic meaning that might hint at an answer. Eigenvalue comes from the German word *eigen*, simply meaning "own," in the sense of the phrase, "I have my *own* principles to uphold and refuse to use emacs." In that sense, eigenvalue simply means a value that is intrinsic to the linear map. In a sense that can be made rigorous, the importance of the study of eigenvalues and eigenvectors is analogous to the importance of the roots of a polynomial to the study of polynomials. Knowing the roots of a polynomial allows you to write the polynomial in a simpler form, and "read off" information about the polynomial from the simpler representation. So it is with eigenvalues and eigenvectors.

We'll start by proving intrinsic-ness; the eigenvalues of a matrix are independent of the choice of basis. Let A be the matrix representation of a linear map $f : \mathbb{R}^n \to \mathbb{R}^n$, written with respect to the standard basis. Let U be a change of basis matrix. That is, the columns of U are the new basis vectors, and if we were to write f with respect to the new basis, its matrix would be $B = UAU^{-1}$. Recall, in words, this matrix converts the input to the standard basis via U^{-1} (the inverse of U), then applies A, then converts the output back to the new basis using U. Now we can state the theorem.

Theorem 12.2. *Let A be a matrix and U be a change of basis matrix, with $B = UAU^{-1}$. Let $v \in \mathbb{R}^n$ be an eigenvector for A with eigenvalue λ. Then $v' = Uv$ is an eigenvector for B that also has eigenvalue λ.*

Proof. We need to show that $BUv = \lambda Uv$. To do this, expand $B = UAU^{-1}$ and apply algebra.[2] In what follows, I_n is the n-by-n identity matrix, i.e., the representation of the function $I(v) = v$ that is the same for every basis.

$$(UAU^{-1})(Uv) = UA(U^{-1}U)v = UAI_nv = U(Av) = U(\lambda v) = \lambda Uv.$$

□

So while (the coordinates of) eigenvectors are not preserved across different bases, the eigenvalues are. A technical way to say this is that eigenvalues of a linear map f are *invariant properties* of f. Invariance means that the property doesn't change under some prespecified family of transformations. In this case, eigenvalues are invariant under the

---

[2] I hopefully assured you in Chapter 10 that basic algebra operations such as regrouping parentheses are legal in matrix algebra, without requiring a detailed and painful derivation of that fact. Such work belongs in textbooks, and we have more exciting things to do here.

operation of changing a basis. Invariance is a natural property to require for something which purports to reveal the divine secrets of a linear map.

This is also related to our earlier discussion in Chapter 8 of the well-definition of the limit. We're saying that the eigenvalues of a linear map don't depend on the arbitrary choices you make to represent them in the nice computational setting of matrix algebra. However, this time it's a bit different because we didn't intentionally bake basis-invariance into the definition. If you sumbled across a matrix-vector equation like $Av = 2v$ in the wild, perhaps while modeling some physical system, it might not occur to you that the number 2 is a special property of the system.

The point is that the invariance of eigenvalues is thought of as a discovered behavior of the definition. Someone once discovered it was useful to look at $Av = \lambda v$ for matrices, and then later discovered invariance, and *then* wrote down Definition 12.1. On the other hand, the definition of a limit had an explicit invariance goal in shaping it.

This notion of invariance is a strong "smell" in mathematics. As we stated, eigenvalues are invariant under the operation of a change of basis. This has something to do with why they are so useful. Invariant objects are signs toward the soul of mathematics. We'll return to the study of invariants when we study hyperbolic geometry in Chapter 16.

But moreover, an eigenvector v of A has a different sort of "invariance" under the operation of left-multiplication by A. That is, if you ignore scaling—or rescale v to a unit vector before and after left multiplying by A—then A sends v to itself. This is why we say that the eigenvectors span the "best axes" in which to view A, because A sends any vector on the axis to another vector within the same line. They exhibit maximal invariance when the linear map is applied to them. And for the limited scope of this chapter, the set of all eigenvalues and eigenvectors of a linear map allows one to represent the entire map in terms of these invariant, independent pieces.

This is the best high-level intuition I can give without getting too deep in the math. Before we do, let's see a compelling example of why eigenvalues are so interesting and complex for specific matrices called *adjacency matrices*. In the next section we won't prove any of the theorems we state.

12.1 Eigenvalues of Graphs

Let $G = (V, E)$ be an undirected graph, the same sort we studied in Chapter 6. There is a natural matrix we can associate with G, defined as follows.

Definition 12.3. Let $G = (V, E)$ be a graph and $V = \{v_1, \ldots, v_n\}$ (i.e., pick an ordering of the n vertices of G). Define the *adjacency matrix of* G, denoted $A(G)$, as the $n \times n$ matrix whose i, j entry is 1 if $(v_i, v_j) \in E$ and 0 otherwise.

In the exercises, you will write down a description of this matrix as a linear map and interpret what it means in graph-theoretic terms. In particular, each of the standard basis vectors $e_i = (0, \ldots, 0, 1, 0, \ldots, 0)$ can be thought of as identifying the i-th vertex v_i of G. Figure 12.1 is an example graph and its adjacency matrix. We call a graph *bipartite* if

$$G \qquad\qquad A(G)$$

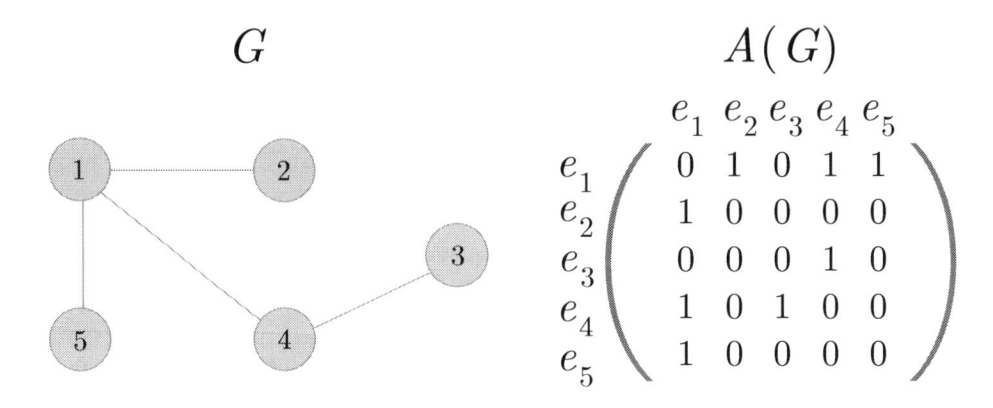

$$\begin{array}{c} \ e_1\ e_2\ e_3\ e_4\ e_5 \\ \begin{array}{c} e_1 \\ e_2 \\ e_3 \\ e_4 \\ e_5 \end{array} \begin{pmatrix} 0 & 1 & 0 & 1 & 1 \\ 1 & 0 & 0 & 0 & 0 \\ 0 & 0 & 0 & 1 & 0 \\ 1 & 0 & 1 & 0 & 0 \\ 1 & 0 & 0 & 0 & 0 \end{pmatrix} \end{array}$$

Figure 12.1: An example of a graph and its adjacency matrix

its vertices can be partitioned into two parts in such a way that all edges cross from one part to the other. The graph G in Figure 12.1 is bipartite because it can be partitioned into $\{1, 3\}$ and $\{2, 4, 5\}$, and no edges go between these two sets.

Bipartite graphs are common in applications, because they naturally encode networks in which there are two classes of things, where things within a class don't relate to each other. For example: students and teachers, with edges being class membership; wholesale factories and distributors, with edges being shipments; or files and users, with edges being access logs. Problems that can be intractable on general graphs can turn out to be easy to solve on bipartite graphs, which is a compelling reason to study them.

Now here is a fantastic theorem that we won't prove. Let $A(G)$ be the adjacency matrix of a (not-necessarily bipartite) graph G. Let λ_1 be the largest eigenvalue, λ_2 the second largest, etc., so that λ_n is the smallest. Note that these eigenvalues may be negative. Also note that, while it is true that adjacency matrices have n eigenvalues, to see why we'll need the theory built up in this chapter (Propositions 12.11 and 12.14).

Theorem 12.4. *Let G be a connected graph. Then G is bipartite if and only if $\lambda_1 = -\lambda_n$.*

This is just one of the many ways that the eigenvalues of the adjacency matrix of G encode information about G. In hindsight, it's obvious that some relationship should exist: there is a systematic way to get from the graph G to the eigenvalues. What's surprising is that they encode such natural and useful information about G, which might otherwise require designing an algorithm to discover.

Here is another theorem, which I will paraphrase slightly to hide the nitty-gritty details. It says that the eigenvector for the second-largest eigenvalue of the adjacency matrix encodes information about tightly-knit clusters of vertices in a graph. In fact, it encodes this information *better* than statistics the following concrete setting.

Let $G = (V, E)$ be a graph constructed by the following process: for each pair of vertices $v_i, v_j \in V$, flip a fair coin. If heads, make (v_i, v_j) an edge of E. Otherwise, skip that edge. You can prove that this process produces all possible graphs with equal likelihood, so the output is simply called a *random graph*.[3]

One can show (though we will not) that for a random graph, with overwhelming probability the densest cluster of vertices will have almost exactly $2\log(n)$ vertices in it. It's also widely believed that no efficient algorithm can reliably find the densest cluster.

So to make this cluster-finding problem easier, after creating the graph in this random way, pick a random subset of vertices of size $\sqrt{n}$, and connect all remaining edges among those vertices. We'll call the chosen subset a *planted clique*. In general, a *clique* is a subset of vertices with a complete set of edges among them. It's a subgraph that forms the complete graph K_m for some m. You might expect that such a dense cluster of vertices would be detectable, simply by being a statistical anomaly. Maybe you could just count up how many edges are on each vertex, looking at the ones that are unusually large, to find the planted clique. I won't prove so here, but this method provably fails. It requires the planted clique to have size $\sqrt{n \log n}$ or bigger. Instead, the following algorithm succeeds:

Theorem 12.5. *Let v be an eigenvector for λ_2, the second largest eigenvalue of the adjacency matrix of G, a random graph on n vertices with a planted clique of size $\sqrt{n}$. The following algorithm recovers the vertices of the planted clique with high probability:*

1. *Recall that the indices of v correspond to vertices of G, and select $\sqrt{n}$ such vertices whose corresponding entries in v are the largest in absolute value. Call this set T.*

2. *Output the set of vertices of G that are adjacent to at least $3/4$ of the vertices in T.*

This is a result that is quite recent by mathematics standards. It was proved in 1998 by Alon et al. No method is known to exist that can reliably find a smaller planted clique, and moreover it can be *proved* that methods that only use statistics about the graph cannot find a smaller clique.[4] All of this is to say, eigenvalues of the adjacency matrix don't just encode information about G, in certain settings they do so in an *optimal way*. The specific area of math studying how and when eigenvalues are useful in encapsulating information about graphs is called *spectral graph theory*. The general idea of using eigenvalues and eigenvectors of matrices derived from a graph to find dense clusters is called *spectral clustering*, and there are many variations.

12.2 Limiting the Scope: Symmetric Matrices

By now I hope I have convinced you that eigenvectors and eigenvalues, together often called an *eigensystem*, encode useful information about linear maps, and the underlying data those linear maps represent.

---

[3] More specifically, it's called an Erdős-Rényi random graph, and the output is a draw from the uniform distribution over graphs with n vertices.

[4] In the sense that they require an exponential number of samples to be correct with good probability. See Feldman et al. 2012, "Statistical Algorithms and a Lower Bound for Detecting Planted Clique."

However, we still have little understanding about why eigensystems reveal such valuable information. The briefest possible answer might be formulated as "eigenvectors, scaled by their eigenvalues, provide the most natural coordinate system in which to view a linear map."

A stronger intuition is difficult to explain without a longer expedition into the theory than we have time in these pages. One reason it's hard is that a linear map f on $\mathbb{R}^n$ might have eigenvalues that are complex numbers instead of real numbers, and eigenvectors with complex entries. Much like the possibly complex roots of a single-variable polynomial, having complex eigenvalues means you have fewer real eigenvalues. More importantly, if you're not comfortable with the geometry of complex numbers, you will have difficulty interpreting how they relate to a linear map for vectors of real numbers. This book skips complex numbers, so we will not be able to give a complete picture.

The second reason is that eigenvalues can occur with multiplicity in two different ways. The differences, and how they manifest in the behavior of the map, are nuanced and not needed for our application, though we do mention some pointers in Section 12.5.

Luckily, there is a nice way to avoid dealing with complex numbers and multiplicity while still seeing the lion's share of eigenvalue power in practice. That is the following theorem:

Theorem 12.6. *Let $f : \mathbb{R}^n \to \mathbb{R}^n$ be a linear map and let A be its associated matrix. If A is symmetric, meaning $A[i, j] = A[j, i]$ for every i, j, then A has n real eigenvalues and eigenvectors.*

A useful notation when working with symmetric matrices is that of the *transpose*. Define by A^T the matrix whose i, j entry is $A[j, i]$. That is, you take A, and flip it along the top-left-to-bottom-right diagonal, and you get A^T. With this notation, saying A is symmetric is saying that $A = A^T$. Here's an example of a symmetric matrix.

$$\begin{pmatrix} 1 & 2 & 3 & 4 \\ 2 & 5 & 6 & 7 \\ 3 & 6 & 8 & 9 \\ 4 & 7 & 9 & -1 \end{pmatrix}$$

In Chapter 10 I promised you that every operation on a matrix corresponds to an operation on a linear map. This is also true for the matrix transpose. If f is a linear map and A is a matrix representation, then A^T corresponds to some linear map f^T that's related to f. However, the operation itself is difficult to describe without a lot of extra notation and definitions. We'll revisit those ideas in the Chapter Notes, but here we'll directly prove the important takeaway of that discussion: symmetric matrices play nicely with the inner product.

First, one can verify that the standard inner product definition results in $\langle Ax, y \rangle = \langle x, A^T y \rangle$ for all x, y. This is often written as $\langle Ax, y \rangle = x^T A^T y$. One considers vectors "single-column matrices," notes that in this perspective $\langle x, y \rangle = x^T y$, and then you get

$$\langle Ax, y \rangle = (Ax)^T y = x^T A^T y = \langle x, A^T y \rangle.$$

And so with symmetry you get a simplified formula $\langle Ax, y \rangle = \langle x, Ay \rangle$. What's special is that symmetric matrices can be defined by this property.

Theorem 12.7. *Let A be a symmetric real-valued $n \times n$ matrix, and let $\langle -, - \rangle$ denote[5] the standard dot product of real vectors. Then A is symmetric if and only if $\langle Ax, y \rangle = \langle x, Ay \rangle$ for every pair of vectors $x, y \in \mathbb{R}^n$.*

Proof. Symmetry gives the forward direction of the "if and only if," since $\langle x, A^T y \rangle = \langle x, Ay \rangle$. For the reverse direction, suppose that $\langle Ax, y \rangle = \langle x, Ay \rangle$ for all x, y. Let $a_1, \ldots, a_n$ be the columns of A, and apply this fact to the vectors $x = e_i, y = e_j$ (the standard basis vectors with a 1 in positions i and j, respectively). We have

$$\langle A e_i, e_j \rangle = \langle a_i, e_j \rangle = A[j, i]$$

And we can do the same thing with A on the other side, by assumption:

$$\langle e_i, A e_j \rangle = \langle e_i, a_j \rangle = A[i, j]$$

Since $\langle A e_i, e_j \rangle = \langle e_i, A e_j \rangle$, we get $A[i, j] = A[j, i]$, implying A is symmetric. $\square$

We will use symmetry to prove that every symmetric matrix with real-valued entries has a real eigenvalue. This is the central lemma needed to prove Theorem 12.6. Funnily, we've spent so long preaching the virtues of eigenvalues, we haven't even considered the basic question of their existence!

Lemma 12.8. *Let A be a symmetric real-valued matrix. Then A has a real eigenvalue.*

Proof. Let x be a unit vector which maximizes the norm $\|Ax\|$, and let $c = \|Ax\|$. Then $Ax = cy$ for some unit vector y. By the maximality of x we know that $\|Ay\| \leq c$. If $y = x$ then we are done (in real life, this happens most of the time).

If $y \neq x$ then we can show that $x + y$ is an eigenvector with eigenvalue c. After the proof we'll explain as a side note why it makes sense in hindsight to consider $x + y$. Now notice that

$$\langle x, Ay \rangle = \langle Ax, y \rangle = \langle cy, y \rangle = c.$$

The first equality is due to Theorem 12.7, the second is the definition of y, and the third is because the inner product is linear and y is a unit vector.

The crucial observation is that $\langle x, Ay \rangle$ is the (signed) length of the projection of Ay onto the unit vector x. Projecting a vector onto a unit vector can only make the first

---

[5] The notation $\langle -, - \rangle$ is used to signify that the function will be expressed in this nonstandard "pairing" notation. If the inputs are $v, w \in V \times V$, the interpretation is to substitute the dashes with the inputs in order, i.e. $\langle v, w \rangle$.

vector shorter. You should have some intuitive sense that this is true after our analysis—particularly the pictures—in Chapter 10. We leave a rigorous proof for the exercises. As a consequence, $c = \langle x, Ay \rangle \leq \|Ay\|$.

Note that $\|Ay\| \leq \|Ax\| \leq c$, since x maximizes $\|Ax\|$. This, combined with the fact that $\langle x, Ay \rangle = c$, gives us

$$c = \langle x, Ay \rangle \leq \|Ay\| \leq c$$

Since c is on either end of this inequality, all of the quantities must be equal! Indeed, the only way for the projection of Ay onto x to have the same length as Ay is for Ay to be in the span of x already. To summarize, we have proved that $Ax = cy$ and $Ay = cx$.

The final observation is simply that $A(x + y) = Ax + Ay = cy + cx$, and so c is an eigenvalue for $x + y$.

□

To fulfill my promise: $x + y$ is a natural choice of eigenvector because it's on the line "halfway" between x and y. Indeed, it's in the span of the vector $(x + y)/2$, which is a more suggestive way to say the "average" of x and y. Symmetry was our guide: A sends x to the span of y and vice versa. The seasoned linear algebraist would guess—and prove shortly thereafter—that the symmetry extends to the whole plane spanning $\{x, y\}$. Since the behavior of any linear map (on this subspace) only depends on its behavior on the basis (of the subspace), we deduce that A behaves as a reflection, flipping the entire plane span$\{x, y\}$. And every reflection in a plane has a line of symmetry, which in this case is through $x + y$.

In any case, it's clear that the inner product is starting to take center stage, and so we should study it in more detail.

12.3 Inner Products

In order to express one very useful aspect of eigenvectors, we must revisit the discussion from Chapter 10 about the inner product. In general, a vector space only has a limited amount of geometry you can describe. However, if you specify an inner product for a vector space, you can describe angles, lengths, and more. The inner product is *imposed* on a vector space, in the same way that a style guide is *imposed* on a programmer: to give structure to (or elucidate structure in) the underlying space. The standard inner product on $\mathbb{R}^n$ is defined by the formula

$$\langle x, y \rangle = \sum_{i=1}^{n} x_i y_i.$$

This formula is intimately connected with geometry. It can be used to compute the angle between two vectors (via $\cos \theta = \langle x, y \rangle / (|x| \cdot |y|)$), and its value is the signed length of the projection of one argument onto the other (scaled by the lengths of the vectors).

The Power of a Generalized Inner Product

Over the years mathematicians have extracted the generic properties of this formula that conjure up its geometric magic. The result is a distilled definition of an inner product.

Definition 12.9. Let V be a vector space with scalars in $\mathbb{R}$. An *inner product* for V is a function $\langle -, - \rangle : V \times V \to \mathbb{R}$ with the following properties:

1. *Symmetric:* For every $v, w \in V$ swapping the order of the inputs doesn't change the inner product, i.e. $\langle v, w \rangle = \langle w, v \rangle$.

2. *Bi-linear:* If you fix any input to a constant $v \in V$ then the restricted function, considered as a map $V \to \mathbb{R}$, is linear. I.e., if we fix the second input $\langle -, w \rangle$, then $\langle cv, w \rangle = c \langle v, w \rangle$ for all $c \in \mathbb{R}$, and likewise $\langle u+v, w \rangle = \langle u, w \rangle + \langle v, w \rangle$. Likewise for fixing the first input.

3. *Nonnegative norms:* For every $v \in V$, the inner product with itself is nonnegative, i.e. $\langle v, v \rangle \geq 0$. This is called the *squared norm* of v. Moreover, we require that the only vector with norm zero is the zero vector.

A vector space V and a specific inner product $\langle -, - \rangle$ are together called an *inner product space*.

In Chapter 10 we proved Theorem 10.17 that every finite-dimensional vector space is isomorphic to $\mathbb{R}^n$. It turns out there's a similar theorem for finite-dimensional inner product spaces. That is, if you are in finite dimensions then every inner product space is isomorphic to $\mathbb{R}^n$ with the usual sum-of-squares inner product. The notion of isomorphism is more complicated here, because it needs to preserve the inner product. See the exercises for more details. This allows us to justify using the standard inner product and $\mathbb{R}^n$ for applications that lack a more principled choice.

More generally, the abstract definition of an inner product becomes more useful and interesting when you're dealing with infinite-dimensional vector spaces. We won't cover this in depth in this book, but a quick aside may pique your interest. The gold standard example of an interesting inner product space is the space of functions of a single real variable $f : \mathbb{R} \to \mathbb{R}$ whose square has a finite integral.[6] Call this space $L^2(\mathbb{R})$, or just L^2 for short (the exponent reminds us we're squaring):

$$L^2(\mathbb{R}) = \left\{ f : \mathbb{R} \to \mathbb{R} \ \bigg| \ \int_{-\infty}^{\infty} f(x)^2 dx \text{ is finite} \right\}$$

A typical example of where these functions occur in real life is as sound waves. L^2 forms a vector space. Addition is the point-wise addition of functions $(f + g)(x) =$

---

[6] We won't cover integration in this book, but you don't need to know (or remember) how to integrate functions to follow along. In all of what follows, the integral $\int_{-\infty}^{\infty} f(x)dx$ is a number that represents the signed area in between $f(x)$ and the x-axis. The meta-motivation for inner products is well-worth any notational discomfort.

$f(x) + g(x)$, and with the requisite calculus one can prove that the sum of two square-integrable functions is square-integrable. The case is similar for the other required vector space properties. And finally, the jewel in the crown, the inner product is

$$\langle f, g \rangle = \int_{-\infty}^{\infty} f(x)g(x)dx.$$

This inner product space—which actually satisfies some additional properties that make it into a so-called *Hilbert space*—is different from vector spaces we've seen so far. In particular, in $\mathbb{R}^n$ there's a "default" basis in which we express vectors without realizing it: the standard basis. L^2 has no obvious basis. From on our discussion of Taylor series in Chapter 8, we know that polynomials can approximate functions in the limit. One might hope that polynomials form a basis of this space, perhaps $\{1, x, x^2 \ldots\}$. But actually none of these functions are even *in* L^2! And moreover, many functions in L^2 aren't differentiable everywhere, so Taylor series can run into trouble.

As it happens, there are many interesting and useful bases for this space. For example, the following basis is called the Hermite basis:[7]

$$\{e^{-x^2/2}, xe^{-x^2/2}, x^2e^{-x^2/2}, \ldots, x^n e^{-x^2/2}, \ldots\} = \{x^k e^{-x^2/2} : k \in \mathbb{N}\}$$

But proving this is a basis is not trivial! There are other useful bases as well. The *Fourier basis*, a staple of the signal-processing world and electrical engineering, is the set of *complex exponentials* $\{e^{2\pi i k x} : k \in \mathbb{Z}\}$. Since we're not officially covering complex numbers in this book, think of this basis as the set of all sine and cosine functions with all possible periods.

These bases are difficult to discover. But even when we have one, how in the name of Grace Hopper can one even *write* a function in such a basis? You can't set up a system of equations because there's no decent starting basis! Not to mention it'd be an infinite system of infinitely long equations.

Using the inner product, and some work to modify the basis to make it geometrically amenable, the process of writing a function with respect to one of these (modified) bases reduces to computing an inner product. Once again, we translate an intuitive but hard mathematical concept into a more computationally friendly language. This should impress upon you the importance of the inner product. Not only does it endow a vector space with new, geometric measurements; it also makes computing basis representations possible where it might otherwise not be. A powerful revelation indeed.

In the rest this chapter, except for the application, the inner product will be considered abstractly, as we study its generic properties and how it relates to eigenvectors. We'll also see how the inner product relates to simplifying the computation of expressing a vector in terms of a basis.

---

[7] More specifically, the Hermite basis is what happens when you apply Gram-Schmidt to orthogonalize and normalize this basis, which we'll see later in this chapter.

Properties of an Inner Product

Definition 12.9 implies some easy consequences. Here are two examples.

Proposition 12.10. *Let $\mathbf{0}$ be the zero vector of V, and 0 the real number zero. Then $\langle v, w \rangle = 0$ for every $w \in V$, if and only if $v = \mathbf{0}$.*

Proof. For the forward direction, if $\langle v, w \rangle = 0$ for every w, then fix $w = v$. The defining properties of an inner product require $v = \mathbf{0}$. For the reverse direction, fix any w and note that $f(v) = \langle v, w \rangle$ is a linear map. Linear maps preserve the zero vector, so $f(\mathbf{0}) = 0$. □

In the exercises you will prove some other basic facts about inner products, but here is one too important to relegate to the end of the chapter.

Proposition 12.11. *Let A be real-valued symmetric matrix. Let v, w be eigenvectors of A with corresponding eigenvalues $\lambda \neq \mu$, respectively. Then $\langle v, w \rangle = 0$.*

Proof. By the symmetry of A:

$$\langle \lambda v, w \rangle = \langle Av, w \rangle = \langle v, A^T w \rangle = \langle v, Aw \rangle = \langle v, \mu w \rangle$$

Since this is an inner product, we can pull out the scalar multiples on the far left and right-hand sides to get $\lambda \langle v, w \rangle = \mu \langle v, w \rangle$. The only way for this equation to be true in spite of $\lambda \neq \mu$ is if $\langle v, w \rangle = 0$. □

Another way to say it is that if two eigenvectors are *not* orthogonal, then they must have the same corresponding eigenvalue (this is the contrapositive statement[8]).

As we proved in Chapter 10, the standard inner product on $\mathbb{R}^n$ allows one to compute angles, and more specifically to determine when two vectors are perpendicular to each other. In a generic inner product space, perpendicularity is undefined, and so we define it by generalizing what we proved in $\mathbb{R}^n$. Perpendicularity and length get new names.

Definition 12.12. *Two vectors $u, v \in V$ in an inner product space are called* orthogonal *if $\langle u, v \rangle = 0$.*

Definition 12.13. *The* norm *of a vector $v \in V$ is the quantity $\|v\| = \sqrt{\langle v, v \rangle}$. Without a square root, it's called the* square norm.

Vectors with norm 1 are called *unit vectors*.

Most of the facts about perpendicularity and projection we proved for $\mathbb{R}^n$ actually don't depend on the definition of the standard inner product. They can be re-proved using any inner product, because the key ingredients from those proofs were extracted into the definition of an inner product. Next we'll show that orthogonal vectors can be used to build up a basis.

---

[8] If "p implies q" is true, then it is equivalently true that "not q implies not p." The latter is called the *contrapositive* form of the former.

Proposition 12.14. *Any set of nonzero vectors $\{v_1, \ldots, v_k\}$ which are pairwise orthogonal are linearly independent.*

Proof. Let $\{v_1, \ldots, v_m\}$ be as in the statement of the proposition, and suppose $c_1v_1 + \cdots + c_mv_m = 0$. To show linear independence, recall, we need to show that all the $c_i = 0$. Fix any i. To show c_i is zero, inspect $\langle c_1v_1 + \cdots + c_mv_m, v_i \rangle$, which is zero because the first argument is zero by assumption. By the linearity of the inner product, this splits up as $\sum_{k=1}^{m} c_k \langle v_k, v_i \rangle$. All of these are zero except $c_i \langle v_i, v_i \rangle$, implying $0 = c_i \langle v_i, v_i \rangle$. Then either $v_i = 0$ (ruled out by assumption) or $c_i = 0$. The same argument applies to evey c_j. $\square$

This explains part of a comment we made earlier about adjacency matrices. We said that an adjacency matrix has n eigenvalues and eigenvectors. It has at most n distinct eigenvalues because each one corresponds to an eigenvector, and together these eigenvectors would form a basis. Such a basis can be *at most* as big as the dimension of the space, which for adjacency matrices was n, the number of vertices. The reason it's *exactly* n is because an adjacency matrix is symmetric, which is the hypothesis for this chapter's crowning result, Theorem 12.24.

12.4 Orthonormal Bases

Bases consisting of orthogonal vectors are glittering treasures for computation. They make it easy to write a vector in terms of that basis.

Let V be an inner product space, and suppose that $\{v_1, \ldots, v_n\}$ is a basis for V, where every v_i is a unit vector and $\langle v_i, v_j \rangle = 0$ for every $i \neq j$. Such a basis is called an *orthonormal basis*. The "ortho" is because each pair is orthogonal, and "normal" because each vector is a unit vector (normalized). Having such a basis allows you to compute the basis representation of any vector using inner products.

Proposition 12.15. *Let $\{v_1, \ldots, v_n\}$ be an orthonormal basis for V, and let $x \in V$. Then x can be written as*

$$x = \langle x, v_1 \rangle v_1 + \cdots + \langle x, v_n \rangle v_n$$

That is, the coefficient of the basis vector v_i is $\langle x, v_i \rangle$.

Proof. Fix any basis vector v_i and let $x = c_1v_1 + \cdots + c_nv_n$ where c_j are the (unknown) coefficients of x's representation with respect to the basis. Then

$$\begin{aligned}\langle x, v_i \rangle &= \langle c_1v_1 + \cdots + c_nv_n, v_i \rangle \\ &= c_1 \langle v_1, v_i \rangle + \cdots + c_n \langle v_n, v_i \rangle \\ &= c_1 \cdot 0 + \cdots + c_{i-1} \cdot 0 + \underbrace{c_i \cdot 1}_{i\text{-th term}} + c_{i+1} \cdot 0 + \cdots + c_n \cdot 0 \\ &= c_i.\end{aligned}$$

And so the inner product gives us exactly the coefficient we wanted.

□

As we've discussed, the naive approach to computing the basis representation of a vector $x \in \mathbb{R}^n$ with respect to a basis $\{v_i\}$ would be to set up the system of linear equations $Ay = x$, where the columns of A are the v_i, and solve for y using a technique like Gaussian elimination. As it turns out, Gaussian elimination takes cubic runtime in the worst case (cubic in n, the dimension of the vector space).

However, with an orthonormal basis all you need to do is compute n inner products. The standard inner product only takes n multiplications and n additions, meaning the entire decomposition only takes time n^2. This is a huge improvement if, suppose, you could compute an orthonormal basis once and use it to compute basis representations many more times, as opposed to doing Gaussian elimination for each vector you wanted to represent in the target basis. It's also worth noting that in practice there's often a natural ordering on a basis, so that the first vectors in the basis contribute "most significantly" to the space, and one can approximate a basis representation using a constant-sized subset of the basis. For our physics application the eigenvalues will determine the ordering.

But beyond that, in a space like L^2 where there's no natural starting basis, this gives us a feasible way to compute basis representations: just compute the inner product! In L^2 you simply integrate.[9]

Going back to finite dimensions, the next important property of an orthonormal basis is that the change of basis matrix (the matrix with the basis vectors as columns) is easy to invert.

Proposition 12.16. *Let $\{v_1, \ldots, v_n\}$ be an orthonormal basis for V. Let B be the change of basis matrix, with the v_i as columns. Then $B^T = B^{-1}$.*

Proof. We can prove this directly by showing that $B^T B$ is the identity matrix, i.e., the matrix 1_n with 1s on the diagonal and zeros elsewhere. Indeed, the entries of $B^T B$ encode all pairwise inner products of the vectors in the basis. The i, j entry of $B^T B$ is the inner product $\langle v_i, v_j \rangle$, which is 1 if and only if $i = j$, and zero otherwise.

□

One may wonder if it's also necessary to show $BB^T = 1_n$ in order to conclude that B^T is a proper inverse of B. A direct proof hits an immediate barrier, because the inner products don't line up as they did above. It turns out this barrier is a mirage. By pure set theory, namely Proposition 4.12 from Chapter 4, a one-sided inverse of a bijection is automatically a two-sided inverse. And, of course, all change of basis matrices are bijections.

This has an almost startling consequence:

Proposition 12.17. *If the columns of A form an orthonormal basis, then so do the rows of A.*

---

[9] Integration is not always computationally easy, but you choose the orthonormal basis so that it is.

Proof. Let $B = A^T$ then B satisfies $B^T B = 1_n$, which as we saw above encodes all the pairwise inner products of columns of B, i.e., rows of A. Since orthogonal vectors are linearly independent (Proposition 12.14), the columns of B form a basis. □

However, if we wanted to prove this without all the set theory hijinks, we could have done so by proving $(A^T)^{-1} = (A^{-1})^T$. You will do this in the exercises.

One natural question you might ask is how to find an orthonormal basis. For finite dimensional inner product spaces there's an algorithmic method, and the method is called the *Gram-Schmidt process*. It falls short of an algorithm by not defining how to do one important step. First, a definition:

Definition 12.18. Let V be an inner product space and $W \subset V$ a subspace with a basis $B = \{w_1, \ldots, w_k\}$. Let v be a vector, and define the *projection of v onto the subspace W*, denoted by $\text{proj}_W(v)$, as follows:

$$\text{proj}_W(v) = \sum_{w_i \in B} \text{proj}_{w_i}(v)$$

The projection of v onto a subspace is the natural geometric generalization of projecting onto a vector. Projecting onto a subspace is the same thing as projecting onto each axis of any basis of that subspace and adding up the results. And just like the one-vector version, $v - \text{proj}_W(v)$ is the part of v that lies perpendicular to the subspace W in the sense that it's perpendicular to every vector in W.

The Gram-Schmidt process operates as follows to build up an orthonormal basis for an n-dimensional inner product space (or subspace).

1. Let $S_0 = \{\}$ be the empty set. S_i will contain the basis built up so far at step i.

2. For $i = 1, \ldots, n$:

    a) Let v be any vector not in the span of S_{i-1}.

    b) Let $v' = v - \text{proj}_{S_{i-1}}(v)$ (get the perpendicular part), or $v' = v$ if $i = 1$.

    c) Let $S_i = S_{i-1} \cup \{v'/\|v'\|\}$ (add normalized v' to the partial basis).

3. Output S_n.

The Gram-Schmidt process doesn't dictate how to find a vector not in the span of a given set, but using that as a subroutine, the rest is well-defined arithmetic. The proof that the result is an orthonormal basis is a simple exercise in induction. The same algorithm allows one to start from a given basis (possibly of a subspace), and transform it into an orthonormal basis with the same span. For this variant, if you have a subspace basis $\{v_1, \ldots, v_k\}$, and you want to know what new vector to choose at step i, you can simply choose v_i.

As a side note, this algorithm is generally not considered "production ready," because it suffers from numerical instability. Most industry-strength linear algebra libraries use one of a few different techniques based on linear algebra primitives (such as Householder reflections and the famed Cholesky decomposition) that have been fine-tuned and optimized for speed and stability.

12.5 Computing Eigenvalues

Our ultimate goal is to come up with an orthonormal basis of eigenvectors. This will combine the computational ease of orthogonality with the deep secrets revealed by eigenvalues. To appreciate the result Theorem 12.24, we should investigate why finding a basis of eigenvectors might be hard.

For instance, we established existence of at least one eigenvalue-eigenvector pair, but can we say anything about uniqueness? Given a linear map A with eigenvector v and corresponding eigenvalue λ, it is obvious that every vector in $\text{span}(v)$ is also an eigenvector for λ. But is it possible that some independent vector is also an eigenvector for λ? A simple example says yes: take the map $f : \mathbb{R}^3 \to \mathbb{R}^3$ sending $(a, b, c) \mapsto (a, b, 0)$, a projection onto the degree-two subspace spanned by $(1, 0, 0)$ and $(0, 1, 0)$. Both $(1, 0, 0)$ and $(0, 1, 0)$ are eigenvectors for the eigenvalue $\lambda = 1$, and so are all linear combinations. The story of an eigenvalue stretches beyond finding a single eigenvector.

Another reason why the analysis of eigenvalues is that zero can be an eigenvalue. The eigenvectors with eigenvalue zero span the preimage of the zero vector.

Definition 12.19. Let $f : V \to W$ be a linear map. Define the *kernel* of f, denoted $\ker(f)$ to be the set of $v \in V$ with $f(v) = 0$.

If you believe that finding roots of single-variable polynomials is hard, you might also be convinced that finding "roots" of linear maps is hard. In fact, you'll prove in an exercise that computing eigenvalues of linear maps is at least as hard as computing roots of polynomials. And as we'll see below, all eigenvalues can be expressed in terms of kernels.

As a quick exercise, prove that the kernel of a linear map is a subspace of V. Rephrasing the above, the eigenvalues of f corresponding to the eigenvalue $\lambda = 0$ are exactly the kernel of f. Also recall that I denotes the identity map $I(x) = x$, with corresponding matrix I_n for n-dimensions.

Proposition 12.20. *Let $f : V \to W$ be a linear map. Then $v \in V$ is an eigenvector corresponding to eigenvalue λ if any only if $v \in \ker(f - \lambda I)$. By $f - \lambda I$ we mean the map $x \mapsto f(x) - \lambda x$.*

Proof. Indeed, $f(v) = \lambda v$ if and only if $f(v) - \lambda v = 0$.

□

We saw an example of a simple map $(a, b, c) \mapsto (a, b, 0)$ that has a two-dimensional eigenspace for the eigenvalue 1. The matrix for this is

$$A = \begin{pmatrix} 1 & 0 & 0 \\ 0 & 1 & 0 \\ 0 & 0 & 0 \end{pmatrix}$$

And we can inspect the matrix $A - \lambda I_3$ to compute the remaining eigenvalues.

$$A - \lambda I_3 = \begin{pmatrix} 1-\lambda & 0 & 0 \\ 0 & 1-\lambda & 0 \\ 0 & 0 & -\lambda \end{pmatrix}$$

A vector (a, b, c) in the kernel of this map (for some unknown λ) must satisfy $a(1-\lambda) = 0$ and $b(1-\lambda) = 0$ and $-\lambda c = 0$. The third equality implies either $\lambda = 0$ or $c = 0$. In the former case, $a = b = 0$ and we get $(0, 0, 1)$ as an eigenvector for $\lambda = 0$. In the latter case, we're left with the same two-dimensional eigenspace for $\lambda = 1$.

Here's a more interesting example, the matrix for the map $(a, b, c) \mapsto (a, a+b, a+b+c)$.

$$B = \begin{pmatrix} 1 & 1 & 1 \\ 0 & 1 & 1 \\ 0 & 0 & 1 \end{pmatrix}$$

This matrix clearly has one eigenvector, $(1, 0, 0)$ for the eigenvalue $\lambda = 1$. But what about other potential eigenvectors? Indeed, we're looking for the kernel of $B - I_3$, which is

$$B - I_3 = \begin{pmatrix} 0 & 1 & 1 \\ 0 & 0 & 1 \\ 0 & 0 & 0 \end{pmatrix}$$

Aside from the span of $(1, 0, 0)$, there are no zeroes. And moreover, $B - \lambda I_3$ has only the trivial kernel $\{0\}$ (set up the system of three equations and verify this).

When an eigenvalue has multiple independent eigenvectors, we get a viscerally interpretable kind of "multiplicity," which goes by the name *geometric multiplicity*.

Definition 12.21. Let $f : V \to V$ be a linear map. The *geometric multiplicity* of an eigenvalue λ for f is the dimension of the space of eigenvectors for that eigenvalue, i.e., the dimension of $\ker(f - \lambda I)$ as a subspace of V.

For for the matrix A above, the eigenvalue 1 has geometric multiplicity 2, but for B the multiplicity is only 1.

There's another, more subtle kind of multiplicity called *algebraic multiplicity*, which I personally don't know how to motivate from "first principles." Specifically, the most common definition uses the definition of the determinant of a matrix (as a polynomial). An alternative way to define it is as follows.

Definition 12.22. The *algebraic multiplicity* of an eigenvalue λ for f is the largest integer m for which $\ker((f - \lambda I)^m)$ is strictly larger than $\ker((f - \lambda I)^{m-1})$.

From this definition, we can see that the algebraic multiplicities of $\lambda = 1$ are different for A and B above. Taking successive powers of $B - I_3$ gives first $(0, 1, 0)$ and then $(0, 0, 1)$ in the kernels, while the algebraic multiplicity for A is just 1.

These two types of multiplicity work together to give a characterization of any linear map in terms of so-called Jordan blocks. These are square sub-matrices with λ on the diagonal and 1's on the adjacent diagonal. For example for $n = 3$:

$$J_{\lambda,3} = \begin{pmatrix} \lambda & 1 & 0 \\ 0 & \lambda & 1 \\ 0 & 0 & \lambda \end{pmatrix}$$

The *Jordan canonical form* theorem states that for any linear map $V \to V$ there is a basis for V, for which the matrix of that linear map consists entirely of Jordan blocks along the diagonal. There may be more than one Jordan block for a given eigenvalue, but the size and number of blocks are determined by the algebraic and geometric multiplicities of that eigenvalue, respectively.

All of this is to note two things: it's possible to compute all of the eigenvalues and eigenvectors for a linear map, and these, along with some auxiliary data (some of which I've left out from this text), do in fact give a complete characterization of the map. However, it's a more nuanced characterization, and one whose benefits are not as easily displayed as when you have an orthonormal basis of eigenvectors. The Jordan canonical form is an important theorem that has generalizations and adaptations in other fields of mathematics. You will explore the Jordan canonical form more formally in the exercises.

Finally, as a quick aside, the set of all eigenvalues together with their geometric multiplicities is called the *spectrum* of a linear map.

Definition 12.23. Let $f : V \to W$ be a linear map between vector spaces. Define the *spectrum* of f as the set

$$\text{Spec}(f) = \{(\lambda, \dim \ker(f - \lambda I)) : f(v) = \lambda v \text{ for some nonzero } v \in V\}.$$

It is interesting to note that most scientific uses of the word "spectrum" refer to this mathematical idea, for example the spectrum of wavelengths of light or the spectrum of an atom.

12.6 The Spectral Theorem

While the Jordan canonical form is a complete characterization, if you're lucky enough that the eigenvectors corresponding to the same eigenvalue are orthogonal, life suddenly becomes much easier. In this case, scaling the eigenvectors to unit vectors gives you an orthonormal basis (recall Propositions 12.11 and 12.14). The matrix for such a linear map, when written with respect to that basis, has all its nonzero entries on the diagonal.

$$A = \begin{pmatrix} \lambda_1 & 0 & \cdots & 0 \\ 0 & \lambda_2 & \cdots & 0 \\ \vdots & \vdots & \ddots & \vdots \\ 0 & 0 & \cdots & \lambda_n \end{pmatrix}$$

And the reverse holds too: if a linear map can be written in this diagonal form, then the basis vectors used must be orthogonal eigenvectors. A linear map that can be written this way for some basis is called *diagonalizable*. What's astounding is that *every* symmetric matrix has an orthonormal basis of eigenvectors. This is the centerpiece theorem of this chapter and the secret ingredient in the physics application to follow.

Theorem 12.24 (The Spectral Theorem). *A real-valued matrix A is symmetric if and only if it has eigenvectors that form an orthonormal basis (i.e., is diagonalizable).*

This theorem requires some nontrivial amount of work, pieces of which we have already proved in this chapter. The easy part, however, is the reverse direction. It uses the fact that $(AB)^T = B^T A^T$.

Proposition 12.25. *A real-valued matrix A with an orthonormal basis of eigenvectors is symmetric.*

Proof. There is a change of basis matrix U, whose columns are the orthonormal basis, for which $A = U^T D U$, for D a diagonal matrix. A diagonal matrix is clearly symmetric, so $A^T = (U^T D U)^T = U^T D^T (U^T)^T = U^T D U = A$, implying A is symmetric.

□

The strategy for the other half of the proof will be by induction on the dimension of the vector space. That is, given the fact that every $(n-1) \times (n-1)$ symmetric matrix has an orthonormal basis of eigenvectors, we'll show that every $n \times n$ symmetric matrix does as well.

Induction suggests we should find one way to "peel off" one dimension in a way that's independent of the rest of the argument. Given A, we'll find an eigenvector v with corresponding eigenvalue λ that will be the first vector in the basis. Then we'll decompose $\mathbb{R}^n$ into two subspaces, a one-dimensional space spanning v, and an $(n-1)$-dimensional space, which we'll apply induction on. In particular, we will be able to rewrite A in a "block" form like so:

$$A \to \begin{pmatrix} \lambda & \mathbf{0} \\ \mathbf{0} & A' \end{pmatrix}$$

In the above, the boldface $\mathbf{0}$ are to denote that zeroes take up the entire "area" implied by the dimensions. If A is an $n \times n$ matrix, and λ is a scalar, then A' is $(n-1) \times (n-1)$ and each boldface zero represents $n-1$ zeroes in the only allowable shape.

Intuitively, what we're doing here is *partially* rewriting the basis in terms of one known eigenvector. Indeed, we have to describe a full basis to get a block decomposition, but as long as whatever process we use to make the basis maintains the symmetry of A', we win. We'll be able to combine the orthonormal basis of A' with v to get a full orthonormal basis for A. The remaining details relate to the algebra of a precise proof, which we'll exhibit now.

Proof. (Finishing the proof of the Spectral Theorem)

Suppose A is a symmetric real-valued $n \times n$ matrix on an n-dimensional vector space $\mathbb{R}^n$. We will show there is an orthonormal basis of eigenvectors of A.

We proceed by induction on n. For $n = 1$ the claim is trivial, because every vector is an eigenvector and every basis is orthogonal. In particular, the linear map corresponding to A must be $f(x) = bx$ for some constant b, and so the unit vector 1 is an eigenvector with eigenvalue b.

Now let $n > 1$, suppose as the inductive hypothesis that every $(n-1) \times (n-1)$ symmetric matrix has an orthonormal basis of eigenvectors, and let A be an $n \times n$ symmetric matrix. We begin by finding any eigenvector v of A, with some associated eigenvalue λ. We know we can do this by Lemma 12.8. Use that v as the first vector in a new basis of $\mathbb{R}^n$.

Construct the rest of this basis as follows. Let W be the subspace of $\mathbb{R}^n$ consisting of all vectors orthogonal to v.[10] Use Gram-Schmidt to choose an orthonormal basis $B' = \{w_2, \ldots, w_n\}$ of W. Joining together, $B = B' \cup \{v\}$ is an orthonormal basis of all of $\mathbb{R}^n$. Note that only v need be an eigenvector; the other vectors in the basis are not necessarily eigenvectors of A, but the whole basis is orthonormal.

Because B is orthonormal, the same argument as Proposition 12.25 implies that A, when written with respect to the basis B, is symmetric. So when we write A with respect to B, the matrix decomposes into blocks (we prove this below):

$$A \xrightarrow{\text{change of basis by } B} \begin{pmatrix} \lambda & \mathbf{0} \\ \mathbf{0} & A' \end{pmatrix} = B^T A B$$

In particular A' is the restriction of A to vectors in the subspace W.

To prove the block form is as we say it is, we just need to reason about the first column of this matrix: if you apply A to v you get λv, which includes none of the other basis vectors. So in the new basis representation you get a column with a λ and zeros elsewhere. As we argued above, this block decomposition is symmetric, so the first row must also have zeros as indicated.

Finally, we can invoke the inductive hypothesis for the matrix A' (which is symmetric because $B^T A B$ is) and the subspace W. I.e., A' has an orthonormal basis of eigenvectors, call it $\{u_2, \ldots, u_n\}$. Then the final basis is $\{v, u_2, \ldots, u_n\}$.

There is one more detail. We defined u_i as an eigenvalue of this sub-matrix A', but can we be sure it's an eigenvalue of the original A? Indeed it is, because of the way we decomposed $\mathbb{R}^n$ into $\text{span}(v)$ and the orthogonal complement W. Specifically, to

---

[10] It is a simple exercise to show that for a fixed nonzero vector v, the set $\{x : \langle x, v \rangle = 0\}$ is a subspace of dimension $n - 1$, and it's called the *orthogonal complement* of v.

compute Ax for any vector x, we write it with respect to the basis, and apply A to each piece. In this case that's $Au_i = \langle u_i, v \rangle v + A'u_i$, and $\langle u_i, v \rangle = 0$. So if u_i is an eigenvector for eigenvalue λ, then $Au_i = A'u_i = \lambda u_i$.[11]

□

12.7 Application: Waves

As you can probably tell from the book to this point, my favorite applications of math are to computer science. Linear algebra is no different. However, it would be intellectually dishonest to omit the influence of linear algebra in physics. Nowhere else does the beauty and utility of eigenvalues shine so bright.

As a demonstration, we consider vibrations (waves) on a string. The analysis we'll perform is a perfect post-hoc motivation for eigenvalues. The string system, with appropriate simplifications, results in a differential equation specified by a symmetric linear map. By the Spectral Theorem, that map has an orthonormal basis of eigenvectors. This allows us to decompose the system into independent components, and results in efficient computation and physical insight. We'll be able to easily compute the long-term behavior of the system—indeed, it will have a formula!—and the eigenvectors will correspond to the "fundamental frequencies" of the vibrating string. In addition to the pictures in this section, there is an interactive demo on the book's website.[12]

The discrete analysis we're about to do also generalizes both in dimension (waves on a surface) and to a continuous setting (the wave equation). While we gave a taste of what linear algebra and eigenvectors look like in infinite dimensions, this application will hopefully motivate further study.

Let's jump right in.

The Setup

Consider the system depicted in Figure 12.2 in which a string is pulled tight through five equally spaced beads. If you pluck the string, it naturally creates a wave that propagates through the string from end to end. Or, if you pluck the middle bead, the string oscillates in a symmetric fashion.

First, we need to write down a formal mathematical model in which we can describe the motion of a bead. We start by defining a function of time that represents an object's position. Ultimately, we'll only care about the *vertical* motion of the beads, but a priori we'll need two dimensions to describe the forces involved.

Let $x : \mathbb{R} \to \mathbb{R}^2$ be a function describing the position of an object at a given time t. In particular, we choose a reference point in the universe to be $(0,0)$ and a basis $\{e_1, e_2\}$ of $\mathbb{R}^2$ for measurement. Then the components of $x(t) = (x_1(t), x_2(t))$ represent the

---

[11] A different argument is to introduce the notion of a *direct sum* of vector spaces. To write a vector space in terms of the direct sum of subspaces (which is what we did here) means that a vector can be written uniquely as a sum of vectors in each subspace. Orthogonal complements always form a direct sum.

[12] pimbook.org

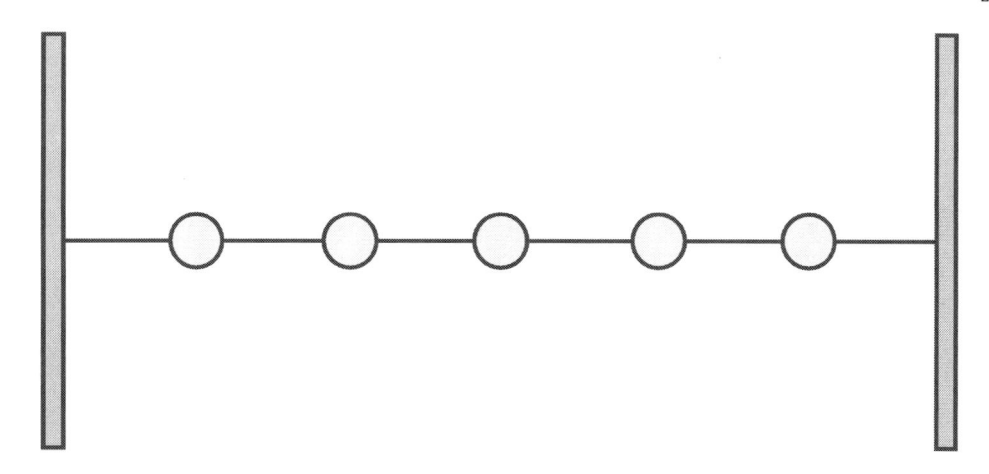

Figure 12.2: A system in which five beads are equidistantly spaced on a taut string.

position of the object, in e_1, e_2 units, respectively, relative to $(0,0)$. The obvious choices of coordinates are the standard basis vectors $(1, 0)$ and $(0, 1)$ representing horizontal and vertical, as aligned with the picture.

Model 12.26. Let $x(t) = (x_1(t), x_2(t))$ be the position of an object at time t. Then its derivative, $x'(t) = (x_1'(t), x_2'(t))$, describes the object's velocity at time t, and the second derivative $x''(t) = (x_1''(t), x_2''(t))$ describes its acceleration at time t.

These should intuitively make sense when thinking of the derivative as a rate of change. Velocity is the rate of change of position, acceleration the rate of change of velocity. As an aside, this kind of vector-valued function that has a 1-dimensional input and a multi-dimensional output is often called a *parametric function*. We'll cover derivatives in more generality in Chapter 14.

We must also describe a mathematical model (one that will suffice for our purposes) for a physical force. Note that while we're doing everything here in two dimensions, the same principles apply to three or more dimensions.

Definition 12.27. A *force* is a function $F : \mathbb{R} \to \mathbb{R}^2$ whose input represents time and whose output is a vector representing the magnitude and direction of the force. Each force is considered as acting on a specific object.

In the formulas below, we're concerned with the force in a particular direction. Indeed, given a force vector $F(t)$ at a specific time t, projecting $F(t)$ onto the appropriate unit vector v gives the component of F in the direction of v. If we choose the basis to align with the vertical direction, the projection is trivial: just look at the second entry of the force vector. But in general you can use projections to get the component of a force in any direction.

In a sense that is not rigorous but part of the mathematical model, forces "act" on objects. By that I mean they are applied to objects and influence their motion. If you

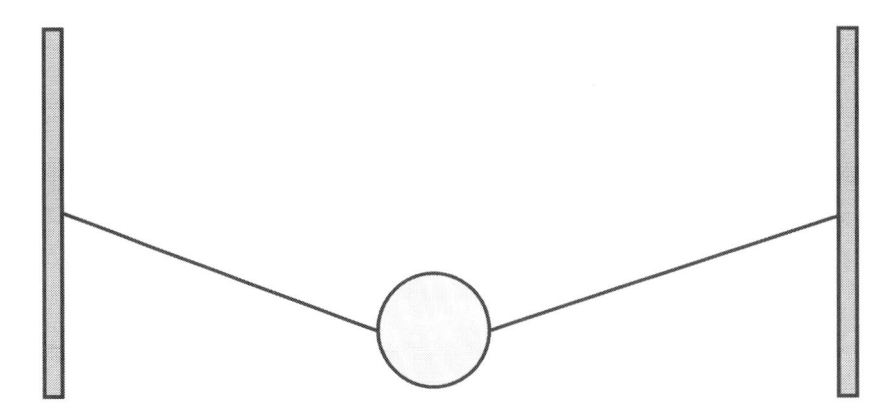

Figure 12.3: A simpler system that has only one bead, displaced from its equilibrium and released.

pluck a string, it moves. The following revolutionary observation allows us to describe exactly *how* forces that act on an object influence their motion.

Model 12.28 (Newton's n-th law for some n). If $F_1, \ldots, F_n$ are forces acting on an object with mass m whose position is described by $x(t)$, then

$$\sum_{i=1}^{n} F(i) = mx''(t)$$

In other words, the sum of the forces applied to an object determines the acceleration of that object. More massive objects need larger forces to move them.

One Bead

Now let's inspect our beaded string in the special case of a single bead in the middle of a string. The bead has been plucked and released, as in Figure 12.3.

Our goal is to model the dynamics of this system as a linear system. At any given time t, we should be able to calculate the acceleration $x''(t)$ of the bead as linear function of its current position. As we'll see that's enough to compute the position $x(t)$ at any time. When we extend the model to include all five beads, it will depend linearly on the positions of multiple beads.

We'll make a whole host of unrealistic assumptions to aid us. Let's pretend the string has no mass, the bead has no width, there is no friction or air resistance, and let's do away with gravity. More generously, we assume that all of these values are "negligibly small" compared to the forces we care about. These kinds of simplifying assumptions are the physics analogue of what mathematicians do when they encounter a hard problem: keep stripping out the difficult parts until you can solve it. If you simplify the problem in the right way, you'll be analyzing just the aspects of the problem that you really care about. After solving it, having hopefully gained useful intuition in the process, you can replace each removed bit and use your newfound intuition to find a solution of the harder

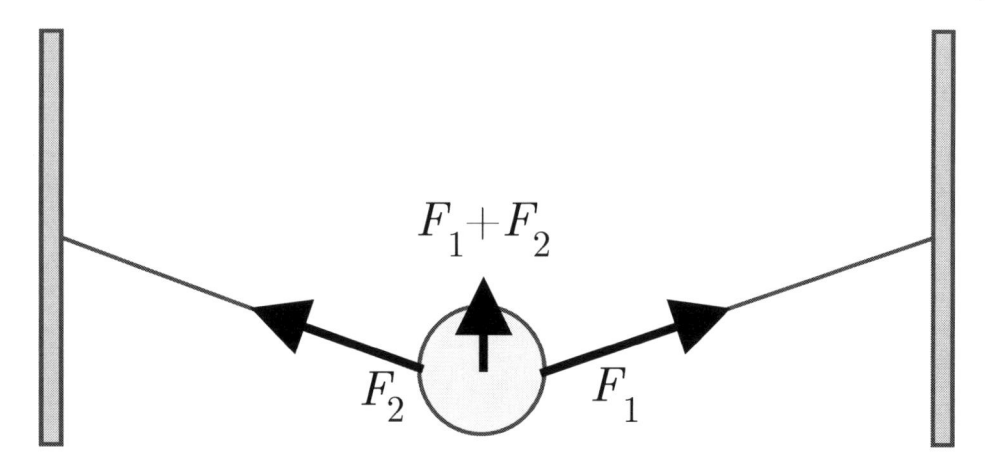

Figure 12.4: The forces pull in opposite directions toward the wall, and together sum to a vertical force.

problem. Or, if you cannot, you can see how the simpler solution breaks with the new assumption, and thus understand *why* the full problem is hard to solve. This process is by no means as easy as it sounds, but it's a powerful guide.

The above assumptions are minor, but there are two crucial assumptions that we have to discuss in more detail. First, we assume the string is not stretched too far. This allows us to use a Taylor series approximation for the sine and tangent of a small angle. Second, assume the string is already stretched tightly when the beads are plucked. This is what allows us to ignore the horizontal motion of the bead. We'll discuss these in more detail when we employ them.

Once we've eliminated gravity and its cohort, there are only two forces acting on the bead: the force of tension in the string on the left and right sides of the bead. When the bead is pulled downward, the string is stretched longer than its resting length, and the bonds between the string's atoms create a force that "pulls" the string back to its normal length. Luckily, tension is well understood. The standard model is Hooke's law.

Model 12.29 (Hooke's law). The force of tension in an elastic string that has been stretched from its resting length by a distance $d \geq 0$ is $-Td$, where T is a constant depending on the material of the string. This model only applies for a sufficiently small d that does not exceed a limit (which again depends on the material in the string).

If the string is tied to a surface and you pull away from the surface, even at an angle, the force is directed back along the string toward the surface. This gives our bead two forces as in Figure 12.4.

Since we assumed the bead has no width (or, if you will, the forces act on the center of mass of the bead), the tails of these vectors are the same point, and when we sum them we get the net force pulling the bead upward.

In our system the string is taut, and we'll suppose it's stretched to begin with. Call $2l$ the natural length of the string (so that l is the length of one of the two halves), T the

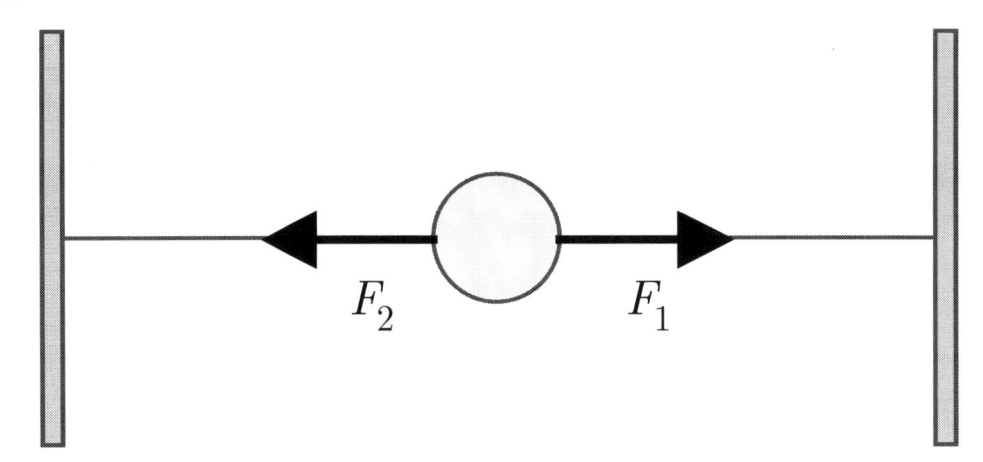

Figure 12.5: At rest, the forces sum to the zero vector.

tension constant, and $2l_{\text{init}}$ the length the string is initially pulled to when the system is at rest. In that case, the two forces on the bead have magnitude $T(l_{\text{init}} - l)$ and face in opposite directions. The bead does not move.

Let's focus on the right hand side of the bead (the left side is symmetric) in Figure 12.6. Choose the resting point of the bead, when the string is completely straight, to be $(0, 0)$. Use the standard basis $\{(1, 0), (0, 1)\}$ and let $x(t) = (x_1(t), x_2(t))$ be the displacement of the bead at time t (some arbitrary but small vertical position not at rest). Call $d(t)$ the length of the right string segment at time t, and $F_1(t)$ the force pulling on the bead by the string. The diagram in Figure 12.6 labels these values.

Now we compute. Our choice of basis and the Pythagorean theorem give $d(t) = \sqrt{l_{\text{init}}^2 + x_2(t)^2}$. We construct $F_1(t)$ first by finding a unit vector in the correct direction, then scaling it so its length is the magnitude of the force. That magnitude is $T(d(t) - l)$, according to Hooke's law. The force vector starts at $x(t)$ and points toward $(l_{\text{init}}, 0)$, so we can take $(l_{\text{init}}, 0) - x(t) = (l_{\text{init}}, -x_2(t))$ and normalize it by dividing by $d(t)$. So far we have

$$F_1(t) = T(d(t) - l) \frac{(l_{\text{init}}, -x_2(t))}{d(t)}$$

The magnitude of the vector has a nonlinear part $d(t) - l$ involving $d(t)$, so let's simplify that first. Since the string was initially stretched to length l_{init}, we have $d(t) - l = (d(t) - l_{\text{init}}) + (l_{\text{init}} - l)$, and so the magnitude of the force is

$$T(d(t) - l_{\text{init}}) + T(l_{\text{init}} - l).$$

Conveniently, the right hand term is the magnitude of tension when the system is at rest. For the left hand term, we can use a Taylor series approximation. First we do some simplification.

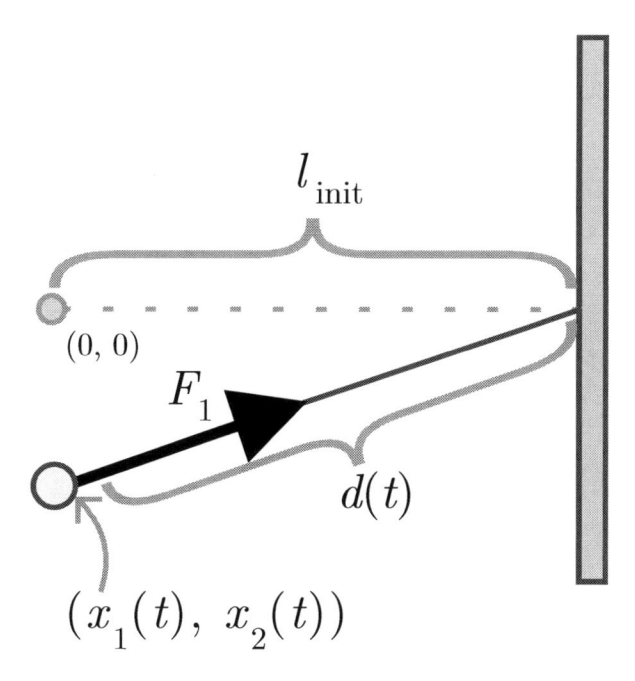

Figure 12.6: The force pulling the bead rightward when the bead is displaced.

$$d(t) = \sqrt{l_{\text{init}}^2 + x_2(t)^2}$$
$$= l_{\text{init}}\sqrt{1 + \left(\frac{x_2(t)}{l_{\text{init}}}\right)^2}$$

Next we compute the Taylor series for $\sqrt{1+z^2}$, substituting $z = x_2(t)/l_{\text{init}}$ at the end. Indeed, the Taylor series is

$$\sqrt{1+z^2} = 1 + \frac{z^2}{2} - \frac{z^4}{8} + \frac{z^6}{16} - \cdots$$

Using the first two terms to approximate, we get $d(t) \approx l_{\text{init}}(1 + \frac{x_2(t)^2}{2l_{\text{init}}^2})$. If we wanted to be more rigorous, we could hide the lower order terms in a big-O notation, but we'll save that for Chapter 15.

Returning to the force of tension, minor algebra gives $T(d(t) - l_{\text{init}}) = T\frac{x_2(t)^2}{2l_{\text{init}}}$. In other words the magnitude of the force of tension in the string is the initial tension, plus a small factor proportional to the square of the deviation.

$$T\frac{x_2(t)^2}{2l_{\text{init}}} + T(l_{\text{init}} - l)$$

The formula above is why we can assume, as most physics texts do without nearly as much fuss as we have displayed here, that the magnitude of tension in the string is constant. This Taylor series approximation is the first assumption showing up in the math: if the initial deviation $x_2(t)$ is small, say much less than 1 unit of measurement, then $x_2(t)^2$ is even smaller and can be ignored, as can all higher powers of $x_2(t)$. Our computation shows that the first power $x_2(t)$ does not show up anywhere in the Taylor series, so if we're committed to simplifying everything to be linear, the Taylor series assures us we're not accidentally ignoring terms we want to preserve.

I personally feel it's important to see how the math justifies the assumptions rather than relying entirely on "physical intuition." Once you state which forces you want to consider—and once you've formalized the mathematical rules governing those forces—the mathematics should stand on its own. In particular, many physics books say that the constant tension assumption rests on the fact that the bead is not displaced very far from rest. Strictly speaking, this is not enough information. What also matters is the relationship between the displacement of the bead and the initial stretch that holds the string taut at rest. The former must contribute an order of magnitude smaller force than the latter to be negligible. The Taylor series revealed this nuance, and further allows us to measure how big a displacement is too big to ignore.[13]

We continue with the assumption, then, that the magnitude of the force of tension in the string is constant over the entire evolution of the system. From this point on we'll use T in place of $T(l_{\text{init}} - l)$ to simplify the formulas (it's all just a constant anyway). Recalling that we formed the unit vector by scaling by $d(t)$, the force on the right string is the vector

$$F_1(t) = T\frac{(l_{\text{init}}, -x_2(t))}{d(t)}$$

Note that while we ignored the $x_2(t)^2$ factor in the magnitude, we haven't yet ignored it in the scaling of the unit vector. That begins now: since the two forces $F_1(t)$ and $F_2(t)$ are symmetric, we only need the components of $F_1(t)$ in the vertical direction. That means we can project $F_1(t)$ onto the vector $(0, 1)$, i.e., isolate the second entry of the vector.

$$F_{\text{vert}}(t) = T(0, -x_2(t)/d(t))$$

And if we expand $d(t) = \sqrt{l_{\text{init}}^2 + x_2(t)^2}$ and ignore $x_2(t)^2$ by setting it to zero, we get $F_{\text{vert}}(t) = (0, -Tx_2(t)/l_{\text{init}})$.[14]

---

[13] In my own confusion writing this section, I verified my suspicions by writing the simulation posted at pimbook.org. Through that exercise an obvious proof dawned on me: if the initial tension is zero (the string is just barely pulled taut), the tension goes from zero to nonzero no matter how small the deviation, a change that cannot be considered constant.

[14] In physics texts you often see the author instead use the cosine formula Theorem 10.19, and the Taylor approximations for $\sin \theta$ and $\tan \theta$. The way we laid it out makes that unnecessary, but we will use those approximations when we generalize to multiple beads.

Now that all our forces are vertical, we can just work with the 1-dimensional picture and see that the sum of the forces on the bead in the vertical direction is $F(t) = -2Tx_2(t)/l_{\text{init}}$. By Newton's law, this dictates the acceleration of the bead, giving

$$mx_2''(t) = -2Tx_2(t)/l_{\text{init}}.$$

Let's simplify the numbers by setting $m = 1$, $l_{\text{init}} = 1$, and $T = 1$, a trick called "choosing units cleverly." Then the formula is $x_2''(t) = -2x_2(t)$. The finish line is in sight. We need one additional, theorem whose proof is left as an investigative exercise. First recall, or learn now, that the derivative of $\sin(x)$ is $\cos(x)$, and the derivative of $\cos(x)$ is $-\sin(x)$, so that the second derivative of $\sin(x)$ is $-\sin(x)$.

Theorem 12.30. *Let $f : \mathbb{R} \to \mathbb{R}$ be a twice differentiable function which satisfies $f''(x) = -f(x)$, and $f(0) = 0$, $f'(0) = 1$. Then $f(x) = \sin(x)$.*

An equation like $f'' = -f$, involving the derivatives of an unknown function, is called a *differential equation*. There is an analogous theorem for the cosine instead using $f(0) = 1$, $f'(0) = 0$. The restrictions on $f(0)$ and $f'(0)$ are called *initial conditions*, and as they change the solution changes. In the case of Theorem 12.30 the solution only changes by constants. In fact, the way these values vary hints at two independent dimensions which provide solutions to $f'' = -f$.

Indeed, the set of solutions to $f'' = -f$ forms a two-dimensional vector space (a subspace of the space of all twice-differentiable functions $\mathbb{R} \to \mathbb{R}$), and $\sin(x)$ and $\cos(x)$ form a basis. As an aside, if we call this vector space U, then the "take a second derivative" function $d : U \to U$ mapping $f \mapsto f''$ is a linear map on U, and the sine and cosine functions are eigenvectors with eigenvalue -1. This hints at the deep truth that sine and cosine are special functions, in part explaining why we should expect a theorem like Theorem 12.30.

So despite how the initial conditions may vary, you know the solution is a linear combination $c_1 \sin(x) + c_2 \cos(x)$. With a bit of algebra, given the initial conditions you can solve for those coefficients based on the initial conditions. We will do this below.

First, we have to wrangle the extra coefficient of $2T$. We can modify the theorem slightly. Note that for a scalar a, the derivative of $\sin(ax)$ is $a\cos(ax)$ (the chain rule, Theorem 8.10), but since we're differentiating twice we have a square in the second derivative $-a^2 \sin(ax)$. I.e., the solution to $x_2''(t) = -2Tx_2(t)$ is a sine or cosine with argument $(\sqrt{2T})t$. Let ω (the Greek letter omega) be $\sqrt{2T}$.

Combining this with the assumption that at time $t = 0$ the bead is displaced by some fixed amount and let go (has zero initial velocity), we get

$$\begin{aligned} x_2(0) &= c_1 \sin(\omega \cdot 0) + c_2 \cos(\omega \cdot 0) & = c_1 \cdot 0 + c_2 \cdot 1 \\ 0 = x_2'(0) &= c_1 \omega \cos(\omega \cdot 0) - c_2 \omega \sin(\omega \cdot 0) & = c_1 \omega \cdot 1 - c_2 \omega \cdot 0 \end{aligned}$$

We can read off the solution as $c_1 = 0, c_2 = x_2(0)$. This means that our lonely bead, plucked and left to wait all this time to learn its destiny, finally has an equation for its

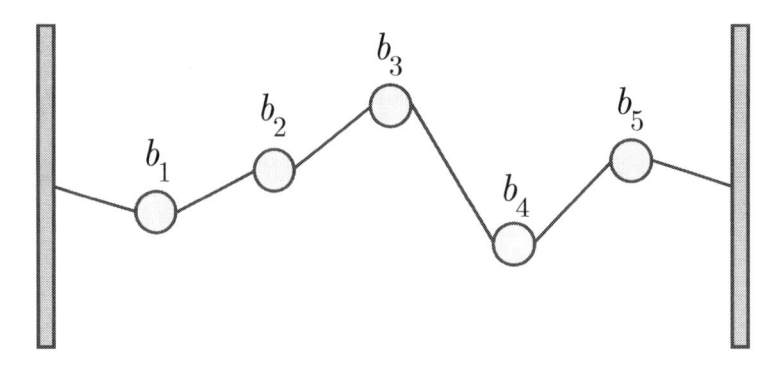

Figure 12.7: Five beads starting from arbitrary initial positions.

motion: $x_2(t) = x_2(0) \cos(t\sqrt{2T})$. It's a smooth cosine with a constant frequency determined by the tension in the string. This is exactly what we expect from a single bead.

Multiple Beads

Now we graduate to multiple beads, shown in Figure 12.7.

Horizontal forces are a new concern. We want to retain our assumption of constant tension in the string. But because the angles are different on different sides of a bead, the fraction of that constant tension pulling the bead left and right can be different, resulting in horizontal motion. We know that the tension in the string will eventually pull the bead back to the center, but we want to feel secure that these violations of our assumptions are minor enough that we can justify ignoring them. We leave it as an exercise to the reader to adapt the setup for a single bead to this scenario, and to use Taylor series approximations to argue that horizontal motion can be ignored.

Since we are ignoring horizontal motion, we'll simplify the notation so that the forces, displacements, velocities, and accelerations are 1-dimensional vectors, i.e., scalars representing vectors pointing in the vertical direction. Let $b_1, \ldots, b_5$ be the beads of mass m_i, and let y_i be the displacement of b_i, with y_i' and y_i'' the velocity and acceleration, as before. The natural resting point of the beads is zero. If we just think about position—and as we saw this completely determines the forces and the acceleration—then the state of this system is a vector $y = (y_1, y_2, y_3, y_4, y_5) \in \mathbb{R}^5$. The forces we're about to compute will form a linear map A mapping $y \mapsto y''$.

Let's now focus on bead b_2 as a generic example, shown in Figure 12.8. In the figure, the gap between b_1 and b_2 is $y_2 - y_1$, and the angle θ_1 is the angle between the string and the horizontal. Likewise for the corresponding data on right hand side of the bead. The tension is a constant T. The projected tension in the vertical direction is $-T\sin(\theta_1) + T\sin(\theta_2)$, with the sign flip because the first is pulling the bead down.[15]

Now we'll use two Taylor series approximations:

---

[15] When b_1 is above b_2, the angle is negative and that reverses the sign: $\sin(-\theta) = -\sin(\theta)$. So the orientations work out nicely.

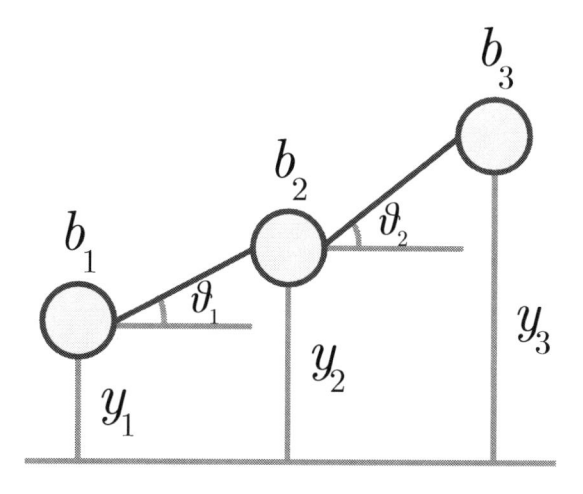

Figure 12.8: A close up of b_2.

$$\sin(\theta) = \theta - \frac{\theta^3}{3!} + \frac{\theta^5}{5!} + \cdots$$

$$\tan(\theta) = \theta + \frac{\theta^3}{3} + \frac{2\theta^5}{15} + \cdots$$

Because the first two terms are equal, and for θ small enough to ignore θ^3 and higher, we can replace $\sin(\theta)$ with $\tan(\theta)$ wherever it occurs. This is the same reasoning as before, because we want to extract the linear aspects of the model. The force on bead b_2 is

$$\begin{aligned} y_2'' m_2 &= F_2(t) \\ &= -T\sin(\theta_1) + T\sin(\theta_2) \\ &= -T\tan(\theta_1) + T\tan(\theta_2) \\ &= -T\frac{y_2 - y_1}{l_{\text{init}}} + T\frac{y_3 - y_2}{l_{\text{init}}} \end{aligned}$$

And rearranging gives

$$\frac{m_2 l_{\text{init}}}{T} y_2'' = y_1 - 2y_2 + y_3$$

Simplify the equation by setting $m_2 = l_{\text{init}} = T = 1$. The forces for the other beads are analogous, with the beads on the end having slightly different formulas as they're attached to the wall on one side. As a whole, the equations are

$$y_1'' = -2y_1 + y_2$$
$$y_2'' = y_1 - 2y_2 + y_3$$
$$y_3'' = y_2 - 2y_3 + y_4$$
$$y_4'' = y_3 - 2y_4 + y_5$$
$$y_5'' = y_4 - 2y_5$$

Rewrite this as a linear map $y'' = Ay$ with

$$A = \begin{pmatrix} -2 & 1 & 0 & 0 & 0 \\ 1 & -2 & 1 & 0 & 0 \\ 0 & 1 & -2 & 1 & 0 \\ 0 & 0 & 1 & -2 & 1 \\ 0 & 0 & 0 & 1 & -2 \end{pmatrix}$$

At last, we turn to eigenvalues. This matrix is symmetric and real valued, and so by Theorem 12.24 it has an orthonormal basis of eigenvectors which A is diagonal with respect to. Let's compute them for this matrix using the Python scientific computing library numpy. Along with Fortran eigenvector computations, numpy wraps fast vector operations for Python.

After defining a helper function that shifts a list to the right or left (omitted for brevity), we define a function that constructs the bead matrix, foreseeing our eventual desire to increase the number of beads.

```
def bead_matrix(dimension=5):
    base = [1, -2, 1] + [0] * (dimension - 3)
    return numpy.array([shift(base, i) for i in range(-1, dimension - 1)])
```

Next we invoke the numpy routine to compute eigenvalues and eigenvectors, and sort the eigenvectors in order of decreasing eigenvalues. For those unfamiliar with numpy, the library uses an internal representation of a matrix with an overloaded index/slicing operator [ ] that accepts tuples as input to select rows, columns, and index subsets in tricky ways.

```
def sorted_eigensystem(matrix, top_k=None):
    top_k = top_k or len(matrix)
    eigenvalues, eigenvectors = numpy.linalg.eig(matrix)

    # sort the eigenvectors by eigenvalue from largest to smallest
    idx = eigenvalues.argsort()[::-1]
    eigenvalues = eigenvalues[idx]
    eigenvectors = eigenvectors[:, idx]

    # return eigenvalues as rows of a matrix instead of columns
    return eigenvalues[:top_k], eigenvectors.T[:top_k]
```

Eigenvalue λ	Eigenvector y_1	y_2	y_3	y_4	y_5
-0.27	0.29	0.50	0.58	0.50	0.29
-1.00	-0.50	-0.50	-0.00	0.50	0.50
-2.00	0.58	-0.00	-0.58	0.00	0.58
-3.00	-0.50	0.50	-0.00	-0.50	0.50
-3.73	-0.29	0.50	-0.58	0.50	-0.29

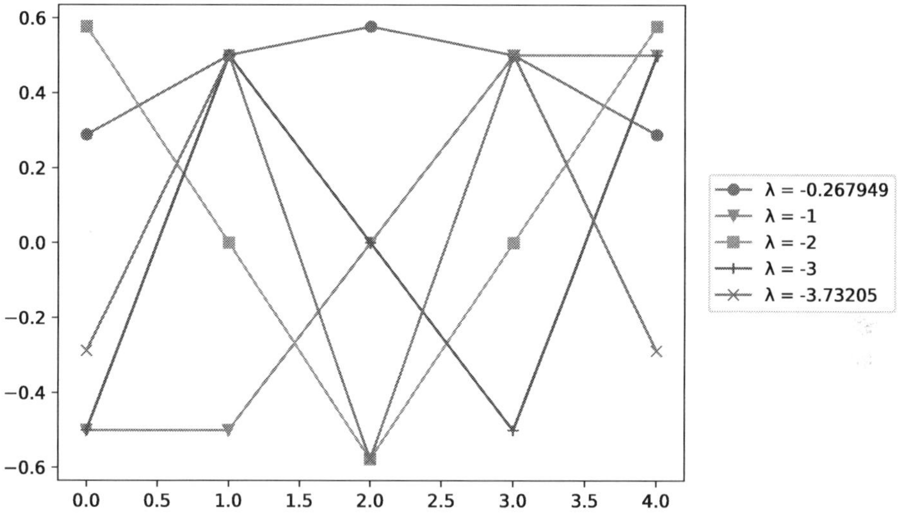

Figure 12.9: The rounded entries of the eigenvectors of the 5-bead system (top) and their plots (bottom).

And, finally, a simple use of the matplotlib library for plotting the eigenvectors. Here our x-axis is the index of the eigenvector being plotted, and the y-axis is the entry at that index. Plotting with five beads gives the plot in Figure 12.9.

In case it's hard to see (there will be a clearer, more obvious diagram at the end of the section), let's inspect it in detail. The top eigenvalue, $\lambda = -0.267\ldots$, corresponds to the eigenvector in the chart above with circular markers. The eigenvector entry starts at 0.29, increases gradually to 0.58, and then back down to 0.29, a sort of quarter-period of a full sine curve. The second largest eigenvalue, $\lambda = -1$ with triangular markers, has an eigenvector starting at -0.5 and increasing up to 0.5, performing a half-period of sorts. The next eigenvector for $\lambda = -2$ performs a single full period, and so on.

Now this is something to behold! The eigenvectors have a structure that mirrors the waves in the vibrating string, and as the corresponding eigenvalue decreases, the "frequency" of the wave plotted by the eigenvector increases. That is, the wave exhibits faster oscillations.

This wave is not a metaphor. If you simulate the beaded string with initial position set to one of these eigenvectors, you'd see a standing wave whose shape is exactly the plot of that eigenvector. In fact, I implemented a demo of this in Javascript, which you can explore for yourself at pimbook.org.[16] The demo is a first-principles simulation of the system, so horizontal forces are not ignored, nor are Taylor series approximations used. Because of this, if you set the initial positions of the beads to be quite large, you'll see irregularities caused by horizontal motion. These are highligthed by how the demo draws the force vector acting on each bead at every instant. It's fun to watch, and it provides a hint as to what assumption allows one to ignore horizontal motion. Indeed, if you set the position to the top eigenvector $100v_1$ (scaled to account for the units being pixels), you can see the same shape as v_1 in the plot above. If you scale it even larger, you can see the horizontal forces come into play. For example, try setting the initial positions to $300v_1 = (87, 150, 174, 150, 87)$.

Let's witness how the formulas work out for the first eigenvector v_1, when the positions start as that eigenvector $y = v_1 \approx (0.29, 0.5, 0.58, 0.5, 0.29)$. In that case each bead's trajectory can be computed independently according to $y'' = Ay = -0.27y$. So the second bead, say, evolves as $y'' = -0.27y$ with initial position $y_2 = 0.5$. This is identical to the single-bead system we solved earlier, and the result is a simple cosine wave with a fixed period and amplitude. The same holds for each bead. The beads in the middle have longer periods and higher amplitudes, as expected.

We have the tools to understand this eigenvector phenomenon beyond concrete computations. As we saw, the eigenvectors of the bead system form an orthonormal basis. The basis vectors are the *independent components* of the joint forces acting on all the beads. What's more, the proof of the Spectral Theorem explains why the eigenvectors have a natural ordering. The way we choose an eigenvector at each step is, according to Lemma 12.8, by maximizing $\|Av\|$ over unit vectors v. In the proof of the Spectral Theorem we then removed that vector, and its span, from consideration for the next vector.[17] So the largest magnitude eigenvalue (in this case the most negative one) is the first one extracted, and that corresponds to the highest frequency. The next eigenvector chosen corresponds to the second largest magnitude eigenvalue, and so on, each having a smaller frequency than the last.

But wait, there's more! Because it's an orthonormal basis of eigenvectors, we can express *any* evolution of this system in terms of the eigenvectors, and do it as simply as taking inner products.

Take, for example, the complex evolution that occurs when you pluck the second bead. Say $y(0) = (0, 0.5, 0, 0, 0)$. The individual beads don't evolve according to a single cosine wave. They jostle in a more haphazard manner. Nevertheless, we can express their trajectory as a sum of five simple cosine waves, one for each eigenvector. Indeed, the following Python snippet performs the decomposition of y (for a concrete, fixed time) in

---

[16] Note the demo is written in ES6 using d3.js, and the implementation is available in the Github repository linked at pimbook.org.

[17] This is suspiciously similar to the singular value decomposition in Chapter 10, though there we focused on the geometric perspective.

terms of the v_i. It uses the simple formula from Proposition 12.15.

```
def decompose(eigenvectors, vector):
    coefficients = {}
    for i in range(len(vector)):
        coefficients[i] = numpy.dot(vector, eigenvectors[i])

    return coefficients
```

With results printed below rounded for legibility, the coefficients for our chosen y can be computed and used to reconstruct the original vector.

```
>>> A = bead_matrix(5)
>>> eigensystem = sorted_eigensystem(A)
>>> eigenvalues, eigenvectors = eigensystem
>>> w = [0, 0.5, 0, 0, 0]
>>> coeffs = decompose(eigensystem, w)
>>> print(coeffs)
{0: 0.25, 1: -0.25, 2: 0, 3: 0.25, 4: 0.25}
>>> numpy.sum([coeffs[i] * eigensystem[1][i] for i in range(5)], axis=0)
array([  0,  5.0e-01,  0,  0,  0])
```

So $y(0) = 0.25v_1 + -0.25v_2 + 0v_3 + 0.25v_4 + 0.25v_5$, and we can compute this sum and pick out any coordinate we want to get the initial position of a particular bead.

Now, in the basis of eigenvectors, we define a new set of variables $z(t) = (z_1(t), \ldots, z_5(t))$. Let $z_i(t)$ be the coefficient of v_i for the representation of $y(t)$ in the basis of eigenvectors. In words, before we were tracking the *position* of the beads as they evolve over time, and now we're tracking the *coefficients of the eigenvectors* as they evolve over time. This is the whole point of the change of basis. In this new representation the differential equation changes to

$$y'' = Ay \implies z'' = Dz$$

Where D is the diagonal matrix of eigenvalues $\lambda_1, \ldots, \lambda_n$ (in any order we please, let's say in decreasing order). Then each coordinate is just like our single-bead case. For example $z_1'' = \lambda_1 z_1$, along with an initial condition $z_1(0) = 0.25$ (as per the decomposition of $y(0)$ above).

We can solve each of these differential equations separately, just as we solved the single-bead equation, and then combine them by converting back to the standard basis of bead positions. The result will give us the trajectory of each bead expressed as a sum of simple cosine waves.

The equations, with initial conditions placed adjacent, are (with some rounding to simplify):

$$z_1'' = -0.27z_1; \quad z_1(0) = 0.25, \quad z_1'(0) = 0$$
$$z_2'' = -z_2; \quad z_2(0) = -0.25, \quad z_2'(0) = 0$$
$$z_3'' = -2z_3; \quad z_3(0) = 0, \quad z_3'(0) = 0$$
$$z_4'' = -3z_4; \quad z_4(0) = 0.25, \quad z_4'(0) = 0$$
$$z_5'' = -3.73z_5; \quad z_5(0) = 0.25, \quad z_5'(0) = 0$$

And the solutions are

$$z_1(t) = 0.25\cos(0.52t)$$
$$z_2(t) = -0.25\cos(t)$$
$$z_3(t) = 0$$
$$z_4(t) = 0.25\cos(1.73t)$$
$$z_5(t) = 0.25\cos(1.93t)$$

Converting back to the bead-position basis, we get

$$y(t) = 0.25\cos(0.52t)v_1 - 0.25\cos(t)v_2 + 0.25\cos(1.73t)v_4 + 0.25\cos(1.93t)v_5$$

Which expanded out coordinate-wise (and again rounded) is

$$y_1(t) = 0.07\cos(0.52t) + 0.125\cos(t) + -0.125\cos(1.73t) + -0.07\cos(1.93t)$$
$$y_2(t) = 0.125\cos(0.52t) + 0.125\cos(t) + 0.125\cos(1.73t) + 0.125\cos(1.93t)$$
$$y_3(t) = 0.145\cos(0.52t) + -0.145\cos(1.93t)$$
$$y_4(t) = 0.125\cos(0.52t) + -0.125\cos(t) + -0.125\cos(1.73t) + -0.125\cos(1.93t)$$
$$y_5(t) = 0.07\cos(0.52t) + -0.125\cos(t) + 0.125\cos(1.73t) + -0.07\cos(1.93t)$$

Fantastic! We started with a tightly coupled system, in which the position and motion of the different beads seem to depend heavily on each other. They do, it's true, but this eigensystem provides a perspective in which their motions can be computed independently! You don't have to know where bead 3 is to compute the future position of bead 2. That's the promise fulfilled by eigenvectors.

Finally, as you may have guessed from the arbitrary choice of five beads, we can generalize this system to any number of beads. If we take even just a hundred beads, and plot the eigenvectors for the top few eigenvalues as we did above, we see smoother, more obvious waves. Figure 12.10 shows this. With such natural shapes of increasing complexity, it makes sense to give a name to these eigenvectors. They're called the *fundamental modes* of the system, and the frequencies of the "sinusoidal curve" of each eigenvector[18] are called the *resonant frequencies* of the system.

---

[18] Or rather, the curves implied to underlie these discrete points.

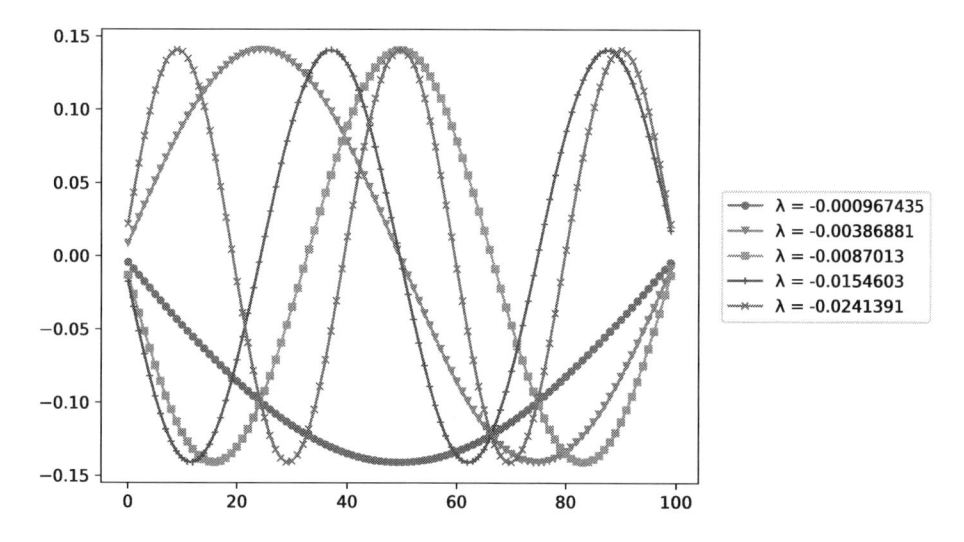

Figure 12.10: The plot of the top five eigenvectors for a hundred-bead system.

If one decreases the distance between beads and increases the number of beads in the limit, the result is the *wave equation*. This is a differential equation (in both time and position along the string) that one can use to track the motion of a traveling wave through a string. See the exercises for more on that. But more importantly for us, the vector space for that continuous model has infinite dimension, it still has a basis of eigenvectors, and they correspond to proper sine curves instead of discrete approximations. In this case, since the "zero-width" beads are now at every position of the string, you can think of them as cross sections of molecules that make up the string itself, with atomic forces playing the role of Hooke's law. These eigenvectors then describe the intrinsic properties of the string itself.

So there you have it. Eigenvectors have revealed the secrets of waves on a string.

12.8 Cultural Review

1. Eigenvalues and eigenvectors often provide the best perspective (basis) with which to study a linear map.

2. An orthonormal basis of eigenvectors allows you to decouple aspects of a complex system that are a priori intertwined, and orthonormality makes computing basis decompositions easy.

3. Invariance is a strong "smell," meaning objects which satisfy an invariance property are probably important, even if you don't know why exactly. In this chapter, it was an eigenvalue being invariant to the choice of basis, and eigenvectors of f being invariant (up to scaling) under the operation of applying f.

4. When trying to solve a complicated problem, a good approach is to simplify the problem as much as possible without losing the essential character of the problem. One can then solve that simplified problem and gain insight. Then gradually add complexity back to the problem and, using the new insights, attempt to solve the harder problem.

12.9 Exercises

12.1 Let V be an n-dimensional inner product space, whose norm is given by the inner product. Prove the following.

1. The only vector with norm zero is the zero vector.
2. The distance function induced by an inner product is nonnegative and symmetric.
3. The distance function induced by an inner product satisfies the triangle inequality. That is, $d(x, y) + d(y, z) \leq d(x, z)$ for all $x, y, z \in V$.

12.2 Prove that a linear map $f : \mathbb{R}^n \to \mathbb{R}^n$ preserves the standard inner product—i.e. $\langle x, y \rangle = \langle f(x), f(y) \rangle$ for all x, y—if and only if its matrix representation A has orthonormal columns with respect to the standard basis. Hint: use the fact that $\langle x, y \rangle = x^T y$.

12.3 Let A be a square matrix with an inverse. Using only the fact that $(BC)^T = C^T B^T$ for two square matrices B, C, prove that $(A^T)^{-1} = (A^{-1})^T$.

12.4 Prove the following basic facts about eigenvalues, eigenvectors, and inner products.

1. Fix a vector y and let $f_y(x) = \langle x, y \rangle$. Prove that if x is restricted to be a unit vector, then $f_y(x)$ is maximized when $x = y$.

2. Let V, W be two n-dimensional inner product spaces with inner products $\langle -, - \rangle_V$ and $\langle -, - \rangle_W$. Define a bijective linear map $f : V \to W$ that is an isomorphism of vector spaces and also satisfies $\langle x, y \rangle_V = \langle f(x), f(y) \rangle_W$ for all $x, y \in V$. Such a map is called an *isometry*. Hint: start by using Gram-Schmidt to choose an orthonormal basis of each vector space.

3. Fix the inner product space $\mathbb{R}^n$ with the standard inner product. Let $A : \mathbb{R}^n \to \mathbb{R}^n$ be a change of basis matrix. Find an example of A for which $\langle x, y \rangle \neq \langle Ax, Ay \rangle$. In other words, an arbitrary change of basis does not preserve the *formula* for the standard inner product. As we saw in the chapter, only an orthogonal change of basis does this. Determine a formula (that depends on the data of A), that shows how to convert inner product calculations in one basis to inner product calculations in another.

12.5 Look up a proof of Theorem 12.30, on the uniqueness of the sine function, that uses Taylor series. The analytical tool required to understand the standard proof is the concept of *absolute convergence*. The central difficulty is that if you're *defining* a function by an infinite series, you have to make sure that series converges with the properties needed to make it a valid Taylor series. Repeat the proof for $\sin(ax)$.

12.6 In Definition 12.3 we defined the adjacency matrix $A(G)$ of a graph $G = (V, E)$. This matrix corresponds to some linear map $f : \mathbb{R}^n \to \mathbb{R}^n$, where $n = |V|$. How would you interpret the vector space V? What is a natural description of the basis of V that we're using to represent $A(G)$? What is a natural (English) description of the linear map f, if you restrict to input vectors whose entries are either 0 or 1? If this is hard to formulate abstractly, write down an example graph on 5 vertices. What happens to your description of f when you allow for non-binary inputs?

12.7 Prove that a connected graph G is bipartite if and only if it contains no cycles of odd length. Write a program to find cycles of odd length, and hence to decide whether a given graph is bipartite.

12.8 Implement the algorithm presented in the chapter to generate a random graph on n vertices with edge probability 1/2, and a planted clique of size k. For the rest of this exercise fix $k = \sqrt{n \log n}$. Determine the average degree of a vertex that is in the plant, and the average degree of a vertex that is not in the plant, and use that to determine a rule for deciding if a vertex is in the clique. Implement this rule for finding planted cliques of size at least $\sqrt{n \log n}$ with high probability, where $n = 1000$.

12.9 As in the previous problem, implement the algorithm in this chapter for finding planted cliques of size $k = 10\sqrt{n}$ in random graphs with $n = 1000$. Use a library such as numpy to compute eigenvalues and eigenvectors for you.

12.10 The *minimal polynomial* of a linear map $f : V \to V$ is the monic polynomial p of smallest degree such that $p(f) = 0$. Since the space of all linear maps $V \to V$ is a vector space, we can interpret a "power" of f^k as the composition of f with itself k times. Likewise, cf is the map $x \mapsto cf(x)$. So $p(f)$ is a linear map $V \to V$, and by $p(f) = 0$ we mean that $p(f)$ is the zero map. Look up a proof that λ is a root of p if and only if λ is an eigenvalue of f.

12.11 We proved that symmetric matrices have a full set of eigenvectors and eigenvalues. In this exercise we will see that to understand eigenvalues of non-symmetric matrices, we must necessarily prove the Fundamental Theorem of Algebra, which we remarked in Exercise 2.12 is quite hard. First prove that r is a root of the polynomial $p(x) = x^n + a_{n-1}x^{n-1} + \cdots + a_1 x + a_0$ if and only if r is an eigenvalue of the matrix

$$A_p = \begin{pmatrix} 0 & 1 & 0 & \cdots & 0 & 0 \\ 0 & 0 & 1 & \cdots & 0 & 0 \\ \vdots & \vdots & \vdots & \ddots & \vdots & \vdots \\ 0 & 0 & 0 & \cdots & 1 & 0 \\ 0 & 0 & 0 & \cdots & 0 & 1 \\ -a_0 & -a_1 & -a_2 & \cdots & -a_{n-2} & -a_{n-1} \end{pmatrix}$$

Notice that this matrix is not symmetric, and that because the roots of a polynomial might necessarily be complex numbers, this implies the eigenvalues of a matrix might also be complex. Walk away from this exercise with a new appreciation for the convenience of symmetric matrices, and the inherent difficulty of writing a generic eigenvalue solver.

12.12 Look up a proof of the theorem that every square matrix can be written in the so-called Jordan canonical form.

12.13 Implement the Gram-Schmidt algorithm using the following method for finding vectors not in the span of a partial basis: choose a vector with random entries between zero and one, repeating until you find one that works. How often does it happen that you have to repeat? Can you give an explanation for this?

12.14 Look up the derivation of the wave equation from Hooke's law for a beaded string (or equivalently, beads on springs) as the distance between adjacent beads tends to zero.

12.15 Look up a proof that the singular values of a non-square real matrix A are the square roots of the eigenvalues of the matrix $A^T A$.

12.16 Generate a "random" symmetric 2000×2000 matrix via the following scheme: pick a random distribution (say, Gaussian normal with a given mean and variance), and let the i, j entry with $i \geq j$ be an independent draw from this distribution. Let the remaining $i < j$ entries be the symmetric mirror. Compute the eigenvalues of this matrix (which are all real) and plot them in a histogram. What does the result look like? How does this shape depend on the parameters of the distribution? On the qualitative choice of distribution?

12.17 At the end of the chapter we converted the eigenvector-coefficient solution to $z'' = Dz$ back to the bead basis by hand. Write a program that, given the initial position of the beads, sets up the independent differential equations in the eigenvector basis, solves those equations, and converts them back to the bead position basis.

12.18 Using Taylor series, find appropriate conditions under which horizontal motion in the 5-bead system can be ignored.

12.19 Generalize our one-dimensional bead system to a two dimensional lattice. That is, fix n and put a bead at each $(i, j) \in \{1, 2, \ldots, n\}^2$, with strings connecting adjacent beads, with fixed walls on each boundary side. Pluck the beads perpendicularly to the lattice. Can you design a symmetryic linear model for this system? If so, what do the eigenvectors look like? If not, what step of the modeling process breaks? What is the fundamental obstacle?

12.20 Consider a one-dimensional "bead system" where instead of the beads physically moving, they are given some initial heat. Adjacent beads transfer heat between them according to a discrete version of the so-called *heat equation*. Find an exposition of the discrete heat equation online that allows you to set up a linear system and solve it for 10 beads. What do the eigenvalues of this system look like?

12.21 PageRank is a ranking algorithm that was a major factor in the Google search engine's domination of the internet search market. The algorithm involves setting up a linear system based on links between webpages, and computing the eigenvector for the largest eigenvalue. Find an exposition of this algorithm and implement it in code. Can you visualize or interpret the eigenvector in a meaningful way?

12.10 Chapter Notes

Transposes and Linear Maps

If $f : V \to W$ is a linear map, and A is a matrix representation of f, how does A^T, the operation of transposing the matrix, correspond to an operation on f? The answer requires some groundwork.

A *linear functional* on a vector space with scalars in $\mathbb{R}$ is a linear map $V \to \mathbb{R}$. That is, it linearly maps vectors to scalars. This is the origin of the name of the subfield of mathematics called "functional analysis," which studies these mappings as a way to study the structure of the (usually infinite dimensional) vector space. We'll stick to finite dimensions. Fix a vector space V over $\mathbb{R}$. The set of all linear functionals on V forms a vector space (using the same point-wise addition and scalar multiplication we saw for L^2). This vector space is called the *dual* vector space of V, and I'll denote it by V^*.

The standard basis $\{e_1, \ldots, e_n\}$ for $\mathbb{R}^n$ corresponds to a *standard dual basis* for the dual space, which we'll denote $\{e_1^*, \ldots, e_n^*\}$. Each e_i^* is the projection onto the i-th coordinate (in the standard basis), i.e. $e_i^*(a_1, \ldots, a_n) = a_i$. This mapping is injective, and in fact every linear functional can be expressed as a linear combination of these dual basis vectors. Hence, $\mathbb{R}^n$ (and, by way of Theorem 10.17, every finite dimensional vector space) is isomorphic to its dual. In particular, they have the same dimension.

This construction works without need for an inner product, but if you have an inner product, you get an obvious way to take a general basis $\{v_1, \ldots, v_n\}$ of V to a dual basis of V^* by mapping v to the function $x \mapsto \langle v, x \rangle$. If the $\{v_i\}$ were an orthonormal basis, this would be the same "coordinate picking" function as we did for the standard basis, due to Proposition 12.15.

Moreover, every linear functional on $\mathbb{R}^n$ can be expressed as the inner product with a *single* vector (not necessarily a basis vector). Expressed in terms of matrices, the linear functional can be written as a $(1 \times n)$-matrix—since it is a linear map from an n-dimensional vector space to a 1-dimensional space. Say we call it $f_v(x) = \langle v, x \rangle$. If you start from the perspective that all vectors are columns, then the matrix representation of f_v is v^T, and the "matrix multiplication" $v^T x$ is a scalar (and also another way to write the inner product, as we saw in this chapter).

Now we finally get to the transpose, which just extends this linear functional picture to a finite number of independent functionals, the outputs of which are grouped together in a vector. Let $f : V \to W$ be a linear map with matrix representation A, an $(m \times n)$-matrix for n-dimensional V and m-dimensional W. Define the *transpose* of f (sometimes called the *adjoint*) as the linear map $f^T : W^* \to V^*$ which takes as input (a linear functional!) $g \in W^*$ and produces as output the linear functional $g \circ f \in V^*$, the composition of the two maps by first applying f and then applying g. And indeed, the matrix representation of f^T with respect to the dual bases for V^*, W^* is A^T.

Since W^* and W are isomorphic, and V^* and V are isomorphic, you may wonder if you can apply this to realize the dual f^T as a map $W \to V$ as well. Indeed you can, and it can even be defined without referring to dual vector spaces at all. Let V, W be inner product spaces and $f : V \to W$ a linear map. Define the transpose $f^T : W \to V$ input-by-input as follows. Let $w \in W$, and define $f^T(w)$ to be the unique vector for which $\langle f(v), w \rangle = \langle v, f^T(w) \rangle$. One needs to prove this is well-defined, but it is. It comes from our discussion about symmetry in Section 12.2 about how in $\mathbb{R}^n$ you get $\langle Ax, y \rangle = x^T A^T y$.

Note that these two definitions of the transpose can only be said to be the same in the case that the vector space has scalars in $\mathbb{R}$. If you allow for complex number scalars, things get a bit trickier.

Chapter 13

Rigor and Formality

> *Mathematics as we practice it is much more formally complete and precise than other sciences, but it is much less formally complete and precise for its content than computer programs. The difference has to do not just with the amount of effort: the kind of effort is qualitatively different. In large computer programs, a tremendous proportion of effort must be spent on myriad compatibility issues: making sure that all definitions are consistent, developing good data structures that have useful but not cumbersome generality, deciding on the right generality for functions, etc. The proportion of energy spent on the working part of a large program, as distinguished from the bookkeeping part, is surprisingly small. Because of compatibility issues that almost inevitably escalate out of hand because the right definitions change as generality and functionality are added, computer programs usually need to be rewritten frequently, often from scratch.*
>
> —William Thurston, "On Proof and Progress in Mathematics"

Programmers who brave mathematical topics often come away wondering why mathematics isn't more like programming. We've discussed some of the issues surrounding this question already in this book, like why mathematicians tend to use brief variable names, and how conventions will differ from source to source. Beneath these relatively superficial concerns is a question about rigor.

Thurston's observations above were as true in the mid 90's as they are over twenty years later. Software is far more rigorous than mathematics, and most of the work in software is about interface and data compatibility—"bookkeeping," as Thurston calls it. This is the kind of work required by the rigor of software. You need to care whether your strings are in ASCII or Unicode, that data is sanitized, that dependent systems are synchronized, because ignoring this will make everything fall apart.

I once took a course on compiler design. The lectures were taught in the architecture building on campus. One day, the architecture students were having a project fair in the building, marveling over their structures and designs. In a lightly mocking tone, the professor observed that software architecture was much more impressive than building architecture. Their buildings wouldn't fall over if they forgot a few nails or slightly changed the materials. But a few misplaced characters in software has caused destruction, financial disaster, and death.

My professor had a point. Regular mayhem is caused by software security lapses, with root causes often related to improper string validation or bad uses of memory copying. Single improperly set bits can cause troves of private data to become public. Financial insecurity is almost synonymous with digital currencies, one particularly relevant example being the 2016 hack of the "Decentralized Autonomous Organization," a sort of hedge fund governed by an Ethereum contract that contained a bug allowing a hacker to withdraw 50 million USD before it was mitigated. The root cause was a bug in the contract allowing an infinite recursion. Multiple (unmanned) space probes, costing hundreds of millions of dollars each, have been destroyed shortly after launch due to coding errors. The Ariane 5 crashed in 1996 because of a bug with integer overflow. The Mariner 1 in 1962 because of a missing hyphen. Finally, in 1991, a bug in the Patriot missile defense system resulted in the death of 28 soldiers at a military base in Saudi Arabia. The bug was an inaccurate calculation of wall-clock time due to a poor choice of rounding. I have little doubt there will be additional deaths[1] caused by lapses and insecurities in self-driving car software, in addition to the damage already caused by accidents (many of which went unreported, according to some 2018 reporting).

These sorts of bugs cause internal debacles at every company with alarming regularity. One consequence is a general feeling among many engineers that "all software is shit." More optimistically, the best engineers work very hard to design interfaces and abstractions that, to the best of software's ability, prevent mistakes. Those who design aircraft control systems do this quite well. Once you've made enough mistakes of your own, you learn a certain air of humility. No matter how smart, even the best engineers get tired, grumpy, overworked, or forgetful—each of which is liable to make them forget a hyphen. Good tools make forgetting the hyphen impossible.

In the subfield of computer science dealing with distributed systems, these issues are exacerbated by the extreme difficulty of even telling whether a system satisfies the guarantees you need it to. A titan of this area is mathematician turned computer scientist Leslie Lamport. Through his work, Lamport essentially defined distributed computing as a field of study. Many of the concepts you have heard of in this area—synchronized clocks, Paxos consensus, mutexes—were invented by Lamport.

Lamport has no particular love of mathematical discourse. In his 1994 essay, "How to Write a Proof," he admits, "Mathematical notation has improved over the past few centuries," but goes on to claim that the style of mathematical proof employed by most of mathematics (including in this book)—mixing prose and formulas in a web of propositions, lemmas, and theorems—is wholly inadequate.

Much of Lamport's seminal work in the last few decades grew out of his frustration with errors in distributed systems papers. As he attests, some researcher would propose (say) a consensus algorithm. It might seem correct at first glance, but inevitably it would contain mistakes—if not be wrong outright. Lamport concludes that *guarantees* about the behavior of distributed systems are particularly hard to establish with the rigor that is

---

[1] I personally attribute the 2018 death of Elaine Herzberg to engineers intentionally disabling safety features and cutting personnel costs than to software bugs.

needed for practical considerations. If you're going to design a new distributed database, you want a much stronger assurance than the assent of some overworked journal referees. Lamport writes,

> *These proofs are seldom deep, but usually have considerable detail. Structured proofs provided a way of coping with this detail. The style was first applied to proofs of ordinary theorems in a paper I wrote with Martín Abadi. He had already written conventional proofs—proofs that were good enough to convince us and, presumably, the referees. Rewriting the proofs in a structured style, we discovered that almost every one had serious mistakes, though the theorems were correct. Any hope that incorrect proofs might not lead to incorrect theorems was destroyed in our next collaboration. Time and again, we would make a conjecture and write a proof sketch on the blackboard—a sketch that could easily have been turned into a convincing conventional proof—only to discover, by trying to write a structured proof, that the conjecture was false. Since then, I have never believed a result without a careful, structured proof. My skepticism has helped avoid numerous errors.*

This is coming from a Turing Award winner, a man considered a luminary of computer science. Even the smartest theorem provers among us make ample mistakes.

Consequently, Lamport designed a proof assistant called TLA+, which he has used to check the correctness of various claims about distributed systems.[2] TLA+ is supposed to prevent you from shooting your own mathematical foot. TLA+ falls in step with a body of work related to automated proof systems. Some systems you may have heard include Coq and Isabelle. Some of these systems claim the ability to prove your theorems for you, but I'll instead focus just on the correctness checking aspects.

So computer scientists like Lamport and software engineers are perturbed by the lack of rigor in mathematics. Each remembers the fresh wounds of catastrophes due to avoidable mistakes. Meanwhile, Lamport and others provide systems like TLA+ that would allow mathematician to achieve much higher certainty in their own results. This raises the question, why don't all mathematicians use automated proof assistants like TLA+? This is a detailed and complex question. I will not be able to answer it justly, but I can provide perspectives built up throughout this book.

We have argued that the elegance of a proof is important. Mathematicians work hard to be able to summarize the core idea of a proof in a few words or a representative picture. Full rigor as the standard for all proofs would arguably strip many proofs of their elegance, increasing the burden of transmitting intuition and insight between humans. The work you put into making an argument automatable is work you could have spent on making math accessible to humans (via additional papers, talks, and working with students). These extra activities already serve as correctness checks, so is there significant added benefit to a formal specification? Lamport's counter is that making it accessible

---

[2] I particularly enjoyed his tutorial video course, which you can find at https://lamport.azurewebsites.net/video/videos.html.

to humans is counterproductive when the result is incorrect. He would also argue that a structured proof *is* easier to understand. One underlying issue Lamport's riposte ignores is that mathematics is a social activity, and formal proof specifications are decidedly antisocial. Good for those who want to ensure planes don't crash, bad for those who want to do mathematics.

Another aspect concerns the priorities and preferences of the subcultures of mathematics. Theory builders might argue that if your proof is too complicated to keep track of—which is why you would want TLA+—it's because your theory has not been built well enough to make the proof trivial. Conversely, problem solvers might complain that proof assistants limit their ability to employ clever constructions. Being able to invoke a result from a disconnected area of math requires you to re-implement that entire field in your new context. Dependency management would turn few-page arguments into thousand-line software libraries.

Both of these attitudes reconverge on Thurston's observation, that the kind of effort that goes into math is categorically different from software. Mathematicians don't *want* to nitpick type errors and missing parentheses. They want to think about ideas at a higher level. Mathematicians have built up so many abstractions over the years specifically to avoid the mundane details that can muddle an idea. One explanation for why TLA+ work so well for distributed systems theorems is that those theorems have relatively few layers of indirection. A handful of bits might represent consensus. On the other hand, in geometry you might think the thought, "this space is very flat, and that should have such-and-such effect." An automated proof assistant will be of no use there, nor will it help you refine the degree to which your hypothesized effect is present. You must lay everything out perfectly formally, even if your definitions haven't been finalized. Then too often you resort to writing and rewriting, and before long you've stopped doing math entirely. Just as Michael Atiyah argues that the proof is the very last step of mathematical inquiry, which implies a proof assistant is useless for the majority of your work.

As most engineers can understand, the degree of rigor to require is a tradeoff with tangible benefits on both sides. Mathematicians opt to let some errors slip through. Over time these errors will eventually be found and reverted or fixed. Since technology rarely goes straight from mathematical publication to space probe control software, the world has enough headway to accommodate it.

Thurston also questions the two assumptions underlying this discussion:

1. *that there is uniform, objective and firmly established theory and practice of mathematical proof,* and
2. *that progress made by mathematicians consists of proving theorems.*

Thurston instead prefers a question more leading to what he feels is the correct answer: "How do mathematicians advance human understanding of mathematics?" Many mathematicians feel unsatisfied by computer-aided proofs because they don't help them *personally* understand the proof. If the core insight can't fit in a single human's head, it might as well be unproved. This is still the attitude of many toward the famous four-color theorem, the shortest proof of which to date involves much brute force case checking by computer. As much as rigor helps one establish correctness, it does not guarantee synthesis and understanding.

Thurston continues,

> *I think that mathematics is one of the most intellectually gratifying of human activities. Because we have a high standard for clear and convincing thinking and because we place a high value on listening to and trying to understand each other, we don't engage in interminable arguments and endless redoing of our mathematics. We are prepared to be convinced by others. Intellectually, mathematics moves very quickly. Entire mathematical landscapes change and change again in amazing ways during a single career. When one considers how hard it is to write a computer program even approaching the intellectual scope of a good mathematical paper, and how much greater time and effort have to be put into it to make it "almost" formally correct, it is preposterous to claim that mathematics as we practice it is anywhere near formally correct.*

Rather, Thurston claims that reliability of mathematical ideas "does not primarily come from mathematicians formally checking formal arguments; it comes from mathematicians thinking carefully and critically about mathematical ideas."

Chapter 14
Multivariable Calculus and Optimization

The world is continuous, but the mind is discrete.

—David Mumford

A large swath of practical applied mathematics revolves around optimization. Financial math optimizes cost and revenue, supply chains optimize routing and allocation of resources, and machine learning optimizes generalization error from training examples. Often these problems are complicated, but can be modeled using multi-input, multi-output functions composed of simple, differentiable pieces. The modeling process is difficult and deserves dedicated books of its own. But once a model is agreed upon by those that will use it, the primary tool to optimize the model is calculus. Often these models are immense, with functions spanning millions of variables and operations. The namesake of calculus, that it is about calculations, hints at a perfect marriage of mathematics and programs.

Thankfully, calculus generalizes quite nicely from one dimension (Chapter 8) to many dimensions. We'll primarily focus on how the derivative generalizes to the so-called *total derivative*, the computation of which reduces to the well-trod problem of computing single-variable derivatives. The magic touch will be a clean definition that isolates the core feature of the derivative, reforging our insights from Chapter 8 into a new foundation for the subject. We'll rely heavily on linear algebra, and discover what we always knew deep in our hearts, that linear algebra is the proper foundation for calculus.

As the application for this chapter, we'll write a neural network from scratch in the style of Google's popular library TensorFlow. In particular, we'll implement a way to decompose an arbitrary function into a so-called *computation graph* of simple operations, and optimize its parameters using a popular technique called gradient descent. We'll apply this to the classic problem of classifying handwritten digits. Along the way, we'll get a whirlwind introduction to the theory and practice of machine learning.

14.1 Generalizing the Derivative

Let's start with our fond memories of single-variable calculus. Recall Definition 8.6 of the derivative of a single-variable function.

Definition 14.1. Let $f : \mathbb{R} \to \mathbb{R}$ be a function. Let $c \in \mathbb{R}$. The *derivative* of f at c, if it exists, is the limit

$$f'(c) = \lim_{x \to c} \frac{f(x) - f(c)}{x - c}$$

On the real line, we defined the symbolic abstraction $x \to c$ to mean "any sequence x_n that converges to c," where we declared the derivative only exists if the limit doesn't depend on the choice of sequence. When we work in $\mathbb{R}^n$ (which, among many other properties, has a nice measure of distance for vectors $d(x, y) = \|x - y\|$) the notion of a convergent sequence generalizes seamlessly. A sequence of vectors $x_1, x_2, \cdots \in \mathbb{R}^n$ converges to $c \in \mathbb{R}^n$ if the sequence $d_n = \|x_n - x\|$ of real numbers converges to zero. Our deeper problem, however, is that despite sequence convergence generalizing, the obvious first attempt to adapt the derivative violates well-definition. Ignoring the obvious type error—one cannot divide a scalar by a vector—the "value" of the derivative would depend on the sequence chosen. That's the "smell" we pointed out in Chapter 8 that makes for a useless definition.

There are many easy examples to demonstrate. For instance: the function $f(x_1, x_2) = -x_2^2$, and the two sequences $x_n = (1 + \frac{1}{n}, 1)$ and $x'_n = (1, 1 + \frac{1}{n})$. Both sequences converge to $(1, 1)$, but because f depends on the second coordinate quadratically, (and doesn't depend on the first coordinate at all!) the direction along which x'_n approaches is steeper than that of x_n. Using the former for "the derivative" would result in something like $\lim_{n \to \infty} \frac{-1+1}{(1/n)} = 0$, while the latter would be $\lim_{n \to \infty} \frac{-1-(2/n)-(1/n^2)+1}{(1/n)} = -2$. This is illustrated in Figure 14.1.

In this brave new world, the underlying idea of "steepness" now inherently depends on direction. This is something one intuitively understands from the natural world; a hiker traverses switchbacks to avoid walking straight up a hill, and a skier skis in an S shape to slow down their descent.[1] In fact, for $f(x_1, x_2) = -x_2^2$, and standing at the point $(1, 1)$, every direction provides a slightly different slope.

This suggests one intuitive way to generalize the one-dimensional definition of the derivative.

Definition 14.2. The *directional derivative* of a function $f : \mathbb{R}^n \to \mathbb{R}$ at a point $c \in \mathbb{R}^n$ in the direction of a unit vector $v \in \mathbb{R}^n$ is the limit

$$\text{Dir}(f, c, v) = \lim_{t \to 0} \frac{f(c + tv) - f(c)}{t}$$

If this limit exists, we say f is *differentiable at c in the direction of v*.

---

[1] I grew up on a hill-covered cattle ranch, and when I was young I noticed the trails traced out by the cows were always nearly flat along the side of the hill. Those massive beasts know how to get from place to place without wasting energy.

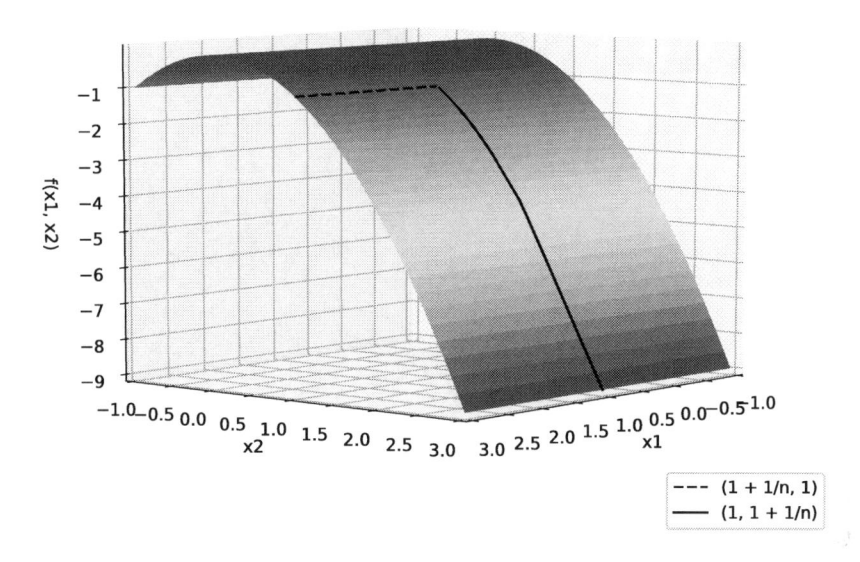

Figure 14.1: The steepness of a surface depends on the direction you look.

So instead of allowing a sequence to approach the point of interest from any direction, we restrict it to the line through the direction v we're interested in. Here we're using $t \to 0$ to denote any sequence $t_n \in \mathbb{R}$ which converges to zero.

This definition has two serious problems. The first is that it's hard to compute. It's not that any individual limit is particularly hard to compute on its own, but that on its face this definition requires us to recompute limits for every direction. With single-variable derivatives, we developed efficient techniques for computing a formulaic derivative. We want a similar mechanism for multivariable derivatives at any point *and in any direction*. We want to compute a formula once, and use that to enable many easy relevant computations later.

The second problem is that it's not strong enough to capture what we *really* want out of a derivative definition. And when I say "we" I mean "mathematicians with centuries of hindsight." This is a bit subtle, but a corkscrew surface shown in Figure 14.2 illustrates the problem.

On this surface at $c = (0,0)$, the directional derivative exists in every direction, but jumps sharply as the direction rotates past the negative x_1 axis. In a mildly technical parlance we have avoided making precise in this book, the directional derivative isn't *continuous* with respect to direction. If I stand at the origin and look directly in the direction of the jump (a ray down the negative x_1-axis), then as my gaze perturbs left and right by any infinitesimally small amount, my view of the steepness of the surface jumps drastically from very steeply negative to very steeply positive. This is bad because it destroys the possibility that a derivative based on the directional derivative can serve

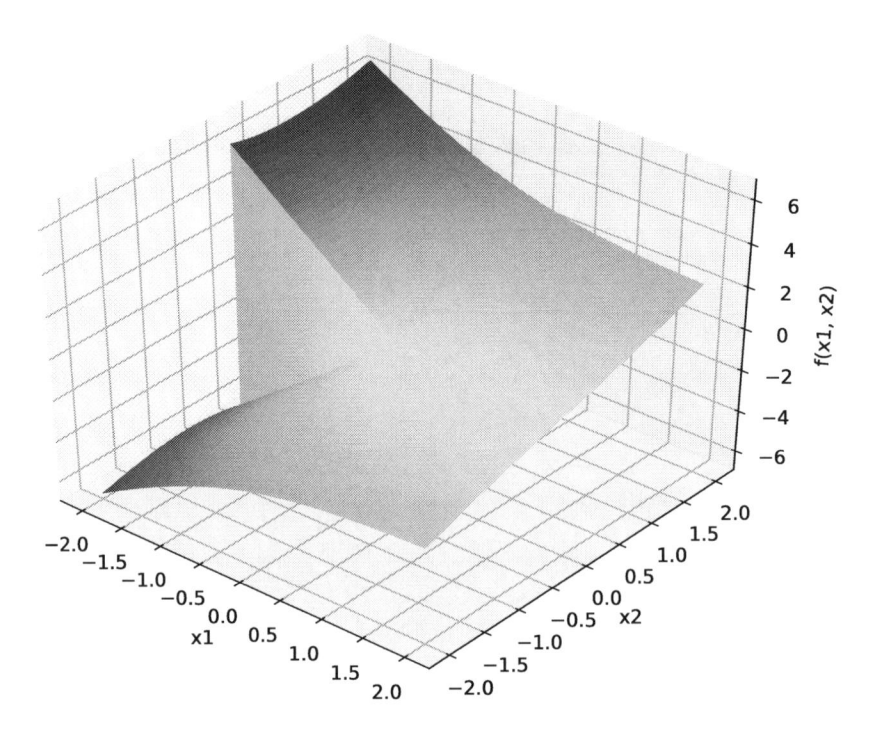

Figure 14.2: A corkscrew function, demonstrating that directional derivatives need not be continuous as the direction changes.

as a global approximation to f near $(0,0)$. It will err egregiously in the vicinity of the jump.

As we'll see soon, a stronger derivative definition avoids these issues. The definition will only apply if the function can be usefully approximated by a linear function. It will provide a linear map representing the whole function, and applying simple linear algebra will produce the one-dimensional derviative in any direction. Since it's linear algebra, we even get the benefit of being able to choose a useful basis, though I haven't yet made it clear what the vector space in question is. That will come as we refine what the right definition of "the" derivative should be.

14.2 Linear Approximations

For dimension 1, the derivative of f had the distinction of providing the most accurate line approximating f at a point. The line through $(c, f(c))$ with slope $f'(c)$ is closer to the graph of f near c than any other line. We proved this in detail in Theorem 8.11.

This approximator is more than just a line. It's a linear map, and now that we have the language of linear algebra we can discuss it. Define by $L_{f,c}$ the linear map $L_{f,c}(z) = f'(c)z$. As input, this linear map takes a (one-dimensional) vector z representing how far

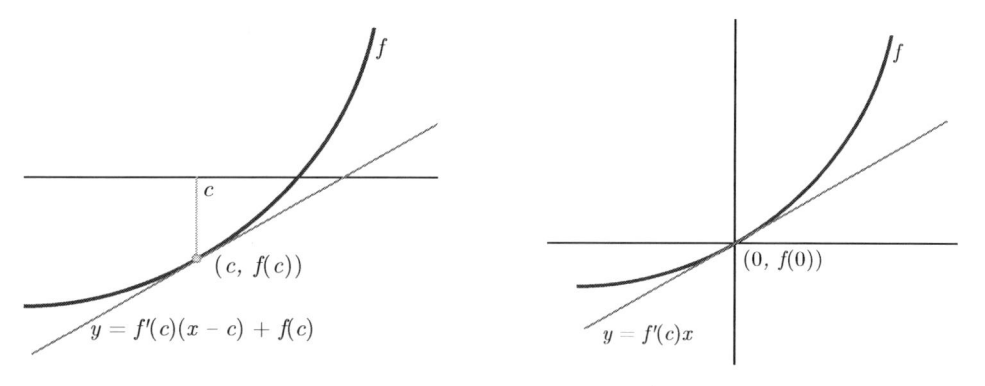

Figure 14.3: Left: a linear approximation without shifting f. Right shifted so that $(c, f(c))$ is at the origin.

one wants to travel away from c. The output is the derivative's approximation of how much f will change as a result. The matrix for $L_{f,c}$ is the single-entry matrix $[f'(c)]$. Moreover, $L_{f,c}(z)$ is exactly the first-degree Taylor polynomial for the version of f that gets translated so that $(c, f(c))$ is at the origin. Figure 14.3 shows the difference.

If you don't like shifting f to the origin, we can define the *affine* linear map (affine just means a translation of a linear map away from the origin), which we'll call a *linear approximation* to f.

Definition 14.3. Let $f : \mathbb{R} \to \mathbb{R}$ be a single-variable differentiable function. Then the linear approximation to f at a point $c \in \mathbb{R}$ is the affine linear map

$$L(c, x) = f'(c)(x - c) + f(c).$$

That is, $L(c, x)$ is the degree-1 Taylor approximation of f at c.

The linear approximator has the following obvious property, which is a restatement of the limit definition of the derivative.

Proposition 14.4. *For any differentiable $f : \mathbb{R} \to \mathbb{R}$ and its linear approximation $L(c, x)$,*

$$\lim_{x \to c} \frac{f(x) - L(c, x)}{x - c} = 0$$

Proof. Split the limit into two pieces:

$$\lim_{x \to c} \frac{f(x) - f(c)}{x - c} - \lim_{x \to c} \frac{f'(c)(x - c)}{x - c} = f'(c) - f'(c) = 0$$

$\square$

I spell this out in such detail because the existence of a linear approximator (an affine linear function satisfying 14.4) becomes a definition for functions $\mathbb{R}^n \to \mathbb{R}$.

Definition 14.5. Let $f : \mathbb{R}^n \to \mathbb{R}$ be a function. We say f has a *total derivative* at a point $c \in \mathbb{R}^n$ if

1. A linear map $A : \mathbb{R}^n \to \mathbb{R}$ exists such that:

2. The affine linear function defined by $L(c, x) = A(x - c) + f(c)$ (which depends on A) satisfies

$$\lim_{x \to c} \frac{f(x) - L(c, x)}{\|x - c\|} = 0$$

If the above both exist, we call L a *linear approximation* of f at c and A a *total derivative* of f at c.

In this definition, we again allow $x \to c$ to mean "any sequence converging to c." Because that's exactly the point! If no proposed linear map works due to a devious choice of approaching sequence, then the function doesn't have the property we want. There is no consistent way to have a linear approximation to f (ignoring how good or bad such an approximation might be). This rules out the confounding corkscrew example; the jump in the directional derivative is the violation of having a linear approximation.

If the definition is satisfied, then near c the function f can be approximated by a linear map A. The term $A(x-c)$ makes the linear map apply to deviations from c. Equivalently, the shift by $x - c$ translates f to the origin to apply A, and the $f(c)$ addition at the end translates back to $(c, f(c))$ afterward, so that f and L can be related to each other in a sensible manner.

One can explain intuitively why the definition of the total derivative avoids the problems of the directional derivative. In two dimensions, the linear approximation defines a plane touching the graph of the surface $z = f(x, y)$ at the point $(c, f(c))$. If the limit above holds, it asserts that no matter the direction of approach, the steepness of f matches the slope of the plane. If f has discontinuous jumps, then the linear approximator can only line up with f on one side of the jump. Figure 14.4 shows an example of the tangent plane to $f(x, y) = -x^2 - y^2$ at $(x, y) = (1, 1)$.

The computational centerpiece of Definition 14.5 is the liner map A. It helps to isolate A to ignore the shifting by c and $f(c)$ in a more principled manner. Let's do this now. We want to make the linear map A the focus of our analysis, and here's how we'll do that. For every point $c \in \mathbb{R}^n$, we "attach" a copy of the vector space denoted $T_f(c) = \mathbb{R}^n$ to $(c, f(c))$, and we call it the *tangent space* of f at c. The tangent space is the set of inputs to the total derivative. Because we view it as "attached" to f at $(c, f(c))$, as in Figure 14.4, we declare the tangent space's origin to be $(c, f(c))$. From that perspective, the linear approximation of f at c is just a linear map $T_f(c) \to \mathbb{R}$, without the shifting by c and $f(c)$.[2]

---

[2] In a manifold, which is a mathematical generalization of $\mathbb{R}^n$ to arbitrary spaces in which calculus can be defined, one "does calculus" entirely in these tangent spaces.

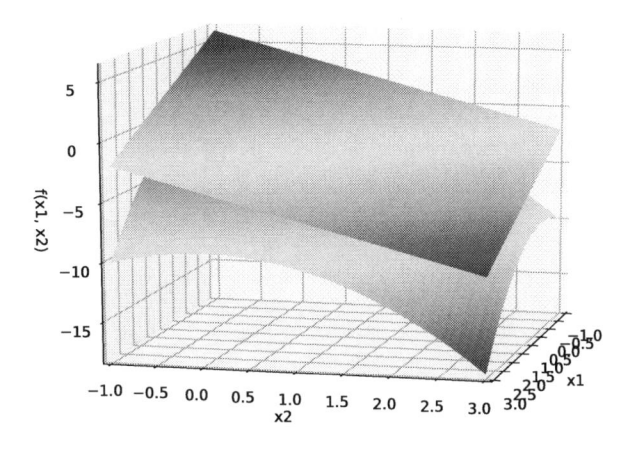

Figure 14.4: The linear subspace defined by the total derivative of f sits tangent to the surface of f at the point the total derivative is evaluated at.

It's worthwhile to do some concrete examples. First in one dimension, then in three. For single-variable functions $f : \mathbb{R} \to \mathbb{R}$, at every point c the tangent space is a one-dimensional vector space. The vectors in the vector space represent left/right deviations of the input of f from c, and the linear map A describes the approximate change in f due to this deviation. As an example, let $f(x) = \sqrt{x+2}$ and consider the point $(c, f(c)) = (2, 4)$. The derivative of f is $1/(2\sqrt{x+2})$, which evaluates to $1/4$ at $c = 2$. Thus, the tangent space $T_f(4)$ is a copy of $\mathbb{R}$, and the total derivative at $c = 2$ is $A(x) = \frac{1}{4}x$. The affine linear map is $L(x) = \frac{1}{4}(x-2) + 4$.

In three dimensions, let $f(x, y, z) = x^2 + (y-1)^3 + (z-2)^4$ and let $c = (3, 2, 1)$. The tangent space $T_f(c) = \mathbb{R}^3$, and so the total derivative $A : \mathbb{R}^3 \to \mathbb{R}$ has three-dimensional inputs. We won't learn how to compute this map from the definition of f until Section 14.4, so for now we give the answer magically; it's the following 1×3 matrix:

$$A = \begin{pmatrix} 6 & 3 & -4 \end{pmatrix}.$$

And as a result

$$\begin{aligned} L(x, y, z) &= A(x-3, y-2, z-1) + f(3, 2, 1) \\ &= 6(x-3) + 3(y-2) - 4(z-1) + 11. \end{aligned}$$

Many elementary calculus books have students compute this ("the equation of the plane tangent to the surface of f") as something of an afterthought, ignoring that it is the conceptual centerpiece of the derivative. Next we turn to some questions of consistency of the definition of the total derivative.

Proposition 14.6. *The total derivative is unique. That is, for any linear map $B : \mathbb{R}^n \to \mathbb{R}$, let L_B be the "linear approximator" defined by $L_B(c, x) = B(x - c) + f(c)$. Let A be the specific matrix used in the definition of the total derivative. Then for every linear map $B \neq A$, the defining limit of the total derivative is nonzero:*

$$\lim_{x \to c} \frac{f(x) - L_B(c, x)}{\|x - c\|} \neq 0$$

Proof. Let B be an arbitrary linear map different from A. We use the trick of writing $B = A + (B - A)$, which allows us to split the limit above in a useful way:

$$\lim_{x \to c} \frac{f(x) - L_B(c, x)}{\|x - c\|} = \lim_{x \to c} \frac{f(x) - [B(x - c) + f(c)]}{\|x - c\|}$$

$$= \lim_{x \to c} \frac{f(x) - [(A + B - A)(x - c) - f(c)]}{\|x - c\|}$$

$$= \lim_{x \to c} \frac{f(x) - A(x - c) + f(c)}{\|x - c\|} + \lim_{x \to c} \frac{(B - A)(x - c)}{\|x - c\|}$$

In the last line, the first limit is zero by the definition of the total derivative, so we need to show that the second term is nonzero. Since we assume $B \neq A$, there must be some unit vector $v \in \mathbb{R}^n$ for which $(B - A)v \neq 0$. Define the sequence $x_n \to c$ by $x_n = c + (1/n)v$. Then

$$\lim_{x \to c} \frac{(B - A)(x - c)}{\|x - c\|} = \lim_{n \to \infty} \frac{(1/n)(B - A)v}{\|(1/n)v\|} = (B - A)v \neq 0.$$

$\square$

This validates us calling the total derivative *the* total derivative. There is no other linear map that can satisfy the defining property. As such, we can define a more convenient notation for the total derivative.

Definition 14.7. Define the notation $Df(c)$ to mean the total derivative matrix A of f at the point c.

A quick note on notation, D is a mapping from functions to functions, but the way it's written it looks like c is an argument to a function called "Df". To be formal one might attempt to curry arguments. $D(f)(c)$ is a concrete matrix of real numbers, and $D(f)$ is a function that takes as input a point c and produces a matrix as output. Mathematicians often drop the parentheses to reduce clutter, and even the evaluation at c if this is clear from context. One might also subscript the c as in Df_c, or use a pipe that usually means "evaluated at," as in $Df|_{x=c}$. We will stick to $Df(c)$, as it achieves a happy middle: just think of the total derivative of f as being named Df.

Now that we've established some consistency, and understand that the total derivative is a linear map at its core, we can dive into the details of how we compute. To make this process cleaner, we first deviate to generalize the derivative to functions $\mathbb{R}^n \to \mathbb{R}^m$.

14.3 Multivariable Functions and the Chain Rule

Comfort with linear algebra makes converting relevant definitions of single-output functions to multiple-output functions trivial. A function $f : \mathbb{R}^n \to \mathbb{R}^m$ consumes a vector $x = (x_1, \ldots, x_n)$ as before, and produces as output a vector

$$f(x) = (f_1(x), f_2(x), \ldots, f_m(x)).$$

Each $f_i : \mathbb{R}^n \to \mathbb{R}$ stands on its own as a function. Moreover, if one defines $\pi_j : \mathbb{R}^m \to \mathbb{R}$ to be the function that extracts the j-th coordinate of its input, then $f_i = \pi_i \circ f$. Mathematicians tend to call the function "extract the i-th coordinate" a *projection* onto the i-th coordinate. Because indeed, it's exactly the linear-algebraic projection onto the i-th basis vector. This is also why you'll see π used as a function, since π is the Greek "p," and "p" stands for projection.

The definition of the derivative is nearly identical, but now all codomains are $\mathbb{R}^m$ and the limit numerator has a vector norm. The diff between Definitions 14.5 and 14.8 is literally four characters (two m's in $\mathbb{R}^m$ and two $\|$'s).

Definition 14.8. Let $f : \mathbb{R}^n \to \mathbb{R}^m$ be a function. We say f has a *total derivative* at a point $c \in \mathbb{R}^n$ if

1. A linear map $A : \mathbb{R}^n \to \mathbb{R}^m$ exists such that:

2. The affine linear function defined by $L(c, x) = A(x - c) + f(c)$ (which depends on A) satisfies

$$\lim_{x \to c} \frac{\|f(x) - L(c, x)\|}{\|x - c\|} = 0$$

We again denote the linear map A as $Df(c)$.

Proposition 14.6 on uniqueness can be rewritten almost verbatim for Definition 14.8. In most of the rest of this chapter, we'll restrict to the special case $m = 1$. However, the chain rule—a singularly powerful and beautiful tool that will guide our proofs and application—shines most brightly in arbitrary dimensions. It says that the derivative of a function composition is the product of their total derivative matrices.

Definition 14.9. Let $f : \mathbb{R}^n \to \mathbb{R}^m$ and $g : \mathbb{R}^m \to \mathbb{R}^k$ be two functions with total derivatives, and let $Df(c)$ be the total derivative matrix of f at $c \in \mathbb{R}^n$ and $Dg(f(c))$ the matrix for g at $f(c) \in \mathbb{R}^m$. Suppose that Df and Dg are represented in the same basis for $\mathbb{R}^m$. Then the total derivative of $g \circ f$ at $c \in \mathbb{R}^n$ is the matrix product $Dg(f(c))Df(c)$.

Restated with fewer parentheses, if F is the total derivative matrix of f at c, and G is the total derivative of g at $f(c)$, then the total derivative of $g \circ f$ at c is GF.

This tidy theorem will be the foundation of our neural network application. It is not far off to say that all you need to train a neural network is "the chain rule with caching." However, we'll delegate the proof—it's admittedly technical and dull—to later in the chapter. We'll return to this in more depth in Section 14.9 when we define computation graphs for our neural network.

The chain rule is an extremely useful tool, and despite being abstract, it lands us within arms reach of our ultimate goal—easy derivative computations. In the next section, we'll see how finding these complicated matrices reduces to computing a handful of directional derivatives.

14.4 Computing the Total Derivative

Back to single-output functions, recall the total derivative at a point c is a linear map $A : \mathbb{R}^n \to \mathbb{R}$, where the domain represents deviations from c. If we want to compute a matrix representation, the obvious thing to do is choose a basis for which A is easy to compute. We'll do this, and arrive at a matrix representation for A (depending on c), by computing a small number of directional derivatives. First we'll show that the total derivative is closely related to directional derivatives.

Theorem 14.10. *Let $f : \mathbb{R}^n \to \mathbb{R}$ be a function with a total derivative at a point $c \in \mathbb{R}^n$. Let $\{v_1, \ldots, v_n\}$ be an orthonormal basis for $\mathbb{R}^n$, and recall that $\mathrm{Dir}(f, c, v)$ is the directional derivative of f at c in the direction of v. The matrix representation of the total derivative of f with respect to the basis $\{v_1, \ldots, v_n\}$ is the $1 \times n$ matrix*

$$\begin{pmatrix} \mathrm{Dir}(f, c, v_1) & \mathrm{Dir}(f, c, v_2) & \cdots & \mathrm{Dir}(f, c, v_n) \end{pmatrix}$$

Proof. The proof is a clever use of the chain rule. We prove it first for v_1 and the first component, but the same proof will hold if v_1 is replaced with any v_i. Fix a small $\varepsilon > 0$. Define by $g : [-\varepsilon, \varepsilon] \to \mathbb{R}^n$ the map $t \mapsto c + tv_1$. Then define $h(t) = f(g(t))$. We chose a "small ε" to ensure h is defined, which it only is if t is sufficiently close to c, but the proof doesn't depend crucially on the value of ε.

$$h : \mathbb{R} \xrightarrow{t \mapsto c + tv_1} \mathbb{R}^n \xrightarrow{f} \mathbb{R}$$

Note that h is a single-variable function $\mathbb{R} \to \mathbb{R}$, and that $h'(0) = Dh(0)$ is[3] exactly $\mathrm{Dir}(f, c, v_1)$, by definition of the directional derivative. Now we apply the chain rule to h, and we get that $Dh(0) = Df(c)Dg(0)$. As $Df(c)$ is a $1 \times n$ matrix, call $z_1, \ldots, z_n$ the unknown entries of $Df(c)$, written with respect to the basis $\{v_1, \ldots, v_n\}$. Also note that $Dg(0)$ can be written as an $n \times 1$ matrix with respect to the same basis (for the codomain of g):

---

[3] I'm implicitly identifying $h'(0)$ with the 1×1 matrix $Dh(0)$.

$$Dg(0) = \begin{pmatrix} 1 \\ 0 \\ \vdots \\ 0 \end{pmatrix}$$

The form of $Dg(0)$ is trivial: $t \mapsto tv_1$ has no coefficient of any other v_i but v_1. Combining these, $Dh(0) = Df(c) \cdot Dg(0)$ is the 1×1 matrix $[z_1]$, proving $z_1 = \text{Dir}(f, c, v_1)$. Doing this for each v_i instead of v_1 establishes the theorem.

□

The same proof can be adapted for functions $\mathbb{R}^n \to \mathbb{R}^m$, which we will explore in the exercises.

Theorem 14.10 provides two pieces of insight. The first is that the directional derivative wasn't so far off from the "right" definition. For "nicely behaved" functions, the total derivative and the directional derivatives agree. There's even a theorem that relates the two: if the directional derivative is continuous with respect to the choice of direction, then the directional derivative matrix from Theorem 14.10 is the total derivative.[4] That theorem implies that our initial counterexample (with a jump as the direction rotates) is the only serious obstacle to exclusively using directional derivatives for computation. This theorem is important enough that it deserves offsetting, despite our negligence in providing a proof, as one can say something slightly stronger.

Theorem 14.11. *Let $f : \mathbb{R}^n \to \mathbb{R}$ be a function and $c \in \mathbb{R}^n$ a point, and $\{v_1, \ldots, v_n\}$ an orthonormal basis. Suppose that for every basis vector v_i the directional derivative $\text{Dir}(f, c, v_i)$ is locally continuous in c, then f has a total derivative given by the matrix in Theorem 14.10.*

See the exercises for a deeper dive.

The second insight is that we can compute *any* directional derivative easily by first computing a small number of directional derivatives—one for each basis vector—and then simply projecting onto the direction of our choice. Recall that projecting one vector onto another is equivalent to taking an inner product, or, for projecting onto the subspace spanned by multiple vectors, to computing a matrix multiplication. When the codomain is $\mathbb{R}$, matrix multiplication just the standard inner product.

Speaking in terms of general bases is fine, and on occasion you'll find derivatives are easier to compute with a clever change of coordinates. However, it's usually easiest to use the same, simple basis: each basis vector is the standard basis vector for $\mathbb{R}^n$, and is denoted dx_i. This vector represents a change in a single input variable while leaving all others constant. If you have names for your variables, like $f(x, y, z) = x^2 y + \cos(z)$, then you would use dx, dy, and dz. When we do examples, we'll stick to using x_i and dx_i.

---

[4] The only proof I know of involves the mean value theorem, which we are not going to cover in this book. It's one of those subtle, technical theorems that happens to show up as a core technique for a lot of proofs. An exercise will recommend you investigate, but we won't explicitly use it.

The standard basis is so useful because it allows one to define an easy computational rule of thumb. For a directional derivative for basis vector dx_2, you may consider all variables except x_2 to be constants, and then apply the same rules for single-variable derivatives the function considered just as a function of x_2. If it helps, you can imagine a "curried" function $f(x_1, x_2, x_3) = f(x_1, x_3)(x_2)$, the former part of which closes over the fixed choices of values for x_1, x_3. The values of x_1, x_3 are fixed, but unknown at the time of derivative computation, and what's left is a single-variable function of x_2. As an example with $f(x_1, x_2, x_3) = x_1^2 x_2 + \cos(x_3)$, we have $\text{Dir}(f, c, dx_1) = 2c_1 c_2 + \cos(c_3)$. You will prove the mathematical validity of this rule in the exercises, but I suspect most readers have seen it and used it before.

The directional derivative along a standard basis vector—i.e., with respect to a single variable—has a special name: the *partial derivative* with respect to that variable. This is denoted using the ∂ sign (which I have always spoken "partial," but it sometimes called "boundary") as $\partial f/\partial x_2$, which is read, "the partial derivative of f with respect to x_2." In the same way that single variable derivatives f' are typically written in the same variables as f (i.e., using x instead of c), the example above can be written as $\partial f/\partial x_1 = 2x_1 x_2 + \cos(x_3)$. One refers to the *operation* of taking a partial derivative with respect to x by the function named $\frac{\partial}{\partial x}$, with the juxtaposition of the f in the numerator taking place of the standard parenthetical function application. Mathematicians have built up a hodgepodge of notations throughout history for this. In part, it's because parentheses are slow to write on a chalkboard—though they are easy for computers to parse, every new lisp programmer discovers they're hard for humans to read unless formatted just so. In part, it's because mathematicians don't always want to think of derivatives as functions. Sometimes they want to highlight a different aspect, such as the vector structure. A mess of lisp-y parentheses would not fit nicely in an inner product or summation.

When your chosen basis is the standard basis for each variable, the resulting total derivative matrix Df is called the *gradient* of f, and it's denoted ∇f. The symbol ∇ is often spoken "grad," and officially called a "nabla." We'll discuss the gradient in more detail below, because the gradient has some nice geometric properties that help in doing optimization.

An example gradient for the function $f(x_1, x_2, x_3) = x_1^2 x_2 + \cos(x_3)$ is as follows. Below I will write the matrix generically in the sense that it works for any choice of $c = (x_1, x_2, x_3)$, in the same way that when writing a single-variable derivative one uses the same variable before and after taking the derivative.

$$\nabla f = \begin{pmatrix} 2x_1 x_2 & x_1^2 & -\sin(x_3) \end{pmatrix}$$

With this, we can compute the directional derivative in the direction of a vector $v = (1, -1, 2)$ by applying the linear map ∇f.

$$\text{Dir}(f, x, v) = (\nabla f)(v) = \langle \nabla f, v \rangle$$

$$= \begin{pmatrix} 2x_1x_2 & x_1^2 & -\sin(x_3) \end{pmatrix} \cdot \begin{pmatrix} 1 \\ -1 \\ 2 \end{pmatrix}$$

$$= 2x_1x_2 - x_1^2 - 2\sin(x_3)$$

As (x_1, x_2, x_3) varies, this expression tracks the derivative of f in the direction of $(1, -1, 2)$ evaluated at (x_1, x_2, x_3). One can also slice it the other way, fixing a position to arrive at an expression that tracks the derivative of f at a specific position as the direction varies. Doing this for f above at $x = (1, 2, \pi/2)$, leaving the unit vector $v = (v_1, v_2, v_3)$ unspecified, we get

$$(\nabla f)(v)\,|_{(x_1,x_2,x_3)=(1,2,\pi/2)}$$

$$= \begin{pmatrix} 2x_1x_2 & x_1^2 & -\sin(x_3) \end{pmatrix} \cdot \begin{pmatrix} v_1 \\ v_2 \\ v_3 \end{pmatrix}$$

$$= \begin{pmatrix} 4 & 1 & -1 \end{pmatrix} \cdot \begin{pmatrix} v_1 \\ v_2 \\ v_3 \end{pmatrix}$$

$$= 4v_1 + v_2 + -v_3$$

Any way you slice it, the value we want is just one inner product away!

Many authors don't write the gradient as a vector in this way. Instead, they denote the basis vectors as dx_i, and the gradient is written as a single linear combination of these basis vectors. For the example f we've been using, it would be

$$\nabla f = 2x_1x_2\,dx_1 + x_1^2\,dx_2 - \sin(x_3)\,dx_3$$

This notation has the advantage that you can use it while still hating linear algebra: this is just the inner product written out before choosing values for v_1, v_2, v_3, i.e., the coefficients of dx_1, dx_2, dx_3 in the vector v to evaluate. It also helps you keep in mind that dx_i are meant to represent deviations of x_i from the point being evaluated. Sometimes they're written as a "delta", Δx_i or δx_i, since delta is commonly used to represent a change.[5] On the other hand, since it uses the symbols dx_i, it's easy to confuse the meaning with d/dx_i. We learned to love linear algebra. We'll stick to the vector notation.

Looking back, we now have exactly what we wanted: a way to compute directional derivatives as easily as taking single-variable derivatives. And now that we have a handle

---

[5] I find it curious how "delta" is used as a synonym for "difference" or "change" by executives in discussions that otherwise lack precision. Perhaps they studied math and incorporated that into their natural speech, or perhaps their faux-technical jargon impresses and confounds their enemies. I have certainly seen instances of both.

on the basic definition, we can study the geometry of the gradient to see how it enables optimization.

Henceforth, when we say "differentiable function" we mean a function with a total derivative, we'll assume all functions are differentiable, and we'll seamlessly swap between total derivatives, directional derivatives, linear maps, and matrices.

14.5 The Geometry of the Gradient

Take the gradient ∇f of a differentiable function $f : \mathbb{R}^n \to \mathbb{R}$, and evaluate it at a concrete point $x \in \mathbb{R}^n$, as we did at the end of Section 14.4. The result is an $n \times 1$ matrix whose entries are all concrete numbers, but since we're working with 1-dimensional outputs, the total derivative is also a vector. This vector represents the linear map $\mathbb{R}^n \to \mathbb{R}$ whose input is a "direction to look in" and whose output is how steep the derivative is in that direction. Since ∇f is derived from f, it's natural to ask how the geometry of ∇f relates to the shape of f.

The answer reveals itself easily with a strong grasp of the projection function from linear algebra. Recall the function $\text{proj}_v(w)$, which projects a vector w onto a unit vector v. We studied this in Chapters 10 and 12, and there we noted some interesting facts. Let's recall them here. Let v be a unit vector and w an arbitrary vector of the same dimension.

1. The standard inner product $\langle w, v \rangle$ is the *signed* length of $\text{proj}_v(w)$. The sign is positive if the result of the projection points in the same direction as v and negative if it points opposite to v.

2. If you project w onto v, and v is not on the same line as w, then $\|\text{proj}_v(w)\| < \|w\|$.

3. An alternate formula for $\langle v, w \rangle$ is $\|v\|\|w\|\cos(\theta)$, where θ is the angle between v and w. In the case that $\|v\| = 1$, the formula is $\|w\|\cos(\theta)$.

All of these point to the same general insight, which is a theorem with a famous name.

Theorem 14.12 (The Cauchy-Schwarz Inequality). *Let $v, w \in \mathbb{R}^n$ be vectors, and $\langle v, w \rangle$ the standard inner product. Then $|\langle v, w \rangle| \leq \|v\|\|w\|$, with equality holding if and only if v and w are linearly dependent.*

The Cauchy-Schwarz inequality has many, many proofs. I'll just share one that uses the cosine formula above to emphasize the geometry. You'll do a different proof in the exercises, and I'll gush over it in the Chapter Notes.

Proof. From $\langle v, w \rangle = \|v\|\|w\|\cos(\theta)$, and since $-1 \leq \cos(\theta) \leq 1$, it follows that $|\langle v, w \rangle| \leq \|v\|\|w\|$.

Because $\cos(\theta)$ repeats after $\theta = 2\pi$, we can restrict our attention to $0 \leq \theta < 2\pi$. For this range, $\cos(\theta) = 1$ if and only if $\theta = 0$, and $\cos(\theta) = -1$ if and only if $\theta = \pi$. For all other values, $\cos(\theta) < 1$. This proves the "if and only if" part of the theorem, because when $\cos(\theta) = \pm 1$, the two vectors lie on the same line, and hence are linearly dependent.

□

The details of this proof show more than the statement. Since the directional derivative is a projection of the gradient ∇f onto a unit vector v—i.e., $\langle (\nabla f)(x), v \rangle$—if you want to maximize the directional derivative, v should point in the same direction as $(\nabla f)(x)$. Said a different way, the gradient $(\nabla f)(x)$ points in the steepest possible direction.

Theorem 14.13. *For every differentiable function $f : \mathbb{R}^n \to \mathbb{R}$ and every point $x \in \mathbb{R}^n$, the gradient $(\nabla f)(x)$ points in the direction of steepest ascent of f at x.*

One is tempted to think this theorem is amazing (it is), but in fact it is not. With linear algebra we've created the perfect conditions for this theorem to be not only true, but trivial. We get further splendors for free: the direction of steepest ascent at c and the *level curve* of f (the set of constant-height inputs $\{(x, f(x)) : f(x) = f(c)\}$, like the topographic altitude lines on a map) are perpendicular to each other. This is simply because if v is a direction on the level curve, then the height of f doesn't change in that direction, so $0 = D(f, c, v) = \langle \nabla f, v \rangle$, and such inner products occur when two vectors are perpendicular.

Since many things in life and science can be modeled using functions $\mathbb{R}^n \to \mathbb{R}$, a common desire is to find an input $x \in \mathbb{R}^n$ which maximizes or minimizes such a function. For the sake of discussion, let's suppose we're looking for minima. Even when a mathematical model f exists for a phenomenon, minimizing it might be algebraically intractable for a variety of reasons. For example, it might involve functions that are difficult to separate, such as trigonometric functions and threshold functions. Alternatively, it might simply be so large as to avoid any human analysis whatsoever, as is often the case with a neural network that has millions of parameters related to labeled data. The rest of this chapter is devoted to understanding how to tackle such situations, and the core idea is to "follow" the direction indicated by the gradient.

14.6 Optimizing Multivariable Functions

Now we'll use the geometry of the gradient to derive a popular technique for optimizing functions $\mathbb{R}^n \to \mathbb{R}$. First, we review the situation for single-variable functions. In Chapter 9 we outlined the steps to solve a one-dimensional minimization problem, which I'll repeat here:

- Define your function $f : \mathbb{R} \to \mathbb{R}$ whose input x you control, and whose output you'd like to minimize. Select a range of interest $a \leq x \leq b$.

- Compute the values $a \leq x \leq b$ for which $f'(x) = 0$ or $f'(x)$ is undefined. These are called *critical points*.

- The optimal input x is the minimum value of $f(x)$ where x is among the critical points, or $x = a$ or $x = b$.

For multivariable inputs, you might reasonably expect an analogous technique to work: look at all the points x for which $(\nabla f)(x)$ is the zero vector, and check them all for optimality. Unfortunately the story is more complicated. There are still critical points—those values x for which $(\nabla f)(x)$ is the zero vector or undefined—but it's not as simple to enumerate them all and check which is the largest.

Take, for example, the function $f(x, y) = x^2 + y^2 + 2xy$. Its gradient is $(2x + 2y, 2y + 2x)$. Equating this to the zero vector results in an infinite family of solutions given by $x + y = 0$. In other words, while one-dimensional functions can be reduced to a discrete set of points to check, the solution to $\nabla f = 0$ can be a complicated surface. Even if you restrict just to polynomial equations life is still hard. There is an entire field of math, called *algebraic geometry*, dedicated to understanding the geometry of so-called *varieties*. A *variety* is the formal term for the space of solutions to a set of polynomial equations. The study of varieties is interesting and nuanced, beyond what can fit in this humble volume. Suffice it to say that understanding the shape of varieties from their defining formulas is not trivial, so we generally shouldn't expect to enumerate the zeros of the gradient.

If the equations are simple enough, one can apply a classical technique called *Lagrange multipliers* to compute optima. This was a central workhorse of a lot of pre-computer-era optimization. In general, Lagrange multipliers fail to help in almost every modern application, so we relegate it to the exercises. We'll instead focus on a more general algorithmic technique that works best when the function you're optimizing is intractable for pen-and-paper analysis. The technique is called *gradient descent*, and in modern times it has grown into a huge field of study.

Gradient descent (or gradient ascent, if you're maximizing) works as follows. Given f, start at a random point x_0. Iteratively evaluate the gradient $(\nabla f)(x_i)$, which points in the direction of steepest *ascent* of f, and set $x_{i+1} = x_i - \varepsilon(\nabla f)(x_i)$, where ε is some small scalar. The subtraction is the focus: you "take a small step" in the *opposite* direction of the gradient to get closer to a minimum of f. So long as the gradient is a reasonable enough approximator of f at each x_i, each $f(x_{i+1})$ is smaller than the $f(x_i)$ before it. Repeat this over and over again, and you should find a minimum of some sort.[6]

Gradient ascent intuitively makes sense, but there are a few confounding details that trick this algorithm into stopping before it reaches a minimum. The devil lies in the details of the stopping condition: if we're at a minimum, the gradient should definitely be the zero vector (there's no direction of ascent at all, so there's no "steepest" direction), but does it work the other way as well?

Definitely not. However, to get a useful feel for why, we have to correct an injustice from Chapter 8: we never discussed the geometry of the second derivative.

---

[6] Or decrease without bound, but in our application zero will be an absolute lower bound by design.

Curvature for Single Variable Functions

The derivative of a single variable function represents the slope of that function at a given point. One can further ask how higher derivatives (f'', $f^{(3)}$, $f^{(4)}$, etc.) correspond to the geometry of f. It turns out that higher derivatives correspond to certain sorts of curvature.

The second derivative is the example with the most common interpretations and theorems. Let $f : \mathbb{R} \to \mathbb{R}$ be a twice-differentiable function, and f'' its second derivative. Then the sign of $f''(x)$ at a given point x is called the *concavity* of f. Positive concavity implies the function is "curved upward" while negative concavity implies "curved downward." When $f''(x) = 0$, the case is a bit more complicated, but it often corresponds to the case where f is changing from having upward curvature to downward curvature, or vice versa.

Moreover, the magnitude of $f''(x)$ describes the "severity" of the curvature. $f(x) = x^2$ and $f(x) = 5x^2$ have different second derivatives at $x = 0$, and the latter is much more "sharply" curved upward.

It's worth noting that there are definitions of curvature that are much more precise and expressive than the second derivative. In fact, the second derivative has a number of shortcomings. In a concrete sense, it only captures "second-order" curvature of the function. So it sees no curvature in $f(x) = x^4$ at $x = 0$, despite that this function is very obviously concave up. The reason is that close to zero x^4 is also very close to zero, and so it makes the function quite flat in that region. Higher derivatives make up for the second derivative's failure, but as one can see just looking at a finite number of derivatives will never provide the whole story.[7] In other words, everything we'll say about the second derivative (and by extension, the Hessian below) will be a sufficiency test for a max/min, not a necessity test. We start with the presence of a local maximum or minimum.

For the sake of rigor I need to clarify what is meant by a *local* max (analogously, min). When I say any property for f holds *locally* at a point c, I mean that there is an interval (a, b) containing c, such that the property is true when f is restricted to (a, b). (a, b) may be very small if need be. In other words, it you "zoom in" to f at c, then the property is true as far as you can see.

To specifically say a point c, $f(c)$ is a *local minimum* of f means there is an interval (a, b) around c for which $f(c) < f(x)$ for all $x \in (a, b)$. In the example function in Figure 14.6, $f(x) = \frac{1}{2}(x-1)^2(x-4)(x+2)$, a sufficiently small interval around $x = 1$ proves that f has a local max at $(1, 0)$, and likewise a local minimum close to $(3, -10)$.

Now we can prove the theorem that concavity is sufficient to detect a local min/max.

Theorem 14.14. *Let $f : \mathbb{R} \to \mathbb{R}$ be a twice-differentiable function and $c \in \mathbb{R}$ be a value for which $f'(c) = 0$. If $f''(c) < 0$, then f has a local maximum at x. If $f''(c) > 0$, then f has a local minimum at c.*

Proof. The Taylor series is our hammer. Since $f'(c) = 0$, near c we can expand $f(x)$ using a Taylor series that primarily depends on $f''(x)$.

---

[7] As we saw in Chapter 8, there are nonzero functions so flat at a point that *all of their derivatives are zero!*

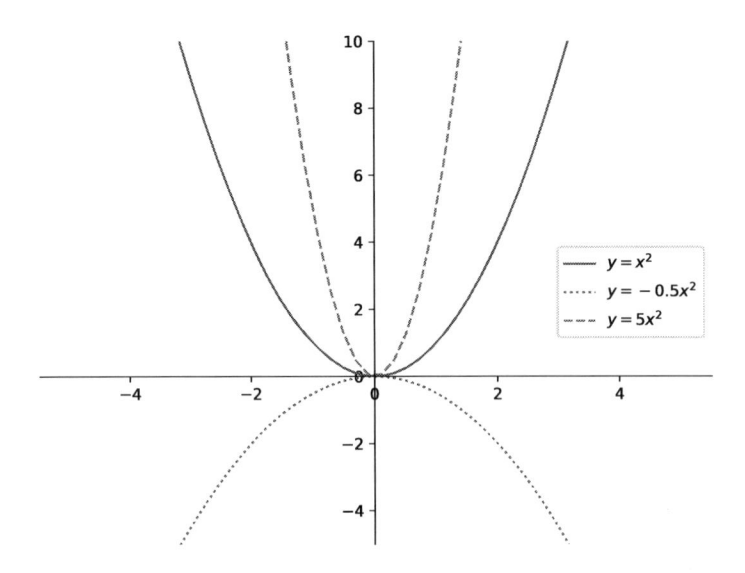

Figure 14.5: Examples of functions with different concavity.

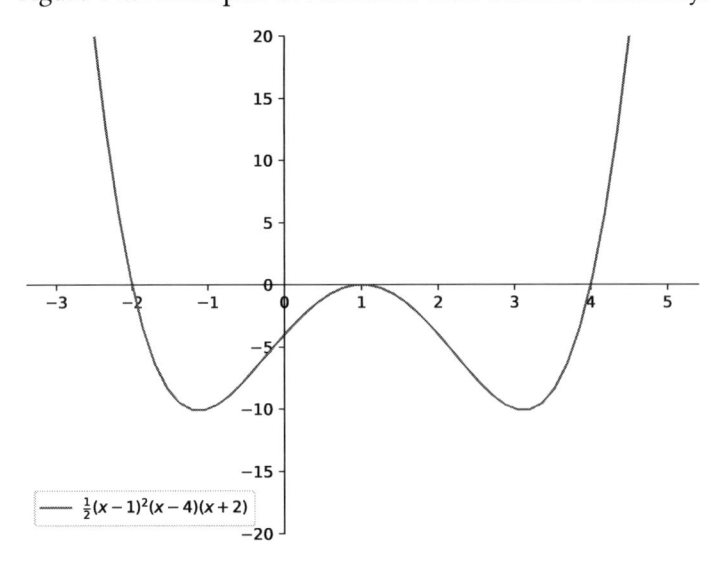

Figure 14.6: An example of a function with a local max at $x = 1$.

$$f(x) = f(c) + \frac{f''(c)}{2}(x-c)^2 + r(x)$$

Here $r(x)$ is the remainder term of the Taylor Theorem (Theorem 8.14). It's a degree-3 polynomial in $x - c$ whose coefficient depends on an evaluation of $f^{(3)}(z)$ at some unknown point $z \in (c, x)$. The most important detail of this is that it's a degree-3 polynomial, but in complete detail, it's

$$r(x) = \frac{f^{(3)}(z)}{6}(x-c)^3 \qquad \text{for some unknown } z \text{ between } c \text{ and } x.$$

We need to argue that because $x - c$ is very small when x is close to c, the value of $(x-c)^3$ is dwarfed by the value of $(x-c)^2$, so that the min/max behavior of f is determined solely by the $(x-c)^2$ term.

Indeed, if you could informally argue that—say, by erasing $r(x)$ with reckless abandon—then $f(x)$ would be a simple, shifted parabola. The sign of $f''(x)$ would dictate whether the curve is concave up or concave down, and the peak would obviously be a min or a max (respectively). To make it more rigorous, we restrict ourselves to a small interval.

Let's suppose that $f''(x) > 0$, so that we need to show $f(c)$ is a local min. In this case we want an interval (a, b) on which $f(c) \leq f(x)$ for all x. Rearranging the formula above,

$$f(c) = f(x) - \frac{f''(c)}{2}(x-c)^2 - r(x).$$

If the term $[-\frac{f''(c)}{2}(x-c)^2 - r(x)]$ is not positive on (a, b), then $f(c) \leq f(x)$. So the theorem will be proved if we can find an interval on which that term is at most zero. Rearranging, we need the following inequality to hold:

$$(x-c)^2 \geq \frac{2f^{(3)}(z)}{6f''(c)}(x-c)^3$$

Since the value of $r(x)$ depends on z (which can be different for different values of x), we can't proceed unless we eliminate the dependence on z. We'll do that by estimating, i.e., replacing $f^{(3)}(z)$ with the max of $f^{(3)}$ over an interval. So start with some fixed interval around c, say $(c - 0.01, c + 0.01)$,[8] and let M be the maximum value of $f^{(3)}(z)/(3f''(c))$ on that interval. I.e., M is the largest value of the coefficient of $(x-c)^3$ in the above inequality that can occur close to c. Then we need to find an interval, perhaps smaller than $(c - 0.01, c + 0.01)$, for which the following (simplified) inequality is true for all x in that interval.

---

[8] All we need is any interval on which f is defined and has no pathological or discontinuous behavior. This is guaranteed to exist because f is differentiable at c. To be completely rigorous one should use $(c - \varepsilon, c + \varepsilon)$ and argue existence of such by continuity/differentiability, but you get the point.

$$(x-c)^2 \geq M(x-c)^3$$

But this is easy! So long as $x \neq c$ we can simplify to see we just need a small enough interval that ensures $(x - c) \leq 1/M$. This will be true of either $(c - 1/M, c + 1/M)$ or $(c - 0.01, c + 0.01)$, whichever is smaller.

□

That was a lot of work to achieve a proof. Recalling our discussion of waves in Chapter 12, the reader might begin to understand why a working physicist would rather erase terms with reckless abandon than wade through the strange existential z's that plague Taylor series. However, as was the case with matrix algebra providing an elegant (though intentionally leaky) abstraction for linear maps, mathematical analyses like these have their own abstractions to aid computation while maintaining rigor. In this case, most programmers are aware of it: big-O notation. We'll display its use in Chapter 15.

When $f''(x) = 0$, we can't conclude anything. f might have a max/min, or it might have neither. One example of having neither is $f(x) = x^3$ at $x = 0$. The function switches concavity from concave down to concave up, but f has no local max or min.

The idea of "local" behavior is a powerful one across mathematics. It is almost always easier to talk about local properties of an object rather than the global structure. A lot of time is spent investigating how a collection of unrelated bits of local information affect a global property. For single variable functions, one incarnation of this is that the local mins and maxes of f—along with a slight amount of extra information—determines the global min/max of f.

One can also think of a directional derivative as a sort of "local" property. It's the derivative when one "only looks" in a certain window, while the total derivative is global. If you can show that each directional derivative is continuous—or even just that the partial derivatives are continuous—then you automatically get the global (total) derivative. You have built global structure out of local pieces. Of course, the total derivative at a point is also a local construct from a different perspective. The total derivative describes the approximate structure of f at a point, and with enough information about the total derivative at every point of f (and a few bits of extra information), you can completely reconstruct f. So there are multiple scales of locality that allow one to discuss local and global properties, and how they relate to each other.

The Hessian

For multivariable functions, locality replaces an interval with an "open ball," i.e., a set $B_r(c)\{x : \|x - c\| < r\}$, which consists of all the points within a given radius of the point in question. The radius takes the place of the length of the interval to say "how local" you're looking.

While there are still local maxes and mins of the obvious sort, there are many ways a local min/max can fail to exist. These are called *saddle points*. The shape of these is quite literal: the surface looks like the saddle of a horse, or the shape of a potato chip, in

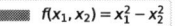

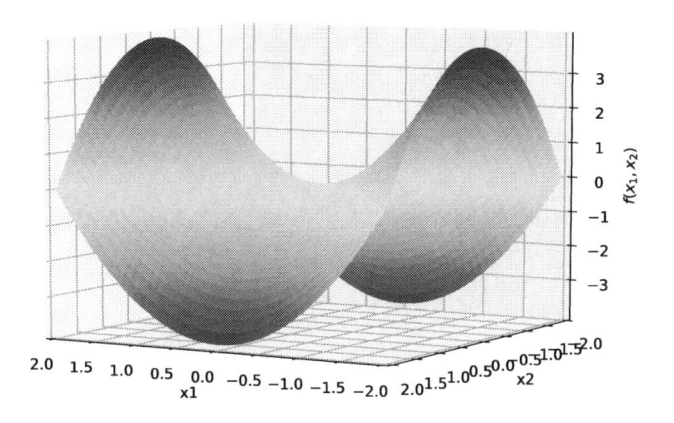

Figure 14.7: An example of a function with a saddle point.

which the curvature goes up along one direction and down along another. A prototypical example of a curve with a saddle point is $f(x, y) = x^2 - y^2$, pictured in Figure 14.7.

With many variables comes many different directions along which curvature can differ. You might imagine a function with 5 variables, each axis giving two choices of up-curvature or down-curvature, for a total of $2^5 = 32$ different kinds of saddles (including the normal max/min). The way to get a handle on these forms is to look at the matrix of all ways to take second derivatives. First we define notation for second derivatives.

Definition 14.15. Let $f : \mathbb{R}^n \to \mathbb{R}$ be a function which has first partial derivatives for every variable (recall, denoted $\partial f / \partial x_i$). The *second (or mixed) partial derivative* with respect to x_i and x_j is the partial derivative of the partial derivative. A compact notation for this is

$$\frac{\partial^2 f}{\partial x_i \partial x_j} = \frac{\partial}{\partial x_j}\left(\frac{\partial f}{\partial x_i}\right)$$

If $i \neq j$, the derivative is called a *mixed partial*. If $i = j$ we write $\partial^2 f / \partial x_i^2$.

Personally I hate this notation, particularly how arbitrarily it's defined so that the "numerator" of the variable names are smushed together. My inner programmer cries out in anguish, because it's breaking algebra and functional notation at the same time by pretending they're the same. Are we taking the squared derivative with respect to a squared variable? Multiplying the top and bottom of a function name separately? Your syntactic sugar is rotting my brain! Alas, the notation is widespread, and the only alternative I know of, $f_{x_i x_j}(x) = \frac{\partial^2 f}{\partial x_i \partial x_j}$, is not all that much better.

One might expect the mixed partials with respect to x_i, x_j and x_j, x_i to be different due to the order of the computation. When f has a total derivative, they turn out to be the same.

Theorem 14.16 (Schwarz's theorem). *Let $f : \mathbb{R}^n \to \mathbb{R}$ be a function. Suppose that all of f's partial derivatives exist and themselves have partial derivatives. Then for every i, j, it holds that $\frac{\partial^2 f}{\partial x_i \partial x_j} = \frac{\partial^2 f}{\partial x_j \partial x_i}$.*

We quote this theorem without proof, but notice that, in addition to reducing our computation duties by a half, it gives a hindsight rationalization for the fraction notation. If the order of partial derivatives doesn't matter, then we need not bother with the functional notation that emphasizes order precedence.

Next we define the Hessian, which is the matrix of mixed partial derivatives of a function.

Definition 14.17. Let $f : \mathbb{R}^n \to \mathbb{R}$ be a function which is twice differentiable. Define the *Hessian* of f, denoted $H(f)$ (or often, when f is fixed, just H), as an $n \times n$ matrix whose i, j entry is $\partial^2 f / \partial x_i \partial x_j$.

$$H = \begin{pmatrix} \frac{\partial^2 f}{\partial x_1^2} & \frac{\partial^2 f}{\partial x_1 \partial x_2} & \cdots & \frac{\partial^2 f}{\partial x_1 \partial x_n} \\ \frac{\partial^2 f}{\partial x_2 \partial x_1} & \frac{\partial^2 f}{\partial x_2^2} & \cdots & \frac{\partial^2 f}{\partial x_2 \partial x_n} \\ \vdots & \vdots & \ddots & \vdots \\ \frac{\partial^2 f}{\partial x_n \partial x_1} & \frac{\partial^2 f}{\partial x_n \partial x_2} & \cdots & \frac{\partial^2 f}{\partial x_n^2} \end{pmatrix}$$

Just like the gradient, $H(f)$ is really a function whose input is a point x in the domain of f, and the output is the matrix $H(f)(x)$. The notation gets even hairier since $H(f)(x)$ is itself a linear map $\mathbb{R}^n \to \mathbb{R}^n$. In an exercise you'll interpret this linear map to make more sense of it.

Because of Schwarz's theorem, any point x we use to make $H(f)$ concrete produces a real symmetric matrix. As we know from Chapter 12, symmetric matrices have an orthonormal basis of real eigenvectors with real eigenvalues, and so we can ask what these eigenvalues tell us about the structure of f local to x. The theorem is a nice generalization of the min/max structure for single variable functions.

Theorem 14.18. *Let $f : \mathbb{R}^n \to \mathbb{R}$ be a function which is twice differentiable, let $x \in \mathbb{R}^n$ be a point, and let H be the Hessian of f at x. If all the eigenvalues of H are positive, then f has a local min at x. If all the eigenvalues are negative, then f has a local max at x. If H has both positive and negative eigenvalues (and no zero eigenvalues), then f has a saddle point at x.*

We'll skip the proof for brevity, but our understanding of eigenvalues and eigenvectors provides a tidy interpretation. The eigenvectors of nonzero eigenvalues correspond to the directions (when looking from x) in which the curvature of f is purely upward or

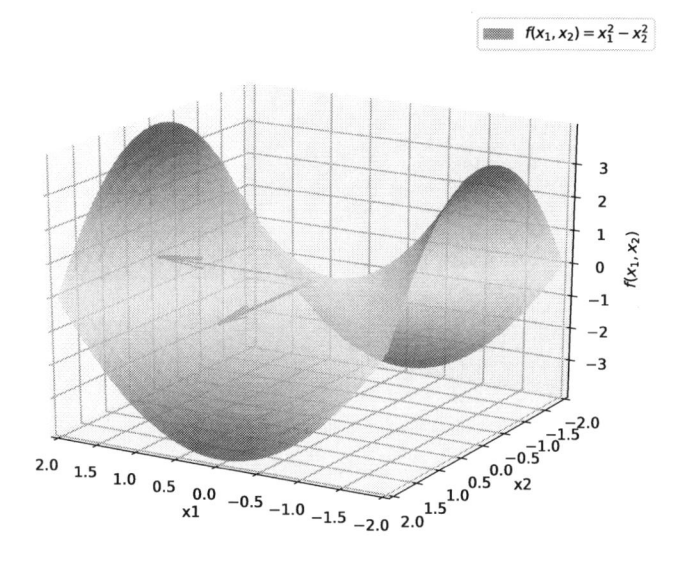

Figure 14.8: A function with a saddle point. The eigenvectors of the Hessian at the saddle point are shown as arrows, and represent the maximally positive and negative curvatures at the saddle point.

downward, and maximally so. In a sense that can be made rigorous, because H has an orthonormal basis of eigenvectors, these curvatures "don't interfere" with each other. If one has an ellipsoidal bowl, the eigenvectors correspond to the "axes" of the bowl. For a saddle point, the eigenvectors are the directions of the saddle that are parallel and perpendicular to the imagined horse's body. This is shown in Figure 14.8.

Of course, all of this breaks down if the sort of curvature we're looking at can't be captured by second derivatives. There might be an eigenvalue of zero, in which case you can't tell if the curvature is positive, negative, or even completely flat.

But this raises a natural question: if the gradient gives you first derivative information, and the Hessian gives you second derivative information, can we get third derivative information and higher? Yes! And can we use these to form a sort of "Taylor series" for multivariable functions? More yes! One difficulty with this topic is the mess of notation. A fourth-derivative-Hessian analogue is a four-dimensional array of numbers. With more dimensions comes more difficulty of notation (or the need for a better abstraction). Nevertheless, we can at least provide the analogue of the Taylor series for the first two terms:

Theorem 14.19. *Let $f : \mathbb{R}^n \to \mathbb{R}$ be a twice differentiable function. Let $x \in \mathbb{R}^n$ be a point and $v \in \mathbb{R}^n$ be a small nonzero vector (a deviation direction from x). Let ∇f be the gradient of f at x, and H the Hessian at x. Then we have the following approximation:*

$$f(x+v) \approx f(x) + \langle \nabla f, v \rangle + \langle Hv, v \rangle$$

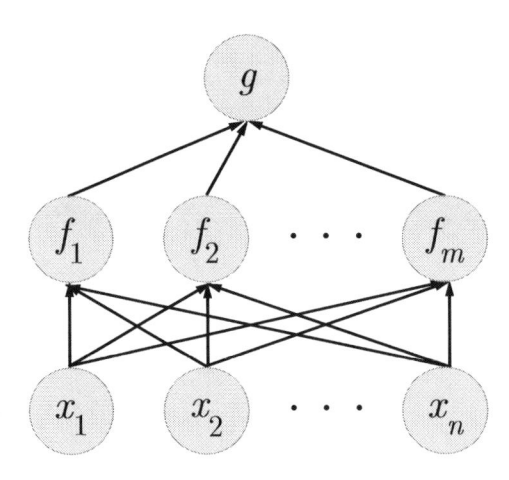

Figure 14.9: The dependence of $g \circ f$ on each x_i contains paths through each of the f_j.

See the exercises for a deeper investigation when $n = 2$.

14.7 The Chain Rule: a Reprise and a Proof

We return again to the chain rule for multivariable functions. Recall the formula for single-variable functions $f, g : \mathbb{R} \to \mathbb{R}$, the chain rule says that the derivative of $f(g(x))$ involves evaluating f' at g, and multiplying the result by g'. I.e., $\frac{d}{dx}(f(g(x))) = f'(g(x))g'(x)$.

Our analogous formula for multivariable functions involves a matrix multiplication:

$$D(g \circ f)(c) = Dg(f(c))Df(c).$$

Let's first think about why this should be harder in principle than the single variable case. Call $x = (x_1, \ldots, x_n)$ the variables input to $f = (f_1, \ldots, f_m)$, a function $\mathbb{R}^n \to \mathbb{R}^m$. The derivative of $g \circ f$ measures how much g depends on changes to each x_i. But while f depends on an input x_i in a straightforward way, g depends on x_i transitively through the possibly many outputs of f. Computing $\partial g / \partial x_i$ should require one to combine the knowledge of $\partial f_j / \partial x_i$ for each j, and that combination might be strange. The function $g \circ f$ is mapped out by a dependency graph like in Figure 14.9, where the arrows $a \to b$ indicate that b depends on a. A similar dependence describes dependence among the partial derivatives.

Luckily the relationship is quite elegant: for one dependent variable you multiply along each branch and sum the results. Doing this for every input variable produces exactly the matrix multiplication that makes up the chain rule. We'll prove a slightly simpler version of the chain rule where g has only one output, which has all the necessary features of the more general proof where $g = (g_1, \ldots, g_k)$ is vector-valued.

Theorem 14.20. *Let $g : \mathbb{R}^m \to \mathbb{R}$ and $h : \mathbb{R}^n \to \mathbb{R}^m$ be differentiable functions. Write $h = (h_1, \ldots, h_m)$, with $h_i = h_i(x)$ for $x = (x_1, \ldots, x_n)$, and $g(y)$ with $y = (y_1, \ldots, y_m)$. Then $g(h(x)) = g(h_1(x), \ldots, h_m(x))$ is differentiable, and the gradient at $c \in \mathbb{R}^n$ is*

$$\frac{\partial g}{\partial x_1}(c) = \sum_{i=1}^m \frac{\partial g}{\partial h_i}(h(c)) \cdot \frac{\partial h_i}{\partial x_1}(c)$$

The other components of the gradient are defined by replacing x_1 with x_j.

Proof. For clarity, in this proof the boldface **v** will denote a vector of numbers or functions (a function with multiple outputs). Denote by $\mathbf{h}(\mathbf{x}) = (h_1(\mathbf{x}), \ldots, h_m(\mathbf{x}))$, so that we can conveniently abbreviate $g(h_1(\mathbf{x}), \ldots, h_m(\mathbf{x}))$ as $g(\mathbf{h}(\mathbf{x}))$. Let H be the matrix representation of the total derivative of **h**,

$$H = \begin{pmatrix} - & H_1 & - \\ - & H_2 & - \\ & \vdots & \\ - & H_m & - \end{pmatrix}$$

Let G be the matrix representation of the total derivative of g (i.e., ∇g). The claimed total derivative matrix for $g(\mathbf{h}(\mathbf{x}))$ is the matrix multiplication GH. This results in the formula claimed by the theorem. We need to show that GH satisfies the linear approximation condition for $g(\mathbf{h}(\mathbf{x}))$, i.e., that

$$\lim_{\mathbf{x} \to \mathbf{c}} \frac{g(\mathbf{h}(\mathbf{x})) - g(\mathbf{h}(\mathbf{c})) - GH(\mathbf{x} - \mathbf{c})}{\|\mathbf{x} - \mathbf{c}\|} = 0$$

We start with a convenient change of variables. Define $\mathbf{t} = \mathbf{x} - \mathbf{c}$, and then we can see that the limit above can equivalently be written in terms of a vector **t** as $\mathbf{t} \to \mathbf{0}$.

$$\lim_{\mathbf{t} \to \mathbf{0}} \frac{g(\mathbf{h}(\mathbf{c} + \mathbf{t})) - g(\mathbf{h}(\mathbf{c})) - GH(\mathbf{t})}{\|\mathbf{t}\|}$$

Now we define two functions that track the error of the linear approximators. More specifically, the first function represents the error of H as a linear approximator of **h** at **c**, and the second is the error of G as a linear approximator of g at $\mathbf{h}(\mathbf{c})$.

$$\mathbf{err}_H(\mathbf{t}) = \mathbf{h}(\mathbf{c} + \mathbf{t}) - \mathbf{h}(\mathbf{c}) - H(\mathbf{t})$$
$$\mathrm{err}_G(\mathbf{s}) = g(\mathbf{h}(\mathbf{c}) + \mathbf{s}) - g(\mathbf{h}(\mathbf{c})) - G(\mathbf{s})$$

Note that in err_G, the vector **s** is in the domain of g, while in $\mathbf{err}_H$ the vector **t** is in the domain of the h_i. We can use these formulas to simplify the limit above. Substitute for $\mathbf{h}(\mathbf{c} + \mathbf{t})$ a rearrangement of the definition of $\mathbf{err}_H$, getting

$$\lim_{\mathbf{t} \to \mathbf{0}} \frac{g\Big(\mathbf{h}(\mathbf{c}) + H(\mathbf{t}) + \mathbf{err}_H(\mathbf{t})\Big) - g(\mathbf{h}(\mathbf{c})) - GH(\mathbf{t})}{\|\mathbf{t}\|}$$

Define $\mathbf{s} = H(\mathbf{t}) + \text{err}_H(\mathbf{t})$, so that we can substitute $g(\mathbf{h}(\mathbf{c}) + \mathbf{s})$ using a rewriting of the definition of err_G.

$$\lim_{\mathbf{t} \to 0} \frac{g(\mathbf{h}(\mathbf{c})) + G(\mathbf{s}) + \text{err}_G(\mathbf{s}) - g(\mathbf{h}(\mathbf{c})) - GH(\mathbf{t})}{\|\mathbf{t}\|}$$

Expand $\mathbf{s}$, apply linearity of G, and cancel opposite terms, to reduce the limit to

$$\lim_{\mathbf{t} \to 0} \frac{G(\text{err}_H(\mathbf{t})) + \text{err}_G(\mathbf{s})}{\|\mathbf{t}\|}.$$

To show this limit is zero, we split it into two pieces. The first is

$$\lim_{\mathbf{t} \to 0} \frac{G(\text{err}_H(\mathbf{t}))}{\|\mathbf{t}\|}.$$

Note that because G is a gradient, $G(\text{err}_H(\mathbf{t}))$ is an inner product—the projection of $\text{err}_H(\mathbf{t})$ onto a fixed vector, $\nabla g(\mathbf{c})$. The Cauchy-Schwarz inequality informs us that the norm of $G(\text{err}_H(\mathbf{t}))$ is bounded from above by $C \text{err}_H(\mathbf{t})$, where $C = \|\nabla g(\mathbf{c})\|$ is constant. So the limit above is

$$C \lim_{\mathbf{t} \to 0} \frac{\text{err}_H(\mathbf{t})}{\|\mathbf{t}\|} = 0.$$

This goes to zero because (by the definition of err_H) it's the defining property of the total derivative of H. It remains to show the second part is zero:

$$\lim_{\mathbf{t} \to 0} \frac{\text{err}_G(\mathbf{s})}{\|\mathbf{t}\|}$$

We would like to bound this limit from above by a different limit we can more easily prove goes to zero. Indeed, if there were a constant B for which

$$\frac{\text{err}_G(\mathbf{s})}{\|\mathbf{t}\|} \leq \frac{\text{err}_G(\mathbf{s})}{B\|\mathbf{s}\|}$$

Then we'd be done: $\mathbf{s} \to \mathbf{0}$ if and only if $\mathbf{t} \to \mathbf{0}$, due to how $\mathbf{s}$ is defined in terms of $\mathbf{t}$, and $\text{err}_G(\mathbf{s})/\|\mathbf{s}\| \to 0$ again by the definition of the total derivative of g. Expanding $\mathbf{s} = H(\mathbf{t}) + \text{err}_H(\mathbf{t})$ and again expanding $\text{err}_H(\mathbf{t})$, the needed B occurs when

$$\frac{1}{B} \geq \left\| \frac{\mathbf{h}(\mathbf{c} + \mathbf{t}) - \mathbf{h}(\mathbf{c})}{\|\mathbf{t}\|} \right\|$$

The quantity on the right hand side is a directional derivative of $\mathbf{h}$ (rather, a vector of directional derivatives), and for sufficiently small $\mathbf{t}$, the quantity is no larger than twice the largest possible directional derivative, i.e., $2\|(\nabla h_1(\mathbf{c}), \ldots, \nabla h_m(\mathbf{c}))\|$. Choose B so that $1/B$ is larger than this quantity, and the proof is complete.

□

This was the most difficult proof in this book. And it's easy to get lost in it. We started from a relatable premise: find a formula for the chain rule for multivariable functions. To prove our formula worked, we reduced progressively trickier and more specialized arguments, boiling down to an arbitrary-seeming upper bound of a haphazard limit of an error term of a linear approximation.

To be sure, the steps in this proof were not obvious. One has to take a bit of a leap of faith to guess that GH was the right formula (though it is the simplest and most elegant option), and then jump from an obtuse limit to the realization that, if one writes everything in terms of error terms, the hard parts (g composed with **h**) will cancel out. Suffice it to say that this proof was distilled from hard work and many examples, and it leaves a taste of mystery in the mouth. Until, that is, one dives deeper into the general subfield of mathematics known as "analysis," where arguments like this one are practiced until they become relatively routine. One gains the nose for what sorts of quantities should yield their secrets to a well-chosen upper bound. Contrast this to subjects like linear algebra and abstract algebra (Chapter 16), in which pieces largely tend to fit together in a structured manner that—in my opinion—tends to appeal to programmers in a way that analysis doesn't. Another demonstration of subcultures in mathematics.

14.8 Gradient Descent: an Optimization Hammer

As we mentioned, the Hessian provides a sufficient condition to determine if a point is a local min: the gradient is zero and all the eigenvalues of the Hessian are positive. There are two caveats to this. First, the Hessian is expensive to compute. It's size is the square of the size of the gradient. Second, a provable optimum is something of a luxury. Most optimization problems benefit just well enough—a sort of 80% of the gain from 20% of the work—from being able to progressively improve existing solutions. Gradient descent does precisely this, and allows you to easily trade off solution quality for runtime.

Informally, recall gradient descent is the process: "go slowly in the opposite direction of the gradient until the gradient is zero." More formally, choose a stopping threshold $\varepsilon > 0$ and a learning rate $\eta > 0$, and loop as follows.

1. Start at some position $x = x_0$ (often a randomly chosen starting point).

2. While $\|(\nabla f)(x)\| > \varepsilon$:

    a) Update $x = x - \eta(\nabla f)(x)$.

3. Output x.

This algorithm can be fast or slow depending on the choice of the starting point and the smoothness of f. If x lands in a bowl, it will quickly find the bottom. If x starts on a plateau of f, it will never improve. For this reason, one often runs multiple copies of this loop, and outputs the most optimal run. If the inputs are chosen randomly, there's a good chance one avoids the avoidable plateaus.

The bottleneck of gradient descent is computing the gradient. When f is complicated, such as in a neural network, efficient use of the chain rule is the primary tool for making gradient computations manageable. The rest of this chapter is dedicated to doing exactly that.

One might wonder, if Hessian gives more information about the curvature of f, why not use the Hessian in determining the next step to take. You can! But unfortunately, since the Hessian is often an order of magnitude more difficult to compute than the gradient—and the gradient *already* requires mountains of engineering to get right—it's simply not feasible to do so. And, as you'll get to explore in the exercises, there are alternative techniques that allow one to "accelerate" gradient descent in a principled fashion without the Hessian.

14.9 Gradients of Computation Graphs

Because the chain rule is an enormous formula, there are some appropriate abbreviations. One often omits the function evaluations, so that one can see the alternating pattern of numerators and denominators:

$$\frac{\partial g}{\partial x_1} = \sum_{i=1}^{k} \frac{\partial g}{\partial h_i} \frac{\partial h_i}{\partial x_1}.$$

In more generality, one will often have a function which depends on some input parameter transitively through many layers of functions. To compute these often requires a long "chain" of partial derivatives. Ignoring that there are sums involved (or assuming only one dependency branch), you'd get chains like this:

$$\frac{\partial f}{\partial x} = \frac{\partial f}{\partial g} \frac{\partial g}{\partial h} \frac{\partial h}{\partial i} \frac{\partial i}{\partial j} \frac{\partial j}{\partial x}.$$

So if you were doing an on-paper analysis of some complex function, you'd generally break it up into parts. More useful for software is to observe that the terms in such a big product can be grouped and re-grouped arbitrarily. For example, if you've already computed $\frac{\partial g}{\partial j}$, then to get $\frac{\partial f}{\partial x}$ you need only compute the missing terms

$$\frac{\partial f}{\partial x} = \frac{\partial f}{\partial g} \frac{\partial g}{\partial j} \frac{\partial j}{\partial x}.$$

This allows one to use caching to avoid recomputing derivatives over and over again. That's especially useful when there are many dependency branches. In fact, as we'll realize concretely when we build a neural network, the concept of derivatives with branching dependencies is core to training neural networks. To prepare for that, we'll describe the abstract idea of a computation graph and reiterate how the chain rule is computed reursively through such a network.

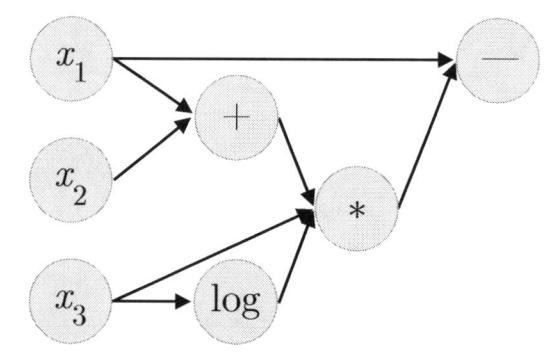

Figure 14.10: A computation graph. Each node N is an input or some mathematical operation on the outputs of dependent nodes feeding into N.

Definition 14.21. Let $G : \mathbb{R}^n \to \mathbb{R}$ be a function. A *computation graph* for G is a directed, acyclic graph[9] (V, E) with the following properties.

1. There is a set of n vertices identified as *input vertices*.

2. Each non-input vertex $v \in V$ has an associated integer $k_v \in \mathbb{N}$ (the number of inputs) and a function $f_v : \mathbb{R}^{k_v} \to \mathbb{R}$.

3. Each non-input vertex v has exactly k_v directed edges with target v.

4. There is exactly one vertex $v \in V$ with no outgoing edges designated as the *output vertex*.

If there's an edge (v, w), we say that v is an *argument* to w and that w *depends on* v.

A computation graph represents the computation of G by first picking operations at each vertex, then specifying the dependencies of those operations, and adding vertices for the input. "Evaluating" a computation graph at a particular input is the obvious computational process of setting "values" for the input vertices, and following the operations of the graph to produce an output. Such a graph is a circuit in which each "gate" corresponds to the function of your choice.

For us, the operations f_v at each vertex will always be differentiable (with one caveat), and hence G will be differentiable, though the definition of a computation graph doesn't require differentiability.

Now we'll reiterate the chain rule for an arbitrary computation graph. Saw we have a programmatic representation of a computation graph for G, and somewhere deep in the graph is a vertex with operation $f(a_1, \ldots, a_k)$. We want to compute a partial derivative

---

[9] Recall, a directed edge $e = (v, w)$ is said to have *source* v and *target* w, and represents a dependency of w on v. A graph is acyclic if it contains no cycles, i.e., no circular dependencies.

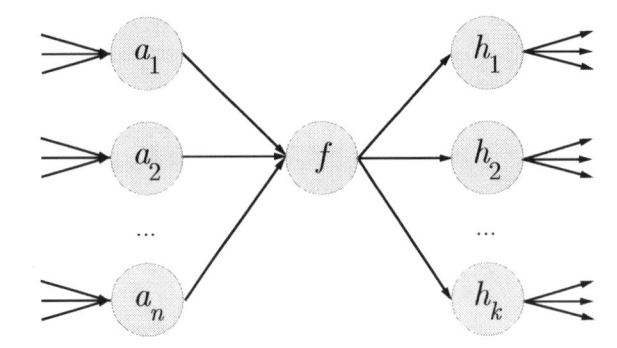

Figure 14.11: A generic node of a computation graph. Node f has many inputs, its output feeds into many nodes, and each of its inputs and outputs may also have many inputs and outputs.

of G with respect to an input variable that may be even deeper than f. Using the chain rule, we'll describe the algorithm for computing the derivative generically at any vertex and then apply induction/recursion. More specifically, at vertex f we'll compute $\partial G/\partial f$ and multiply it by $\partial f/\partial a_i$ to get $\partial G/\partial a_i$.

So given a vertex with operation $f(a_1, \ldots, a_n)$, argument vertices $a_1, \ldots, a_n$, and whose output is depended on by vertices $h_1, \ldots, h_k$. We're interested in computing $\partial G/\partial a_1$ (with the other arguments a_i to f being analogous). This is illustrated in Figure 14.11.

We know $\partial f/\partial a_1$ by assumption, having designed the graph so the gradient ∇f_v of each vertex v is easy to compute. By induction, for each output vertex h_j we can compute $\partial G/\partial h_j$. Then apply the chain rule:

$$\partial G/\partial f = \sum_{i=1}^{k} \frac{\partial G}{\partial h_i} \cdot \frac{\partial h_i}{\partial f}.$$

Once we have that, each $\partial G/\partial a_i = (\partial G/\partial f) \cdot (\partial f/\partial a_i)$, as desired. Note that if a_i has another path to G, they need to be summed.

Because we use the vertices that depend on f as the inductive step, the base case is the output vertex, and there $\partial G/\partial G = 1$. Likewise, the top of the recursive stack are the input vertices, and at the end we'll have $\partial G/\partial x_i$ for all inputs x_i.

As one can easily see, a network with heavily interdependent vertices requires one to cache the intermediate values to avoid recomputing derivatives everywhere. That's exactly the strategy we'll take with our neural network.

14.10 Application: Automatic Differentiation and a Simple Neural Network

Neural networks are extremely popular right now. In the decade between 2010 and 2020, neural networks—specifically "deep" neural networks—have transformed subfields of computer science like computer vision and natural language processing. Neural networks and techniques using them can, with rather high fidelity, identify objects and scenes, translate simple language, and play abstract games of logic like Go. This was enabled, in large part, by the increased availability of cheap compute resources and graphical processing units (GPUs).

Perhaps surprisingly, the mathematical techniques that are used to train these networks are largely the same as they were decades ago. They are all variations on gradient descent, and the specific instance of gradient descent applied to training neural networks is called *backpropagation*.

In this section, we'll implement a neural network from scratch and train it to classify handwritten digits with relatively decent accuracy. Along the way, we'll get a taste for the theory and practice of machine learning.

Machine Learning is All About the Data

Machine learning is the process of using data to design a program that performs some task. A prototypical example is classifying handwritten digits: you want a function which, given as input the pixels of an image of a handwritten digit, produces as output the digit in the picture. To solve such a problem, ignoring issues of engineering maintenance over time, you need a broad recipe of three steps:

1. Collect a large sample of handwritten digits, and clean them up (as all programmers know, we must sanitize our inputs!).

2. Get humans to provide labels for which pictures correspond to which digits.

3. Run a machine learning training algorithm on the labeled data, and get as output a *classifier* that can be used to label new, unseen data.

One usually defines an allowed universe of possible classifiers—say, the class of decision trees that make decisions based on individual pixels—and the training algorithm uses the data to select a decision tree. An example decision tree might ask yes/no questions like, "does pixel $(12, 25)$ have intensity higher than 128?" The answer determines the next question to ask, and eventually the final classification.

A slow, brutish training algorithm might be: generate all possible decision trees in increasing order of size, and select the first one that's consistent with the data.

To get a more pungent whiff, let's jump right into the handwritten digit dataset we'll use in the remainder of this chapter. The dataset is a famous one that goes by the irrelevant acronym MNIST (Modified National Institute of Standards and Technology referring to the institution that created the original dataset). The database consists of 70,000 data

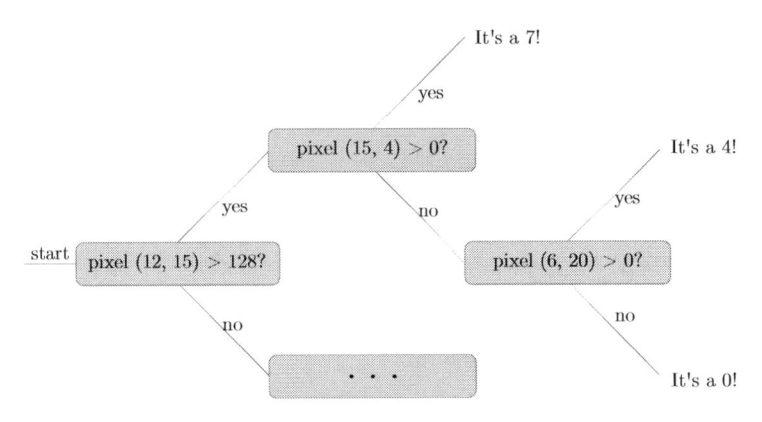

Figure 14.12: An example decision tree classifying an image by looking at specific pixels.

```
0 0 0 0 0 0 0 0 0 0 0 0 0 0 0 0 0 0 0 0 0 0 0 0 0 0 0 0
0 0 0 0 0 0 0 0 0 0 0 0 0 0 0 0 0 0 0 0 0 0 0 0 0 0 0 0
0 0 0 0 0 0 0 0 0 0 0 0 0 0 0 0 0 0 0 0 0 0 0 0 0 0 0 0
0 0 0 0 0 0 0 0 0 0 0 0 0 0 0 0 0 0 0 0 0 0 0 0 0 0 0 0
0 0 0 0 0 0 0 0 0 0 0 0 0 0 0 0 0 0 0 0 0 0 0 0 0 0 0 0
0 0 0 0 0 0 0 0 0 0 0 0 0 0 0 0 0 0 0 0 0 0 0 0 0 0 0 0
0 0 0 0 0 0 0 0 0 0 0 0 0 0 0 0 0 0 0 0 0 0 0 0 0 0 0 0
0 0 0 0 0 0 0 0 0 0 0 0 0 0 115 121 162 253 253 213 0 0 0 0 0 0 0 0
0 0 0 0 0 0 0 0 0 0 0 0 0 63 107 170 251 252 252 252 252 250 214 0 0 0 0 0
0 0 0 0 0 0 0 0 0 25 192 226 226 241 252 253 202 252 252 252 252 252 225 0 0 0 0 0
0 0 0 0 0 0 0 68 223 252 252 252 252 252 39 19 39 65 224 252 252 183 0 0 0 0 0 0
0 0 0 0 0 0 0 186 252 252 252 245 108 53 0 0 0 150 252 252 220 20 0 0 0 0 0 0
0 0 0 0 0 0 70 242 252 252 222 59 0 0 0 0 0 178 252 252 141 0 0 0 0 0 0 0
0 0 0 0 0 0 185 252 252 194 67 0 0 0 0 17 90 240 252 194 67 0 0 0 0 0 0 0
0 0 0 0 0 0 83 205 190 24 0 0 0 0 0 121 252 252 209 24 0 0 0 0 0 0 0 0
0 0 0 0 0 0 0 0 0 0 0 0 0 77 247 252 248 106 0 0 0 0 0 0 0 0 0 0
0 0 0 0 0 0 0 0 0 0 0 0 0 253 252 252 102 0 0 0 0 0 0 0 0 0 0 0
0 0 0 0 0 0 0 0 0 0 0 0 134 255 253 253 39 0 0 0 0 0 0 0 0 0 0 0
0 0 0 0 0 0 0 0 0 0 0 6 183 253 252 107 2 0 0 0 0 0 0 0 0 0 0 0
0 0 0 0 0 0 0 0 0 0 10 102 252 253 163 16 0 0 0 0 0 0 0 0 0 0 0 0
0 0 0 0 0 0 0 0 0 13 168 252 252 110 2 0 0 0 0 0 0 0 0 0 0 0 0 0
0 0 0 0 0 0 0 0 0 41 252 252 217 0 0 0 0 0 0 0 0 0 0 0 0 0 0 0
0 0 0 0 0 0 0 0 40 155 252 214 31 0 0 0 0 0 0 0 0 0 0 0 0 0 0 0
0 0 0 0 0 0 0 0 0 165 252 252 106 0 0 0 0 0 0 0 0 0 0 0 0 0 0 0
0 0 0 0 0 0 0 0 43 179 252 150 39 0 0 0 0 0 0 0 0 0 0 0 0 0 0 0
0 0 0 0 0 0 0 0 137 252 221 39 0 0 0 0 0 0 0 0 0 0 0 0 0 0 0 0
0 0 0 0 0 0 0 0 67 252 79 0 0 0 0 0 0 0 0 0 0 0 0 0 0 0 0 0
0 0 0 0 0 0 0 0 0 0 0 0 0 0 0 0 0 0 0 0 0 0 0 0 0 0 0 0
```

Figure 14.13: A training point for a digit 7 (aligned to make it easier to see).

points, each of which is a 28-by-28 pixel black and white image of a handwritten digit. The digits have been preprocessed in various ways, including resizing, centering, and anti-aliasing. The raw dataset was originally created around 1995, and since 1998 the machine learning researchers Yan LeCun, Corinna Cortes, and Christopher Burges have provided the cleaned copy on LeCun's website.[10] We also include a copy in the code samples for this book, since their version of the dataset has a non-standard encoding scheme.

MNIST is the Petersen graph of machine learning: every technique should first be tested on it as a sanity check. Figure 14.13 shows an example of a training point with label 7, pretty-printed from its raw format as a flat list of 784 ints.

[10] http://yann.lecun.com/exdb/mnist/

The data is split into a training set and a test set, the former having 60,000 examples and the latter 10,000, which are stored in separate files. The separation exists to give a simulation of how well a classifier trained on the training data would perform on "new" data. As such, to get a good quality estimate, it's crucial that the training algorithm uses no information in the test set. We load the data using a helper function, which scales the pixel values from $[0, 255]$ to $[0, 1]$. For our application, we'll simplify the problem a bit to distinguishing between two digits: is it a 1 or a 7? The digit 1 corresponds to a label of 0, and a digit 7 corresponds to a label of 1.

```
def load_1s_and_7s(filename):
    print('Loading data {}...'.format(filename))
    examples = []
    with open(filename, 'r') as infile:
        for line in infile:
            if line[0] in ['1', '7']:
                tokens = [int(x) for x in line.split(',')]
                label = tokens[0]
                example = [x / 255 for x in tokens[1:]] # scale to [0,1]
                if label == 1:
                    examples.append([example, 0])
                elif label == 7:
                    examples.append([example, 1])
    print('Data loaded.')
    return examples
```

But before we go on, I must emphasize that the first two steps in the "machine learning recipe," collecting and cleaning data, are much harder than they appear. A misstep in any part of these processes can cause wild swings in the quality of the output classifier, and getting it right requires clear and strict procedures. We were fortunate enough to have LeCun and his colleagues vet MNIST for us. These prepared datasets are like goods in supermarkets. A shopper doesn't see, appreciate, or viscerally comprehend the amount of work and resources required to rear the cow and grow the almonds, nor even the general form of the pipeline. A common refrain among data scientists and machine learning practitioners is that machine learning is 10% machine learning, and 90% data pipelines.

For example, deciding on the meaning of a label is no simple task. It seems easy for problems like handwritten digits, because it's mostly unambiguous what the true label for a digit is. But for many interesting use cases—detecting fraud/spam, predicting what video a user will enjoy, or determining whether a loan applicant should receive a loan— determining what constitutes a positive or negative label requires serious thought, or worse, the hindsight of a disaster caused by getting it wrong. Harder still are the system-level implications of how a classifier will be used. If a video website deploys a system that naively optimizes for a shallow metric like total time watched, creators will upload superficially longer videos. This wastes everyone's time and hurts the reputation of the site.

Another concern is bias in the training data. Not just statistical bias, which can be a result of errors in data collection on the part of the process designer, but *human* bias beyond one's control. When you collect data on human preferences, it's easy for population majorities to overwhelm less prevalent signals. This happens roughly because machine learning algorithms tend to look for the statistically dominant trends first, and only capture disagreeing trends if the model is complex enough to have both coexist. Think of Chapter 12 in which we studied a physical model by throwing out small order terms. In this context, if those terms corresponded to a coherent group of users, those users would be ignored or actively harmed by the mathematical model.

Even worse, active discrimination can be encoded into training labels. If one trains an algorithm to predict job fitness on a dataset of hiring information, incorporating the reviews of human interviewers can muddy the dataset. You have to be aware that humans, and especially humans in a position of power, can exhibit bias for any number of superficial characteristics that are unrelated to job fitness, most notably that an applicant looks and behaves like the people currently employed. An algorithm trained on this data will learn to mimic the human preferences, which may be unrelated to one's goal.

While mathematics and engineering do weigh in on these problems, it's extremely important to realize that the transition to numbers and equations doesn't magically *avoid* problems like bias and bad process. If anything it obfuscates them from those who aren't fluent in the language. All the user sees is the biscuit that the algorithm decided was appropriate for them to eat. When math is applied to the real world, it serves as a model with assumptions as a foundation. If the assumptions disagree with reality, the levee will break. Riots can literally ensue. We acutely understand this in software: most systems rely on a mess of consistency constraints, some validated explicitly and others not, and when you put garbage data into a software system, you'll get garbage results. So it is for machine learning, which is why it's sometimes called the "high interest credit card of technical debt." These sorts of problems, though interesting and important, are beyond the scope of this book. Instead we'll focus on the "easy" part, actually training an algorithm and producing a classifier.

Learning Models and Hypotheses

In mathematical terms, the process of training a machine learning algorithm starts with defining the domain over your data. Very often the domain is $\mathbb{R}^n$ or $\{0,1\}^n$, so that an input datum is transformed from its natural format, such as an analog image, into a vector of numbers, such as the 4096 pixels in a discrete 64-by-64 digital image. Labels, though they can often have multiple values, will for our purposes be restricted to two options: $\{0,1\}$. For the handwritten digits example, think of this as the classifier for "is the digit a 7 or not?"

With these definitions, a dataset is a set of input-output pairs called *labeled examples*, $S = \{(x, l) : x \in \mathbb{R}^n, l \in \{0,1\}\}$, where x is the example and l is the label. If $f : \mathbb{R}^n \to \{0,1\}$ is the "true" function that labels examples correctly, then $f(x) = l$ for every $(x, l) \in S$.

Next, one defines a so-called *hypothesis class*. This is the universe of all possible output classifiers that a learning algorithm may consider. A useful hypothesis class has natural parameters that vary the behavior of a hypothesis. The learning algorithm learns by selecting parameters based on examples given to it. One of the most common examples, and a building block of neural networks, is the inner product.

Definition 14.22. Fix a dimension $n \in \mathbb{N}$. A *linear threshold function* is a function $L_{w,b} : \mathbb{R}^n \to \{0, 1\}$, parameterized by a vector $w = (w_1, \ldots, w_n) \in \mathbb{R}^n$ called the *weights* and a scalar $b \in \mathbb{R}$ called a *bias*, which is defined as

$$L_{w,b}(x) = \begin{cases} 1 & \text{if } \langle w, x \rangle + b \geq 0 \\ 0 & \text{otherwise.} \end{cases}$$

Linear threshold functions have $n + 1$ parameters: the n weights w and the bias b. The linear threshold function lives up to its name, thanks to the geometry of the inner product. In particular, w defines an $(n-1)$-dimensional vector space $w^\perp = \{v : \langle w, v \rangle = 0\}$, which splits $\mathbb{R}^n$ into two halves.[11] If $b = 0$, then $w^\perp$ passes through the origin, and the inner product $\langle w, x \rangle$ is positive or negative depending on whether x is on the same side of $w^\perp$ as w or the opposite side (respectively). If $b \neq 0$, then $w^\perp$ is shifted away from the origin by a distance of b in the direction of $-w$.

One must also decide how to measure the quality of a proposed classifier. Measures vary depending on the learning model, but in practice it usually boils down to: does the classifier accurately classify the slice of data that has been cordoned off solely for the purpose of evaluation? This special slice of data is the test set. In the exercises, we'll explore a handful of theoretical learning models that give provable guarantees. Though these models are theoretical—for example, they assume the true labels have a particular structure—they serve as the foundation for all principled machine learning models. In these models, if a classifier is accurate on a test set, it will provably generalize to accurately classify new data.

A simple example learning model and problem, which is a building block for many other learning problems,[12] is the following. Given labeled data points chosen randomly from a distribution over $\mathbb{R}^n$ that *can* be separated by a linear threshold function, design an algorithm that finds a "good" threshold function, i.e., one that will generalize well to new examples drawn from the same distribution. We'll explore this more in the exercises.

Summarizing, given a hypothesis class H and a dataset S, a learning algorithm takes as input S and produces as output a hypothesis $h \in H$. We want training algorithms to be efficient and classification to be "correct," where correct means that h should accurately classify the test data.[13]

Neural Networks as Computation Graphs

In Section 14.9 we explored how a differentiable function can be represented as a computation graph of simple operations, each of whose derivative is known. We saw how to

---

[11] For this reason, a linear threshold functions is sometimes called a "halfspace." I can't help but think of a halfspace as a fantasy convention for halflings and half-bloods.
[12] Such as neural networks and the support vector machine.
[13] We're ignoring some concerns related to overfitting, which is an important topic, but beyond the scope of this book.

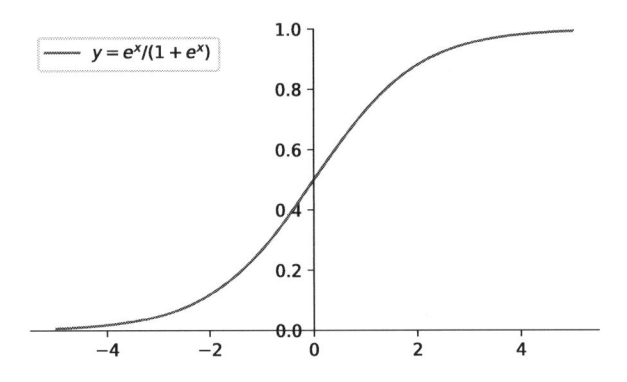

Figure 14.14: A sigmoid function used to introduce nonlinearity into a computation graph.

compute the gradient of a complicated multivariable function by breaking it into pieces and using recursion and caching.

A neural network is exactly this: a massive function composed of simple, differentiable parts, whose output is a real number approximating the desired label of a training example. In Python, our network is an object wrapping the computation graph data structure, and the trained network will evaluate an input and produce a binary label saying whether the input is a 1 (a label of zero) or a 7 (a label of one).[14]

```
network = NeuralNetwork(computation_graph, ...)
network.train(dataset)
network.evaluate(new_example)
```

The most important component operation that is used to build up a neural network is the linear halfspace, the same $L_{w,b}$ of Definition 14.22. We'll call a vertex of the computation graph corresponding to a linear halfspace a *linear node*, and each linear node will have its own distinctly tunable set of parameters, the choice of w and b.

However, there must be more to a neural network than linear nodes. As we know well from linear algebra, a composition of linear functions is still linear. The geometry of the space of handwritten digits is probably more complicated than a linear function can model. That is to say, we need to include operations in our computation graph that transform the input examples in nonlinear ways.

A historically prevalent operation is the sigmoid function, that is, the single-variable function defined by $\sigma(x) = e^x/(1 + e^x)$, with the graph depicted in Figure 14.14.

The sigmoid is clearly nonlinear, nice and differentiable, and its output is confined to $[0, 1]$. You may hear of this operation being compared to the "impulse" of a neuron in a brain, which is why the sigmoid is often called an *activation function*. Though neural

---

[14] The full program is available in the repository linked at pimbook.org.

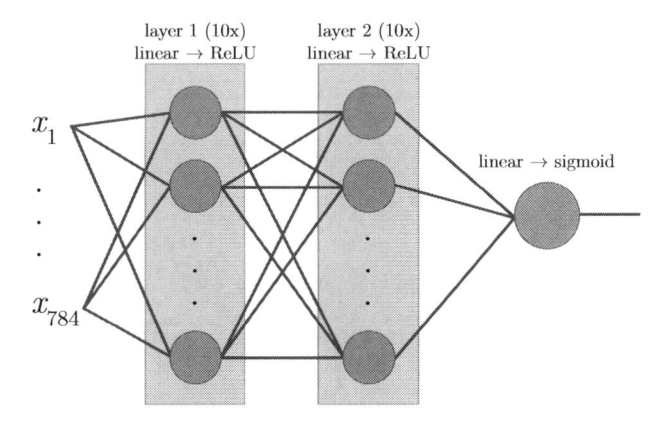

Figure 14.15: A simple neural network architecture for MNIST.

networks are called "neural," the name is mostly an inspiration. Simply put, sigmoids and other activation functions introduce nonlinearity in a useful way.

Typically, one applies the single-input activation function to the output of every linear node. Occasionally, the combined pair of a linear node and its activation function are called a *neuron*. Activation functions usually do not have tunable parameters.

Another important activation function, which is particularly popular in deep learning, is the *rectified linear unit*.

Definition 14.23. The *ReLU* function is the function

$$\text{ReLU}(x) = \begin{cases} x & \text{if } x \geq 0 \\ 0 & \text{otherwise} \end{cases}$$

Equivalently, it can be defined as $\text{ReLU}(x) = \max(0, x)$.

A ReLU needs no plot, as it's simply the function: truncate negative values to zero. The ReLU is particularly interesting because it is *not* differentiable! However, it's only fails to have a derivative at $x = 0$, and in practice one can simply ignore the problem. One nice thing about the ReLU, which is particularly nice when you need lightning-fast computations for training massive networks, is that its evaluation and derivative require only branching comparisons and constants. No exponential math is required.

The network we'll build is architected (quite arbitrarily, as it happens) as depicted in Figure 14.15. The leftmost layer consists of the 784 input nodes, which are inputs to each node of the first layer of 10 linear nodes, each of which has a ReLU activation function. The outputs of the first-layer ReLUs feed as input to a second layer of 10 linear nodes, again with ReLUs, and the output of those goes into a final single linear node with a sigmoid activation.

```
def build_network():
    input_nodes = InputNode.make_input_nodes(28*28)

    first_layer = [LinearNode(input_nodes) for i in range(10)]
    first_layer_relu = [ReluNode(L) for L in first_layer]

    second_layer = [LinearNode(first_layer_relu) for i in range(10)]
    second_layer_relu = [ReluNode(L) for L in second_layer]

    linear_output = LinearNode(second_layer_relu)
    output = SigmoidNode(linear_output)
    error_node = L2ErrorNode(output)
    network = NeuralNetwork(output, input_nodes, error_node=error_node)

    return network
```

The final output of the network is a real number in $[0, 1]$. Labels are binary $\{0, 1\}$, and so we interpret the output as a probability of the label being 1. Then we can say that the label predicted by a network is 1 if the output is at least $1/2$, and 0 otherwise.

You might be wondering how someone comes up with the architecture of a neural network. The answer is that there are some decent heuristics, but in the end its an engineering problem with no clear answers. Our network is quite small, only about 7,500 tunable parameters in all (because it's written in pure Python, training a large network would be prohibitively slow). In real production systems, networks have upwards of millions of parameters, and the process of determining an architecture is more alchemy than science. There is a now-famous 2017 talk by Ali Rahimi in which he criticized what he argued was a loss of rigor in the field. He quoted, for example, how a change to the default rounding mechanism in a popular deep learning library (from "truncate" to "round") caused many researcher's models to break completely, and nobody knew why. The networks still trained, but suddenly failed to learn anything. Rahimi argues that brittle optimization techniques (gradient descent) applied to massively complex and opaque networks create a house of cards, and that theory and rigor can alleviate these problems. Brittle or not, gradient descent on neural networks has proved to be remarkably useful, making some learning problems tractable despite the failure of decades of research into other techniques. So let's continue.

Once we've specified a neural network as a computation graph and obtained a dataset S of labeled examples (x, l), we need to choose a function to optimize. This is often called a *loss function*. For a single labeled example (x, l), it's not so hard to come up with a reasonable loss function. Let f_w be the function computed by the neural network and w the combined vector of all of its parameters. Then define $E(w) = (f_w(x) - l)^2$ as the "error" of a single example. This is just the squared distance of the output of f on an example from that example's label. Note we're not doing any rounding here, so that $f(x) \in [0, 1]$.

If we wanted to convert this to a loss function for an entire training dataset, we could, as $E_{\text{total}}(w) = \frac{1}{|S|} \sum_{(x,l) \in S} (f_w(x) - l)^2$. Then the natural method is to use gradient descent to minimize E_{total}. However, this loss function requires us to loop over the entire training dataset for each step of gradient descent. That is prohibitively slow. Instead, one rather applies what's called *stochastic* gradient descent. In stochastic gradient descent, one chooses an example (x, l) at random, and applies a gradient descent step update to $E(w) = (f_w(x) - l)^2$. Each subsequent gradient step update uses a different, randomly chosen example. The fact that this usually produces a good result is not obvious.[15]

There are many different loss functions, and the loss function we chose above is called the L_2-loss. The name L_2 comes from mathematics, and the number 2 describes the 2's that occur in the formula for the norm: $\|x\|_2 = (\sum_i x_i^2)^{1/2}$. Changing the 2 to, say, a 3 results in an L3 norm, and for a general p these are called L_p norms. You will explore different loss functions in the exercises.

As we outlined in Section 14.9, each vertex of our computation graph needs to know about various derivatives related to the operation computed at that node, and that these values need to be cached to compute a gradient efficiently. Now we'll see one way to manifest that in code. Let's start by defining a generic base node class, representing a generic operation in a computation graph. We'll call the operation computed at that node f, which has arguments $z_1, \ldots, z_m$, and possibly tunable parameters $w_1, \ldots, w_k$.

$$f = f(w_1, \ldots, w_k, z_1, \ldots, z_m)$$

Call the function computed by the entire graph E. The inputs to E are both the normal inputs and all of the tunable parameters at every node. For the sake of having good names, we'll define the *global* derivative of some quantity x to mean $\partial E/\partial x$, while the *local* derivative is $\partial f/\partial x$ (it's local to the node we're currently operating with). These are not standard terms.

Now we define a cache to attach to each node, whose lifetime will be a single step of the gradient descent algorithm.

```
class CachedNodeData(object):
    def __init__(self):
        self.output = None
        self.global_gradient = None
        self.local_gradient = None
        self.local_parameter_gradient = None
        self.global_parameter_gradient = None
```

The attributes are as follows, with each expression evaluated at the current input x and the current choice of tunable parameters.

1. output: a single float, the output of this node.

2. global_gradient: a single float, the value of $\partial E/\partial f$.

3. local_gradient: a list of floats, the values $(\partial f/\partial z_1, \ldots, \partial f/\partial z_m)$; i.e., the components of ∇f that correspond to the arguments of f.

---

[15] There are also compromises: pick a random subset of 100 examples, and compute the average error and gradient for that "mini batch." Variations abound.

4. `local_parameter_gradient`: the same thing as `local_gradient`, but for the components of ∇f corresponding to the tunable parameters of f.

5. `global_parameter_gradient`: the same thing as `local_parameter_gradient`, but for the components of ∇E corresponding to the tunable parameters of f.

Now we define a base class `Node` for the vertices of the computation graph. Its children are `InputNode`, `ConstantNode`, `LinearNode`, `ReluNode`, `SigmoidNode`, and `L2ErrorNode`. Here's an example of how the subclasses of `Node` are used to build a computation graph:

```
input_nodes = [InputNode(i) for i in range(10)]
linear_node_1 = LinearNode(input_nodes)
linear_node_2 = LinearNode(input_nodes)
linear_node_3 = LinearNode(input_nodes)
sigmoid_node_1 = SigmoidNode(linear_node_1)
sigmoid_node_2 = SigmoidNode(linear_node_2)
sigmoid_node_3 = SigmoidNode(linear_node_3)
linear_output = LinearNode([sigmoid_node_1, sigmoid_node_2, sigmoid_node_3])
output = SigmoidNode(linear_output)
error_node = L2ErrorNode(output)

network = NeuralNetwork(output, input_nodes, error_node=error_node, step_size=0.5)
network.train(dataset)
network.evaluate(new_data_point)
```

And now we define `Node` and its subclasses.

```
class Node(object):
    def __init__(self, *arguments):
        # if has_parameters is True, child class must set self.parameters
        self.has_parameters = False
        self.parameters = []
        self.arguments = arguments
        self.successors = []
        self.cache = CachedNodeData()

        # link argument successors to self
        for argument in self.arguments:
            argument.successors.append(self)

        '''Argument nodes z_i will query this node f(z_1, ..., z_k) for f/z_i,
        so we need to keep track of the index for each argument node.'''
        self.argument_to_index = {node: index for (index, node) in enumerate(arguments)}
```

The list of arguments is ordered, so that all inputs and gradients correspond index-wise. We'll define the core methods in `Node` that perform gradient descent training momentarily, but first we have to define what functions the subclasses need to implement. They are:

1. `compute_output`: take as input a list of floats representing the concrete values of the global input to the computation graph (called `inputs`), and produce as output

the output of this node, by recursively calling `output` on the argument nodes and performing an operation to produce an output.

2. `compute_local_gradient`: Take nothing as input and produce as output the list of local gradients $\partial f/\partial z_i$.

3. `compute_local_parameter_gradient`: Take nothing as input and produce as output the local parameter gradient $\partial f/\partial w_i$.

4. `compute_global_parameter_gradient`: Take nothing as input and produce as output the global parameter gradients $\partial E/\partial w_i$.

The example of the linear node illustrates each of these pieces. Let

$$f(w, b, x) = \langle w, x \rangle + b$$
$$= b + \sum_{i=1}^{n} w_i, x_i$$

We model the bias term b by adding an extra input as a `ConstantNode`. We also have a simple `InputNode` for the input to the whole graph.

```python
class ConstantNode(Node):
    def compute_output(self, inputs):
        return 1

class InputNode(Node):
    def __init__(self, input_index):
        super().__init__()
        self.input_index = input_index

    def compute_output(self, inputs):
        return inputs[self.input_index]

    @staticmethod
    def make_input_nodes(count):
        '''A helper function so the user doesn't have to keep track of
           the input indexes.
        '''
        return [InputNode(i) for i in range(count)]
```

Now we can define `LinearNode`. First, we initialize the weights and add a constant node for the bias. In this way, the bias is treated the same as any other input, which makes the formulas convenient.

```
class LinearNode(Node):
    def __init__(self, arguments):
        super().__init__(ConstantNode(), *arguments) # first arg is bias
        self.initialize_weights()
        self.has_parameters = True
        self.parameters = self.weights # name alias

    def initialize_weights(self):
        arglen = len(self.arguments)
        # set the initial weights randomly, according to a heuristic distribution
        weight_bound = 1.0 / math.sqrt(arglen)
        self.weights = [random.uniform(-weight_bound, weight_bound) for _ in
            range(arglen)]
```

A common heuristic to initialize a linear node's weights is to set the weights to be random numbers in between $1/\sqrt{d}$, where d is the number of weights. This aligns with gradient descent: start at a random initial configuration and try to optimize.

The rest of the class consists of the required implementations of the Node interface. The gradients are particularly simple formulas. For $f = \sum_{i=0}^{n} w_i x_i$, we have

$$\frac{\partial f}{\partial x_i} = w_i, \qquad \frac{\partial f}{\partial w_i} = x_i, \qquad \frac{\partial E}{\partial w_i} = \frac{\partial E}{\partial f} \frac{\partial f}{\partial w_i}$$

This turns into code as follows:

```
class LinearNode(Node):
    [...]

    def compute_output(self, inputs):
        return sum(
            w * x.evaluate(inputs)
            for (w, x) in zip(self.weights, self.arguments)
        )

    def compute_local_gradient(self):
        return self.weights

    def compute_local_parameter_gradient(self):
        return [arg.output for arg in self.arguments]

    def compute_global_parameter_gradient(self):
        return [
            self.global_gradient * self.local_parameter_gradient_for_argument(argument)
            for argument in self.arguments
        ]

    def local_parameter_gradient_for_argument(self, argument):
        '''Return the derivative of this node with respect to the weight
        associated with a particular argument.'''
        argument_index = self.argument_to_index[argument]
        return self.local_parameter_gradient[argument_index]
```

The other nodes are defined similarly, with the parameter functions returning empty lists as the LinearNode is the only node with tunable parameters. For each of the four `compute_` methods defined on each child class, we define corresponding methods on the parent class that check the cache and call the subclass methods on cache miss. They all look more or less like this:

```
class Node:
    [...]

    @property
    def local_gradient(self):
        if self.cache.local_gradient is None:
            self.cache.local_gradient = self.compute_local_gradient()
        return self.cache.local_gradient
```

The methods in the child classes use these properties when referring to their arguments, so the values will be lazily evaluated and then cached as needed. Finally, the computation of the global gradient for a node doesn't depend on the formula for that node, so it can be defined in the parent class.

```
class Node:
    [...]

    def compute_global_gradient(self):
        return sum(
            successor.global_gradient * successor.local_gradient_for_argument(self)
            for successor in self.successors)

    def local_gradient_for_argument(self, argument):
        argument_index = self.argument_to_index[argument]
        return self.local_gradient[argument_index]
```

At this point we've enabled the computation of all the gradients we need to do a step of gradient descent.

```
class Node:
    [...]

    def do_gradient_descent_step(self, step_size):
        '''The core gradient step subroutine: compute the gradient for each of
        this node's tunable parameters, step away from the gradient.'''
        if self.has_parameters:
            for i, gradient_entry in enumerate(self.global_parameter_gradient):
                self.parameters[i] -= step_size * gradient_entry
```

Recall, each subclass defines its vector of parameters, and the `global_parameter_gradient` has to line up index by index. Also recall that we're subtracting because we want to minimize the error function E, and ∇E points in the direction of steepest increase of E.

The very last node of the computation graph, which computes the error for a training example, has some extra methods that depend on a training example's label. For the L_2 error, the entire class is:

```
class L2ErrorNode(Node):
    def compute_error(self, inputs, label):
        argument_value = self.arguments[0].evaluate(inputs)
        self.label = label # cache the label
        return (argument_value - label) ** 2

    def compute_local_gradient(self):
        last_input = self.arguments[0].output
        return [2 * (last_input - self.label)]

    def compute_global_gradient(self):
        return 1
```

Now we define a wrapper class `NeuralNetwork` that keeps track of the input and terminal nodes of the computation graph, resets caches, and controls the training of the network. We start with a self-explanatory constructor, and a helper function for applying some function to each node of the computation graph exactly once.

```
class NeuralNetwork(object):
    def __init__(self, terminal_node, input_nodes, error_node=None, step_size=None):
        self.terminal_node = terminal_node
        self.input_nodes = input_nodes
        self.error_node = error_node or L2ErrorNode(self.terminal_node)
        self.step_size = step_size or 1e-2

    def for_each(self, func):
        '''Walk the graph and apply func to each node.'''
        nodes_to_process = set([self.error_node])
        processed = set()

        while nodes_to_process:
            node = nodes_to_process.pop()
            func(node)
            processed.add(node)
            nodes_to_process |= set(node.arguments) - processed
```

The `for_each` function performs a classic graph traversal (whether it's depth-first or breadth-first depends on the semantics of `pop` and `add`, but we only care that each node is visited exactly once). We can use it to reset the caches at every node. We can also trivially define the `evaluate` function and `compute_error` functions as wrappers.

```
class NeuralNetwork(object):
    [...]

    def reset(self):
        def reset_one(node):
            node.cache = CachedNodeData()
        self.for_each(reset_one)

    def evaluate(self, inputs):
        self.reset()
        return self.terminal_node.evaluate(inputs)

    def compute_error(self, inputs, label):
        '''Compute the error for a given labeled example.'''
        self.reset()
        return self.error_node.compute_error(inputs, label)
```

Finally, the training loop. It's as simple as randomly choosing an example, computing the output error for that example, and then calling `do_gradient_descent_step` on each node.

```
class NeuralNetwork(object):
    [...]

    def backpropagation_step(self, inputs, label, step_size=None):
        self.compute_error(inputs, label)
        self.for_each(lambda node: node.do_gradient_descent_step(step_size))

    def train(self, dataset, max_steps=10000):
        '''dataset is a list of pairs ([float], int) where the first entry is
        the input point and the second is the label.'''
        for i in range(max_steps):
            inputs, label = random.choice(dataset)
            self.backpropagation_step(inputs, label, self.step_size)
```

Now let's apply this to the MNIST dataset. First we build our network, with two fully connected layers of `LinearNodes` and `ReluNodes`, with a final `LinearNode` with a `SigmoidNode` output.

```
def build_network():
    input_nodes = InputNode.make_input_nodes(28*28)
    first_layer = [LinearNode(input_nodes) for i in range(10)]
    first_layer_relu = [ReluNode(L) for L in first_layer]
    second_layer = [LinearNode(first_layer_relu) for i in range(10)]
    second_layer_relu = [ReluNode(L) for L in second_layer]

    linear_output = LinearNode(second_layer_relu)
    output = SigmoidNode(linear_output)
    error_node = L2ErrorNode(output)
    return NeuralNetwork(output, input_nodes, error_node=error_node, step_size=0.05)
```

Then we split the training set into batches, separating from each batch a so-called *validation* set, which we use to measure the quality of the training as it progresses. At the end, we evaluate the error on the test set.

```
train = load_1s_and_7s('mnist/mnist_train.csv')
test = load_1s_and_7s('mnist/mnist_test.csv')
network = build_network()
n, epoch_size = len(train), int(len(train) / 10)

for i in range(5):
    shuffle(train)
    validation, train_piece = train[:epoch_size], train[epoch_size:2*epoch_size]
    print("Starting epoch of {} examples with {} validation".format(
        len(train_piece), len(validation)))

    network.train(train_piece, max_steps=len(train_piece))
    print("Finished epoch. Validation error={:.3f}".format(
        network.error_on_dataset(validation)))

print("Test error={:.3f}".format(network.error_on_dataset(test)))
```

During training we see:

```
Starting epoch of 1300 examples with 1300 validation
Finished epoch. Validation error=0.015
Starting epoch of 1300 examples with 1300 validation
Finished epoch. Validation error=0.007
Starting epoch of 1300 examples with 1300 validation
Finished epoch. Validation error=0.007
Starting epoch of 1300 examples with 1300 validation
Finished epoch. Validation error=0.006
Starting epoch of 1300 examples with 1300 validation
Finished epoch. Validation error=0.010
Test error=0.011
```

Which is about 1.1% error. Figure 14.16 shows some examples of classifications of digits after training. To make it easier to display in the book, I've rounded any nonzero values to 0 and 1, though in the full code we provide a helper function show_random_examples that shows the raw pixel values. As you can see, the first two are correct, and the third is incorrect (though the correct classification of that digit is hardly obvious).

Looking closely at the validation error as training progresses, the validation error progressively decreases, but at the end increases from 0.6% to 1%. One possible explanation for this is the phenomenon of *overfitting*. We'll explore it more in the exercises, but a cursory explanation is that as a sufficiently expressive machine learning model continues to be trained, it can learn to encode specific features of the dataset. That is, the longer one trains on the same data, the more the trained model resembles a lookup table. We'll explore this more in the exercises.

So there we have it! A functioning neural network, built as a computational graph of arbitrary operations, with automatic gradient computations.

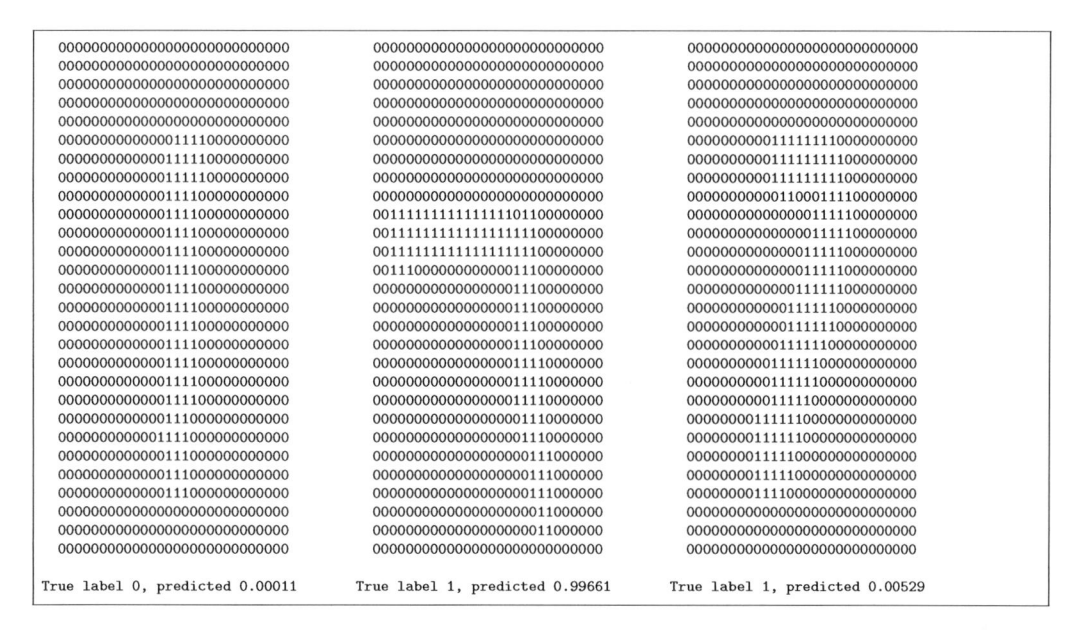

```
0000000000000000000000000000          0000000000000000000000000000          0000000000000000000000000000
0000000000000000000000000000          0000000000000000000000000000          0000000000000000000000000000
0000000000000000000000000000          0000000000000000000000000000          0000000000000000000000000000
0000000000000000000000000000          0000000000000000000000000000          0000000000000000000000000000
0000000000000000000000000000          0000000000000000000000000000          0000000000000000000000000000
0000000000000111100000000000          0000000000000000000000000000          0000000000001111111100000000
0000000000000111110000000000          0000000000000000000000000000          0000000000001111111110000000
0000000000001111100000000000          0000000000000000000000000000          0000000000001111111110000000
0000000000001111000000000000          0000000000000000000000000000          0000000000000110001111000000
0000000000001111000000000000          0011111111111111101100000000          0000000000000000001111000000
0000000000001111000000000000          0011111111111111111100000000          0000000000000000001111000000
0000000000001111000000000000          0011111111111111111100000000          0000000000000000111110000000
0000000000001111000000000000          0011100000000001110000000000          0000000000000001111100000000
0000000000001111000000000000          0000000000000001110000000000          0000000000000011111000000000
0000000000001111000000000000          0000000000000001110000000000          0000000000001111110000000000
0000000000001111000000000000          0000000000000001110000000000          0000000000011111100000000000
0000000000001111000000000000          0000000000000001111000000000          0000000000111111000000000000
0000000000001111000000000000          0000000000000001111000000000          0000000000111110000000000000
0000000000001111000000000000          0000000000000001111000000000          0000000000111110000000000000
0000000000001110000000000000          0000000000000001110000000000          0000000011111000000000000000
0000000000011110000000000000          0000000000000001110000000000          0000000011111100000000000000
0000000000011100000000000000          0000000000000001110000000000          0000000011110000000000000000
0000000000011100000000000000          0000000000000001111000000000          0000000011110000000000000000
0000000000011100000000000000          0000000000000001111000000000          0000000011110000000000000000
0000000000011100000000000000          0000000000000001111000000000          0000000011110000000000000000
0000000000000000000000000000          0000000000000000011000000000          0000000000000000000000000000
0000000000000000000000000000          0000000000000000011000000000          0000000000000000000000000000
0000000000000000000000000000          0000000000000000000000000000          0000000000000000000000000000

True label 0, predicted 0.00011       True label 1, predicted 0.99661       True label 1, predicted 0.00529
```

Figure 14.16: Example predictions of our neural network.

14.11 Cultural Review

- At its core, the derivative is the linear approximation of a function at a point. This view applies to both single- and multivariable settings.

- Local properties—those properties which hold only in a narrow slice around a point of interest—tend to be easier to reason about and compute, and they often inform one about the global properties of an object.

14.12 Exercises

14.1 A function $\mathbb{R}^n \to \mathbb{R}$ is called *continuous* at a point $c \in \mathbb{R}^n$ if for every $\varepsilon > 0$ there exists a $\delta > 0$ such that whenever $\|x - c\| < \delta$ it holds that $|f(x) - f(c)| < \varepsilon$. Using this definition, show that $f(x, y, z) = x^2 + y^2 + z^2$ is continuous at $(0, 0, 0)$, but that $g(x, y, z) = \frac{xyz}{x^2y^2+z}$ is not continuous at $(0, 0, 0)$.

14.2 Prove the analogue of Theorem 14.10 for functions $\mathbb{R}^n \to \mathbb{R}^m$. In that case, if $f = (f_1, \ldots, f_m)$, the total derivative matrix should be:

$$\begin{pmatrix} Df_1(c, v_1) & Df_1(c, v_2) & \cdots & Df_1(c, v_n) \\ Df_2(c, v_1) & Df_2(c, v_2) & \cdots & Df_2(c, v_n) \\ \vdots & \vdots & \ddots & \vdots \\ Df_m(c, v_1) & Df_m(c, v_2) & \cdots & Df_m(c, v_n) \end{pmatrix}$$

Hint: the same proof works, but the construction of the single-variable function to apply the chain rule to is slightly different.

14.3 Look up a proof of the fact that a function $f : \mathbb{R}^n \to \mathbb{R}$ is differentiable (has a total derivative) if all of its partial derivatives exist and are continuous (Theorem 14.11). This theorem relies on a chain of results: the definition of continuity, Rolle's Theorem for single-variable functions, and the Mean Value Theorem for single variable functions. The Mean Value Theorem is one of the most powerful technical tools in the fields of mathematics that deal with continuous functions.

14.4 Find and study a proof of Schwarz's theorem, that mixed partial derivatives of sufficiently nice functions don't depend on the order you take them in. The proof is gritty, but enlightening.

14.5 Prove the first part of the Cauchy-Schwarz inequality for real vectors, that $|\langle v, w \rangle| \leq \|v\|\|w\|$, using basic algebra.

14.6 Prove that the rule for computing partial derivatives by assuming other variables are constant is valid.

14.7 Make sense of the Hessian as a linear map.

14.8 The gradient of a function $\mathbb{R}^n \to \mathbb{R}$ is a vector which points in the direction of steepest ascent of the function, which we investigated via projections. What, if anything, can be said about the direction of steepest ascent of a multi-output function $\mathbb{R}^n \to \mathbb{R}^m$ by inspecting its total derivative matrix?

14.9 Find and understand a statement of Taylor's theorem for two-dimensional functions (with an arbitrary number of approximation terms).

14.10 Perhaps the most famous theoretical machine learning model is called the *Probably Approximately Correct* model (abbreviated PAC). This model formalizes much of modern machine learning. Given a finite set X (the universe of possible inputs), the PAC model involves a probability distribution D over X used both for generating data and evaluating the quality of a hypothesis. A machine learning algorithm gets as input the ability to sample as much data as it wants from D, and its output hypothesis h must have high accuracy on D (hence the name "approximately" in PAC). Since the sampled data is random, the learning algorithm may fail to produce an accurate classifier with small probability. However—and this is the most stringent qualification—in order for a learning algorithm to be considered successful in the PAC model, it must provably succeed *for any* distribution on the data. If the distribution is uniformly random or focused on just a small set of screwy points, a valid "PAC learner" must be able to adapt. Look up the formal definition of the PAC model, find a simple example of a problem that can be PAC-learned, and read a proof that a successful algorithm does the trick.

14.11   Another important learning model involves an algorithm that, rather than passively analyzing data that's given to it (as in the PAC model of the previous exercise), is allowed to formulate queries of a certain type, an "oracle" (a human) answers those queries, and then eventually the algorithm produces a hypothesis. Such a model is often called an "active learning" model. Perhaps the most famous example is *exact learning with membership and equivalence queries.* Look up a formal definition of this model, and learn about its main results and variations.

14.12   Write a program that uses gradient descent to solve the linear threshold function problem from the end of Section 14.10. That is, determine what the appropriate loss function should be, determine a formula for the gradient, and enshrine it in code.

14.13   In this chapter, our gradient descent used a fixed ε as the step size. However, it can often make sense to adjust the rate of descent as the optimization progresses. At the beginning of the descent, larger steps can provide quicker gains toward an optimum. Later, smaller steps help refine a close-to-optimal solution. A popular way to do this due to Yurii Nesterov involves keeping track of a so-called *momentum* term, and adding both the normal gradient descent step plus the momentum term. Research Nesterov's method (Under what conditions does it work? Do these reasonably apply to neural networks?) and adapt the program in this chapter to use it. Measure the improvement in training time.

14.14   Another popular technique for training neural networks is the so-called *minibatch*, where instead of a stochastic update for each example, one groups the examples into batches and computes the average loss for the batch. Research why minibatch is considered a good idea, and augment the program in this chapter to incorporate it. Does it improve the error rate of the learned hypothesis?

14.15   There are many different loss functions for a neural network. Look up a list of the most widely used loss functions, and research their properties.

14.16   One particularly relevant loss function is called *softmax*, because it applies to a vector-valued input. Softmax is typically used to represent the loss of a categorical (1 out of N options) labeling, and it's particularly useful to adapt MNIST from a binary two-digit discriminator to a full ten-digit classifier. Augment the code in this chapter to incorporate softmax, and use this to implement a classifier for the full MNIST dataset.

14.17   *Overfitting* is the phenomenon of a machine learning algorithm "hard-coding" the labels of specific training examples in a way that does not generalize. Imagine a robot that memorizes a lookup table for conversation replies, but then fails to respond to every unexpected query. It could hardly be called learning! Overfitting seeps into neural networks in pernicious ways, such as not properly separating training, validation, and test data. Overusing validation data can also cause some degree of overfitting of tuned parameters. The most common type of overfitting occurs simply when training goes on too long on the same set of examples. Explore the degree to which overfitting occurs in the neural network in this chapter for MNIST by running the training loop for a long time. Try decreasing the size of the training set, and observe the overfitting get worse.

14.18 Space and orientation is particularly useful to computer vision applications. One industry-standard "feature" used in deep neural networks for computer vision is a primitive called *convolution*. Research this new operation, and implement a 4×4 convolution node in the neural network from this chapter. Design an architecture that incorporates convolution, and train MNIST on it. Does the quality improve?

14.13 Chapter Notes

The Cauchy-Schwarz Inequality

If you want to build an appreciation for mathematical proofs, it's hard to find a better focal point than the Cauchy-Schwarz inequality. This theorem has many genuinely different mathematical proofs, each of which generalizes in different ways to different settings. My favorite treatise on the subject is Michael Steele's beautifully written book, "The Cauchy Schwarz Master Class." The book has something for everyone. I have been known to spend airplane flights filling scratch paper with solutions to the cornucopia of genuinely fun exercises the book has to offer.

Interestingly, Hermann Schwarz (whom the inequality is named after) was the first to provide a correct proof of the equality of mixed partial derivatives, Theorem 14.16.

Scaling Neural Networks

Our neural network and computation graph are almost laughably small. And, having written our network in pure Python, training proceeds at a snail's pace. It should be obvious that our toy implementation falls far short of industry-strength deep learning libraries like TensorFlow, even though the underlying concepts of computation graphs are the same. I'd like to lay out a few specific reasons.

Our network for learning (a subset of) MNIST has roughly $7,500$ tunable parameters. Large-scale neural networks can have millions or even billions of tunable parameters. It's no surprise that many additional mathematical and engineering tricks are required to achieve such scale.

One aspect of this is hardware. Top-tier neural networks take advantage of the structure of certain nodes (for example, many nodes are linear) and the typical architecture of a network (nodes grouped in layers) to convert evaluation and gradient computations to matrix multiplications. Once this is done, graphics cards (GPUs) can drastically accelerate the training process. Even more, companies like Google develop custom ASICs (application-specific integrated circuits) that are particularly fast at doing the operations neural networks need for training. One such chip is called a Tensor Processing Unit (TPU). The proliferation of graphics cards and custom hardware has resulted in the ability to train more ambitious models for applications like language translation and playing board games like Go.

However, fancy hardware won't fix issues like overfitting, where a model with billions of parameters essentially becomes a lookup table for the training data and doesn't generalize to new data. To avoid this, experts employ a handful of engineering and architectural tricks. For example, between each layer of linear nodes, one can employ a technique called *dropout*, in which the outputs of random nodes are set to zero. This prevents nodes in subsequent layers from depending on specific arguments in a fragile way. In other words, it promotes redundancy. Such techniques fall under the umbrella of *regularization* methods.

Other techniques are specific to certain application domains. For example, the concept of *convolution* is used widely in networks that process image data. While convolution has a mathematically precise definition, we'll suffice to describe it as applying a "filter" to every 4×4 pixel window of an image. Such techniques allow individual neurons to encode edge detectors. When combined in layers—filters of filters, and so on—the results are nodes that act as quite sophisticated texture and shape detectors.

The individual computational nodes also get much consideration. Historically, the original nonlinear activation node for a linear node was the sigmoid function. However, because the function plateaus for large positive and negative values, training a network that solely uses sigmoid activations can result in prohibitively slow learning. The ReLU function avoids this, but brings its own problems. In particular, when linear weights are randomly initialized as we did, ReLU nodes have an equal chance of being zero or nonzero. When a ReLU activation is zero, that neuron (and all the input work to get to that neuron) is essentially dead. Even if the neuron should contribute to the output of an example, the gradient is zero and so gradient descent can't update it. Other activation functions have been defined and studied to try to get the best of both worlds.

For the reader eager to dive deeper into production-quality neural networks, check out the Keras library. Keras is a layer on top of Google's TensorFlow library that makes implementing neural networks in Python as straightforward as in this post. The designer of Keras also wrote a book, "Deep Learning with Python," which—beyond including a multitude of examples—covers the nitty-gritty engineering details with plenty of references.

Chapter 15

The Argument for Big-O Notation

[Big-O notation] significantly simplifies calculations because it allows us to be sloppy—but in a satisfactorily controlled way. [...] The extra time needed to introduce O notation is amply repaid by the simplifications that occur later.

– Donald Knuth

Big-O notation is a common plight of programmers seeking a job at a top-tier software company. It can feel extremely unfair to be rejected from a job for not being able to rattle off the big-O runtime of an algorithm, despite being able to implement that algorithm on the spot on a whiteboard. It's a loathsome feeling conspicuously detached from the job.

As we've discussed, the bulk of software is bookkeeping, moving and reshaping data to adhere to APIs of various specifications, and doing this in a way that's easy to extend and maintain. The ever-present specter of software is the fickle user who thinks they know what they want, only to change their mind when you finish implementing it.

However, one should try to see the other side of the coin as well. Often an interviewer doesn't particularly care about the exact big-O runtime of an algorithm. They aren't testing your aptitude to recall arbitrary facts and do algebra. They care that you can reason about the behavior of the thing you just wrote on the whiteboard. As we all know, beyond correctness, an important part of software is anticipating how things will break in subtler ways. What kind of data will make the system hog memory? For what sort of usage will a system thrash? Can you guarantee there are no deadlocks? Most importantly, can you be concrete in your analysis?

Among the simplest things one could possibly ask is what part of the algorithm you just wrote is the bottleneck at scale. To do that, you have to walk a fine line between being precise and vague. Define the quantities of interest—whether they're joins in a database query or sending data across a network—and the simplifying assumptions that make it possible to discuss in principle. You also have to sweep an immense amount of complexity under the rug. Maybe you'll ignore problems that could occur due to multithreading, or the overhead of stack frame management incurred by splitting code into functions in just such a way, or even ignore the *benefits* of helpful compiler optimizations and memory locality, when the application doesn't depend on it.

In dealing with this, we weigh the consequences of a double-edged sword. Be too precise and you drown in a sea of details. It becomes impossible to have a discussion with principled arguments and reasonable conclusions. On the other hand, be too vague and you risk invalid conclusions, leading to wasted work and worse software. Like we did with waves on a string in Chapter 12, even if we know we're ignoring certain details, we want to understand the dominant behavior of the system—the aspects we care about—while ignoring the complexities that prevent us from gaining a deeper understanding.

Few tools in computer science help one balance on the tightrope. We have experimental measurements, tests against historical data, and monitoring on live data. But these are tools designed for incrementalism. For most big decisions, such as designing a new database, data structure, operating system, or a truly novel product—as companies like Google, Amazon, Facebook, and Microsoft have done many times—the investment required for a redesign requires strong and principled justification. No users exist yet, nor does any usage data.

Mathematics provides an abstraction that helps one, as Knuth says, be sloppy in a precisely controlled way. The abstraction is big-O notation, along with its cousins little-o, big-Ω and little-ω, and big-Θ.[1] Together they are called *asymptotic notation*. Big-O notation is a language in which to phrase tradeoffs, compare critical resource usage, and measure things that scale.

The key part of that description is *language*. Big-O is a piece of technical mathematics specifically designed to make conversation between humans about messy math easier. It fits that description more obviously and shamelessly than any other bit of math I know. And it's not just about runtime. You can use big-O and its relatives to describe the usage of any constrained resource, be it runtime, space, queries, collisions, errors, or bits sent to a satellite.

Of course, like any tool big-O is not a panacea. Often one needs to peek behind the curtain and optimize and a granular level. Customer attention is a matter of milliseconds. In time-critical engines like text editors and video games, frame rate and response latency are the bottom line. But big-O has the advantage of being able to fit entirely inside your head, unlike tables of measurements. As a language aid, a first approximation, and a start to a conversation, big-O is hard to beat.

So in this short chapter I'll introduce big-O notation, describe some of its history, show how it simplifies some of the calculations in this book, and then describe some of my favorite places where big-O takes center stage.

History and Definition

The original use of big-O notation was by Landau and Bachmann in the 1890's for approximating the accuracy of function approximations at a point. The O notation was chosen because O stands for "Order" (more precisely, the German *Ordnung*). Big-O notation is meant to replace an expression with its order of magnitude. This is doubtlessly

---

[1] big-Ω, little-ω, and big-Θ are defined in terms of big-O and little-o, which we'll make clear.

useful for mathematics, and it was a particularly popular notation in number theory. It was not until mid-century 1900's that big-O found its way to computer science. Donald Knuth opens a 1976 essay with, "Most of us have gotten accustomed to [big-O notation]," and goes on to formalize it and introduce lower-bound analogues.

For understanding function approximations, big-O is relevant to Taylor series. In the language of big-O, $\sin(x)$ being well approximated by x near $x = 0$ is phrased as

$$\sin(x) = x + O(x^3)$$

To explain what this means, recall that the Taylor series for $\sin(x)$ at $x = 0$ is

$$\sin(x) = x - \frac{x^3}{3!} + \frac{x^5}{5!} - \cdots + \frac{(-1)^n x^{2n}}{(2n)!}$$

Big-O says the x^3 terms and smaller are dominated by the x term. What's unspoken here is what "dominates" means. In the analysis of algorithms, "dominates" usually means an upper bound as the size of the input grows larger. But here nothing is growing! Instead, here the big-O notation implies a limit $x \to 0$. I.e., when x shrinks, x^3 vanishes much faster than x. The formal definition is as a limit.

Definition 15.1. Let $a \in \mathbb{R}$ and let $f, g : \mathbb{R} \to \mathbb{R}$ be two functions with $g(x) \neq 0$ on some interval around a. We say $f(x) = O(g(x))$ as $x \to a$ if the limit of their ratios does not diverge.

$$\lim_{x \to a} \left| \frac{f(x)}{g(x)} \right| < \infty$$

The limit notation needs a disambiguation. We're not saying that the limit has to exist. Indeed, $\sin(x)$ does not have a limit, but $\sin(x) = O(1)$. Rather, we simply need that the limit does not grow without bound.

So when we say $f = O(g)$, we mean that g is a sort of upper bound on f under some limit. Usually the limit point a is established once at the beginning of a discussion, or obvious from context (e.g., you're doing a Taylor series at a). In the rare cases one needs to disambiguate, one can use $O_{x \to a}(g(x))$.

Unpacking this definition a bit, consider the special case when the limit exists and is finite. Then there is some constant C for which

$$\lim_{x \to a} \left| \frac{f(x)}{g(x)} \right| = C,$$

and so there is some interval around a so that $|f(x)| \leq (C+1)|g(x)|$. Indeed, $|f(x)| \leq Dg(x)$ for some constant D, so long as x is near the point of interest.

This notation satisfies some straightforward properties that allows one to do algebra with big-O quantities. Their proofs are straightforward from Definition 15.1 and standard properties of limits.

1. $f = O(f)$ for any f.

2. If $f_1 = O(g_1)$ and $f_2 = O(g_2)$, then $f_1 f_2 = O(g_1 g_2)$.

3. If $f_1 = O(g_1)$ and $f_2 = O(g_2)$, then $f_1 + f_2 = O(g_1 + g_2)$

4. $f + f = O(f)$, and moreover $Cf = O(f)$ for any constant C.

Take care, because when we say $f = O(g)$, the symbol $=$ doesn't mean $=$ in the usual sense. For example, it's not symmetric or transitive; $x = O(x^3)$ and $x^2 = O(x^3)$ as $x \to 0$, but $x \neq x^2$. When someone uses big-O notation like $f = O(g)$, it's best to read $=$ as "is," and then the sentence makes sense: "f is (at most) order of g." Moreover, when we include $O(g(x))$ in the context of some larger expression, like $\sin(x) = x + O(x^3)$, what we mean is that $\sin(x) = x + f(x)$ for some $f(x) = O(x^3)$. Fluent use of big-O involves "native support" for this implicit association in your head, which can take some time to get used to.

Continuing with the example of $\sin(x)$, say we wanted an estimate of $\sin(x)\sqrt{1-x^2}$. Recall from Section 12.7 that the Taylor series for $\sqrt{1-x^2}$ is

$$\sqrt{1+x^2} = 1 + \frac{x^2}{2} - \frac{x^4}{8} + \frac{x^6}{16} - \cdots$$

The generic n-th term of $\sqrt{1+x^2}$ is not that easy to write down, so we won't. But we just want to compute an approximation of the product $\sin(x)\sqrt{1+x^2}$ near zero. One thing we could do is compute the Taylor series of the entire thing by hand, computing derivatives for every term. Quite laborious! Another thing we could do is try to reason about the infinite product of their Taylor series. That would still be a lot of work, and without extra prior knowledge, we might question whether it's valid to take a term-by-term product of two infinite series.

Big-O can help. If we decide in advance how many terms we care about, then we can truncate the two series with big-O and we're left with a finite product. Note that if these next computations look strange, it's probably because you're used to seeing big-O as an infinite limit, whereas the big-O used here is a limit as $x \to 0$. In this context, $x^5 = O(x^3)$. We'll see the "usual" version of big-O shortly.

$$\sin(x) = x + O(x^3),$$
$$\sqrt{1+x^2} = 1 + O(x^2),$$
$$\begin{aligned}\sin(x)\sqrt{1+x^2} &= (x + O(x^3))(1 + O(x^2)) \\ &= x + O(x^3) + x \cdot O(x^2) + O(x^3)O(x^2) \\ &= x + O(x^3) + O(x^3) + O(x^3) \\ &= x + O(x^3)\end{aligned}$$

In particular, this makes rigorous the idea that "$(x+$ something small), multiplied by $(1+$ something small), is still $(x+$ something small)." It's the kind of reasoning that one sees in physics books all the time, but instead of using the mathematically valid big-O, they say "we'll ignore this term" or "assume this term is zero." However, being sloppy in this uncontrolled way can result in unforeseeable errors. Missing error terms can get combined in ways that the combination of the error is of the same order of magnitude as the term you care about. With big-O, error terms are still present, but they're present in a way that doesn't complicate calculations. When two terms get combined, you're forced to ask if the combined error is too big. The interface helps prevent careless mistakes. Following one of the major themes of this book, it reduces both the cognitive load of doing algebra, and the cognitive load of keeping track of error terms.

We can extend this notation to infinite limits:

Definition 15.2. Let $f, g : \mathbb{R} \to \mathbb{R}$ be two functions with $g(x) \neq 0$ for all sufficiently large x. We say $f(x) = O(g(x))$ as $x \to \infty$ if the limit of their ratios does not diverge.

$$\lim_{x \to \infty} \left| \frac{f(x)}{g(x)} \right| < \infty$$

With the infinite limit, we're saying $|f(x)| \leq D|g(x)|$ for all sufficiently large x and some constant D. Here and elsewhere in math, "sufficiently large" abbreviates the $\lim_{x \to \infty}$. Some N exists, above which ($x > N$) the property is always true.

Definitions 15.1 and 15.2 have the same name because they satisfy the same properties. However, the hypotheses of these properties are different. For example, $x^2 = O_{x \to 0}(x)$ and $x^3 = O_{x \to 0}(x)$, implying $x^2 + x^3 = O_{x \to 0}(x)$, but for infinite limits the part that fails is $x^2 \neq O_{x \to \infty}(x)$. Instead, $x^2 = O_{x \to \infty}(x^3), x = O_{x \to \infty}(x^3)$, and so $x + x^2 = Ox \to \infty(x^3)$.

Little-o, Omega, and Theta

There is one other important asymptotic notation known as little-o notation. If big-O is phrased as "less than or equal to," then little-o is "much less than." Formally, instead of the defining limit being finite, for little-o the defining limit is zero.

Definition 15.3. Let $f, g : \mathbb{R} \to \mathbb{R}$ be two functions with $g(x) \neq 0$ for all x sufficiently close to $a \in \mathbb{R}$. We say $f(x) = O(g(x))$ as $x \to a$ if the limit of their ratios is zero.

$$\lim_{x \to a} \left| \frac{f(x)}{g(x)} \right| = 0$$

We allow $a = \infty$, in which case the nonzero condition is again "sufficiently large" instead of "sufficiently close to a."

In other words, the function $f(x)$ vanishes compared to $g(x)$. So while $2x^3 = O(x^3)$ as $x \to \infty$, it's not $o(x^3)$. Little-o requires something smaller, for example $x^2 = o(x^3)$. Shaving off any sufficiently large-growing function can also be the difference between big-O and little-o. In particular, as $x \to \infty$ it's true that $x = o(x \log x)$ and even $x = o(x \log(\log(\log(x))))$.

The rest of the asymptotic notation family is defined by relation to big-O and little-o.

Definition 15.4. Let f, g be functions as before.

- Define $f = \Omega(g)$ if $g = O(f)$. This is a big-O "lower bound."
- Define $f = \omega(g)$ if $g = o(f)$. This is a little-o "lower bound."
- Define $f = \Theta(g)$ if $f = O(g)$ and $g = O(f)$. This is an asymptotic "equality."

Little-o in particular has some nice uses simplifying calculus. In particular, we can define the derivative entirely in terms of O notation. Donald Knuth is a champion of this approach.

Definition 15.5. Let $f(x)$ be a function. We say that $f'(x)$ is the derivative of f if (for a parameter $\varepsilon \to 0$)

$$f(x + \varepsilon) = f(x) + f'(x)\varepsilon + o(\varepsilon)$$

As an exercise, prove that this is a restatement of the usual definition of the derivative as a limit.

Part of what makes this version of the derivative definition so elegant is that it puts the core idea of multivariable derivatives—that we care about a linear approximator—front and center. The function f is literally approximated by a linear map $\varepsilon \mapsto f'(x)\varepsilon$ in the formula. All of the cruft about limits is now hidden by the O notation. As an example of its usage, the derivative of x^2 is computed to be $2x$:

$$(x + \varepsilon)^2 = x^2 + 2x\varepsilon + \varepsilon^2 = x^2 + (2x)\varepsilon + o(\varepsilon).$$

Recall the chain rule, Theorem 8.10, which you proved in an exercise and we generalized in Chapter 14. We can prove this theorem using easy calculations.

Theorem 15.6. *The derivative of $f(g(x))$ is $f'(g(x))g'(x)$.*

Proof. Using the definition of differentiability for g,

$$f(g(x + \varepsilon)) = f(g(x) + g'(x)\varepsilon + o(\varepsilon))$$

Define $\eta = g'(x)\varepsilon + o(\varepsilon)$. Note that as $\varepsilon \to 0$ we also have $\eta \to 0$. So we can apply the definition of the derivative to f.

$$\begin{aligned}
f(g(x + \varepsilon)) &= f(g(x) + \eta) \\
&= f(g(x)) + f'(g(x))\eta + o(\eta) \quad \text{(now expand } \eta\text{)} \\
&= f(g(x)) + f'(g(x))g'(x)\varepsilon + [f'(g(x))o(\varepsilon) + o(g'(x)\varepsilon + o(\varepsilon))]
\end{aligned}$$

Note that $f'(g(x))$ and $g'(x)$ are constants relative to the little-o, so the bracketed terms simplify to $o(\varepsilon)$. What's left is the coefficient of ε, which is $f'(g(x))g'(x)$. □

Half of the work in this book is finding computationally friendly representations of interesting conceptual ideas. In this case big-O allowed us to turn *proofs* into easy computation!

Algorithm Analysis

Infinite limit big-O notation is a hallmark of algorithm runtime and space analysis. One cares about the runtime of an algorithm as the input size scales. The prototypical example is sorting. If an input list has n fixed-length integers, then BubbleSort has $O(n^2)$ worst-case runtime, while MergeSort has $O(n \log n)$ worst-case runtime. For this essay we ignore the worst-case/best-case/average-case distinction.

To say anything meaningful about which algorithm is better, we want big-O for two reasons. First, just as the interface for a software system shouldn't depend on the implementation, our analysis of the quality of an algorithm shouldn't depend on the fine-grained details of the implementation. If one decides to structure the algorithm as three functions instead of four, the raw runtime will change; extra steps are taken to push stack frames and handle return values! Of course, many engineers spend a lot of important and valuable time studying the fine-grained runtime of time-critical algorithms. There are experts in loop-unrolling, after all. But big-O isn't meant for those situations; rather, it's meant for the life of the system that comes before fine-tuning. Big-O is a first-responder to the scene. By the time you're fine-tuning, big-O's job is done.

Second, and closely related, the analysis of the quality of the algorithm shouldn't depend on features of the system the code is being run on that are beyond the programmer's control. If you're sensitive to whether your C compiler is run with aggressive or *extremely* aggressive optimization flags, then big-O will not help. But most systems don't ever reach that level of care in their entire lifetime. Big-O allows you to ignore it.

And so we package those details up into a "constant factor" of overhead, which we accept as the penalty for being able to make decisions on principle. As such, given two algorithms with different big-O runtime, the order of magnitude change inside the big-O is our main focus. When we ask, "can this algorithm be solved any faster?" we don't mean can the constant be improved. Rather, we mean can it be solved an *order of magnitude* faster, ignoring constants and runtime for small inputs.

I often hear the complaint, "But what if the constant factor is a billion! Then it's completely useless to use big-O!" Computer scientists are well aware of the possibility that the hidden constant might be absurd. A witty meme, whose origin I can't recall and failed to hunt down, involves the Black Knight of Monty Python and the Holy Grail. This character famously loses his limbs in a sword fight, but refuses to surrender, exclaiming, "It's just a flesh wound!" On this image, the meme superimposes the quote, "It's just a constant factor!" Joking aside, more often than not the constant factors are mere flesh wounds. Constants dominating runtime—i.e., when big-O misleads—is the exception to the rule, and usually a sign of recent, or purely theoretical research. A famous example is the linear-time algorithm for polygon triangulation. This algorithm has a large constant factor, and is so tricky to implement that it has been called "hopeless" by Steve Skiena, the author of "The Algorithm Design Manual."

We've established that big-O can be used to measure things beyond algorithm runtime and space usage, like the quality of an approximation. Indeed, big-O can be used to discuss the usage of *any* constrained resource. For Taylor series the resource is "deviation from the truth," but in computer science there are a whole host of other things that big-O is used to analyze.

- **Communication:** In a distributed system, a common bottleneck is the amount of data that needs to be communicated across servers in order to finish a computation.

- **Randomness:** In cryptography, one can measure the security of a protocol in terms of encryption key size, which is usually proportional to the number of bits of a random seed. High quality random number generators can be slow, and time-sensitive cryptographic applications need to make a tradeoff between security and time.

- **Collisions:** Load balancers have to assign jobs to servers with an extremely high rate of jobs assigned per second. In particular, they almost never have enough time to ask a server how many jobs its processing. Instead, load balancing algorithms use randomness and reason about the expected worst-case load of a server. One can think of collisions of job assignments as a constrained resource a load balancer wants to minimize for the most impacted server.

- **Errors:** In systems where data integrity is important, expensive, and bits are often lost or flipped (such as data being transmitted through space, or on a scratched up disc), one often employs redundancy schemes called *error-correcting codes* that allow one to recover from these errors. Such schemes require one to store additional bits, and so there's a tradeoff between how many additional bits one needs to store and the error tolerance of the scheme.

- **Labeled examples:** Most machine learning systems require labeled training data to produce a classifier. Since compute power is generally cheaper than getting humans to label examples, one major bottleneck on the efficiency of a learning system is access to clean data. Many learning systems are studied under the lens of so-called *query complexity*, which measures access to data. A popular topic these days is also interactive learning, in which a learning system has a "human in the loop" that helps the machine with difficult examples. A human doing work is clearly a bottleneck to an automated system.

- **Regret:** Some machine learning systems involve an explore/exploit tradeoff, where the learning algorithm receives a reward for each action it takes, and would like to find the best actions while still getting a good reward as it searches.[2] The quantity one wants to optimize for is *regret*, the difference between the reward you got and the reward you would have got had you behaved optimally in hindsight.

---

[2] If you're interested in this, a keyword to search for is "bandit learning."

Each of these topics has a rich history of design and analysis, and for each the principles of the discussion revolve around asymptotic analysis. An interactive learning system that takes n pieces of input data but requires $\Omega(n)$ queries to a human to learn can already be determined unscalable, but one that only needs $O(\log(n))$ might work. A load balancer that spreads m jobs over n servers and causes the worst server to have $\Theta(m/n + \sqrt{m})$ jobs is almost certain to crash servers during peak hours compared to one that guarantees $O(m/n + \log n)$.

Big-O is a cognitive tool that allows a human to organize and make sense of a mess of details in a rigorous fashion. It's a tool for high level thinking. Software is full of constrained resources, tradeoffs, and the desire for principled decision making. Fluency in asymptotic language will help you navigate these decisions efficiently and formulate hypotheses that can then be backed up by data.

Chapter 16
Groups

> *We need a super-mathematics in which the operations are as unknown as the quantities they operate on, and a super-mathematician who does not know what he is doing when he performs these operations. Such a super-mathematics is the Theory of Groups.*
>
> – Sir Arthur Eddington

In Chapter 10 we briefly discussed the shift in mathematics from thinking about objects to thinking about transformations between objects. This shift was radical for mathematics and much of physics. It has been less dramatic for programmers, because many ideas that brewed in mathematics for centuries have commonplace analogues in programming. That, and that software matured as a discipline largely after these mathematical revolutions took hold.

Embodying part of this novelty are ideas like programs that transform other programs. You write programs. Compilers are programs that turn your programs into other programs. A program analyzes the quality of a compiler. Programs test the correctness of the compiler analyzer. Software automates the running of the tests of the correctness of the compiler analyzer. And, of course, you use a program to help refactor the programs that automate the running of the tests of the correctness of compiler analyzer. It's programs all the way down.

What's less obvious to a programmer is that studying the class of transformations of an object provides insight into that object. It's as if studying a refactoring tool not only taught you how to write easily-refactorable programs, but also gave you a shortcut to understanding an unfamiliar program! Building up a theory based on transformations is like a slick development framework, which you later learn applies to programs you never anticipated writing. Group theory is a fantastic example of this.

Group theory is the mathematical study of *symmetry*. As we'll see in this chapter, symmetry has algebraic structure. We can work with symmetry in much the same way we do algebra with numbers or matrices. This is why group theory is part of a general area of mathematics called *abstract algebra*.

The original insight of group theory,[1] bringing us full circle to Chapter 2, is that the roots of a single-variable polynomial have symmetric structure. Such structure can be for-

---
[1] By many accounts attributed to a Frenchman named Évariste Galois in the early 1800's

mulated as a group, and used to analyze the properties of a polynomial. Or, as the case may have it, to make general statements about all polynomials. Indeed, as we mentioned in Chapter 8, it can be hard to analytically find the roots of a polynomial of large degree. By "analytically" I mean in the sense of the quadratic formula: a single algebraic expression using elementary operations, involving the coefficients of the polynomial, which one could use to find all the roots. The difficulty of this motivated us to derive and implement Newton's method for numerically finding approximate roots.

We have group theory in part to thank for not wasting our time on the analytical approach. Using group theory one can prove that it's not merely *difficult* to find an algebraic formula for the roots of a generic degree-5 polynomial. It's impossible. We foreshadowed this in Chapter 2 when we discussed existence and uniqueness. This theorem—known as the Abel-Ruffini theorem—is a crown jewel of mathematics. And though this book is too short to do the theorem justice, the modern proof relies heavily on the shift in thought from objects to transformations.

A second perspective on groups is understood easily, almost trivially, from programming. One beautiful aspect of group theory is how it allows one to cleanly compartmentalize the difference between a mathematical object and its representation. The definition of a group serves as an interface or a template class—in the sense of object-oriented programming—and concrete groups are semantically equivalent implementations of this interface in different contexts. True surprises occur when a family of objects that has been studied for a long time is discovered to implement the group interface. Such is the case with elliptic curves of cryptography fame. Any time a field of mathematics has the word "algebraic" prepended to it—such as algebraic geometry or algebraic topology—you automatically know the subject is about finding algebraic structures like groups hidden among seemingly non-algebraic company. When such miracles occur, you can leverage the power of algebra to compute in the cleaner, abstract setting of the algebraic structure.

Michael Atiyah, a famous geometer, once quipped,

Algebra is the offer made by the devil to the mathematician. The devil says: "I will give you this powerful machine, it will answer any question you like. All you need to do is give me your soul: give up geometry and you will have this marvellous machine."

Hermann Weyl echoed a similar idea seventy years earlier: "In these days the angel of topology and the devil of abstract algebra fight for the soul of each individual mathematical domain." While these seem like superstitious warnings to the unsuspecting apprentice of mathematics, the utility of algebra for computation is undeniable. If there's anything to read from these quotes, it's that geometric arguments are considered fashionable, pure, and beautiful by a certain group of influential mathematicians. Subcultures abound.

But you, dear programmer, would never patronize computation as mere contentedness. We know deep in our hearts that computation is beautiful. It deserves to be cherished as an equal to geometry, analysis, logic, and the rest. Algebra deserves our special attention in that, to the extent it destroys geometry, it enables computation.

A   B

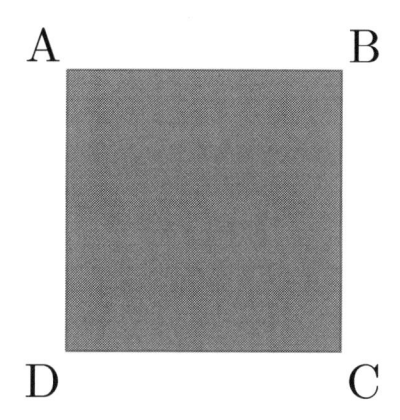

D   C

Figure 16.1: A square with each of its corners labeled.

The most common example of a group—and its raison d'etre—is the set of symmetries of some object. That is to say, a group is nothing if it does not "act" on some set by transforming it in a composable, reversible way. You use groups to elucidate the symmetry in objects of interest. In this final chapter we'll see how the concept manifests itself in Euclidean and hyperbolic geometry, and in the exercises we'll explore groups as they show up in number theory, cryptography, polynomials, graphs, and others.

We'll finish off the chapter, and the book, with a dive into hyperbolic geometry. We'll see how of geometry can be studied via the groups that transform geometric space. Finally, we'll apply what we learned to draw hyperbolic tessellations, of the same sort that M.C. Escher studied to create his art.

16.1 The Geometric Perspective

The simplest approach to understanding groups as objects describing symmetry is with geometry. Picture a square in the plane. We're going to transform this square. To keep track of what we're doing, we label each corner with a letter, as in Figure 16.1.

Now imagine cutting this square out of the plane, doing some kind of rigid physical manipulation, and placing it back into the same hole so that it fills up all the same space. For example, you could rotate the square counterclockwise by a quarter turn, or reflect it across the AC diagonal, or both. These are *rigid motions* of the square. As functions, they are bijections from the square to itself. Moreover, they preserve the distances between all pairs of points. In symbols, let's give coordinates (x, y) to the square. Say the square is the product of two intervals

$$[0, 1] \times [0, 1] = \{(x, y) \in \mathbb{R}^2 : 0 \leq x, y \leq 1\},$$

and call $f(x, y)$ one of the rigid motions described above. Then f has the property that for every pair of points $(x_1, y_1), (x_2, y_2)$, the distance between (x_1, y_1) and (x_2, y_2)

is equal to the distance between $f(x_1, y_1)$ and $f(x_2, y_2)$.

Definition 16.1. Given a set X, a *metric* is a non-negative function $d : X \times X \to \mathbb{R}$ with the following three properties:

- $d(x, y) = 0$ if and only if $x = y$.
- $d(x, y) = d(y, x)$ for all $x, y \in X$.
- The "triangle inequality": $d(x, y) \leq d(x, z) + d(z, y)$ for all $x, y, z \in X$.

These properties make an arbitrary function sensible enough that one could reasonably call it a "distance" function. Of particular interest in the triangle inequality, which says that taking a direct path from x to y is never worse than taking an indirect path through z.

In Chapters 10 and 12 we discussed how the Euclidean inner product gives rise to a distance metric

$$d(x, y) = \|x - y\| = \sqrt{\langle x - y, x - y \rangle}.$$

This metric is the same metric for Euclidean geometry. However, not all metrics arise from an inner product. Our study of hyperbolic geometry will produce a highly non-linear metric, so it's worth teasing apart the two concepts.

Definition 16.2. Let X be a set with a distance function $d : X \times X \to \mathbb{R}$. An *isometry* or *rigid motion* of X is a bijection $f : X \to X$ such that $d(x, y) = d(f(x), f(y))$ for every $x, y \in X$.

Back to our example of the square. Since we labeled the corners, we can track how an isometry affects the corners. And in a sense that will become clear shortly, we *only* care about how it affects the corners. If we denote a counterclockwise quarter-turn by ρ (the Greek lower-case rho) and a flip across the AC diagonal[2] by σ (the Greek lower-case sigma), we can write down a sequence of these operations like

$$\rho\rho\sigma\rho,$$

where we apply the operations in order from right to left. That is, the above operation is "rotate a quarter turn, then flip, then rotate twice more." Figure 16.2 shows how the symmetries transform the square.

To emphasize that we're talking about isometries that fix the square (as it might be viewed inside an ambient space like $\mathbb{R}^2$), we call these isometries *symmetries* of the square. This gives rise to the natural question: what are all of the different symmetries of the

---

[2] This flip is specific to the *initial* position of A and C. As A and C move around, the flip operation is still top-left-corner to bottom-right-corner.

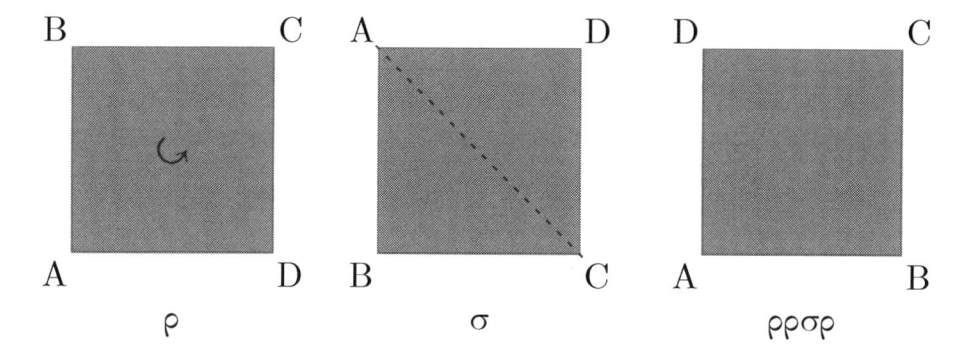

Figure 16.2: Example symmetries of the square.

square? There are infinitely many ways to compose symmetries on paper, but two symmetries created via different methods can result in the same operation.

To study this, we identify some core properties of symmetries.

- The operation where we "do nothing" (the identity function $f(x, y) = (x, y)$) is a symmetry.

- Every symmetry has an opposite symmetry. This follows from isometries being bijections.

- We can compose any two symmetries to get another symmetry.

Two different ways to compose symmetries can result in the same symmetry. Indeed, flipping across the same diagonal twice is the same thing as doing nothing, and rotating four times in the same direction is also the same thing as doing nothing. Note we only consider the relative change of the square compared to how it started. To apply the next rigid motion in a sequence, you need not know how it was previously transformed.

Moreover, a symmetry of the square is completely determined by how it acts on the corners. We sketch a proof. By our requirement that distances are preserved, the corners must also go to corners. Specifically, opposite diagonal corners have a maximal distance between any two points in the square. Their distance can't be achieved except by opposite-corner points. Once the corners are chosen every other point in the square is required to be a certain distance from each corner. And there is a short but not completely trivial proof that three or more circles (whose centers don't form a line) that have a simultaneous intersection point must have *exactly* one such point. Figure 16.3 shows an example.

As an exercise, flesh out this proof sketch in more detail. However, be warned that not all possible labelings of the corners arise from symmetries of the square. Opposite corners of the square cannot be mapped by an isometry to neighboring corners.

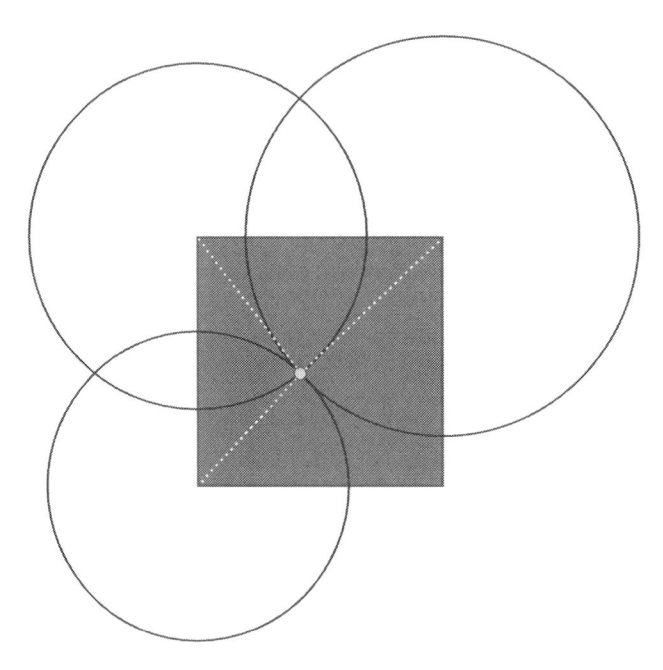

Figure 16.3: The position of a point is uniquely determined by its distance from the three corners.

With a handful of symmetries, such as our ρ and σ from earlier, we can write down compositions of those symmetries, and make equations of symmetries. The following three are some particularly simple ones:

$$\rho^4 = 1$$
$$\sigma^2 = 1$$
$$\rho\sigma\rho = \sigma$$

Where 1 is a placeholder for the identity symmetry. The suggestive algebraic notation hints at our goal: 1 is the multiplicative *identity* satisfying, e.g., $\rho \cdot 1 = \rho$. We even write ρ^{-1} as the quarter-turn in the reverse direction.

These three identities allow us to reduce complicated expressions, such as $\sigma\rho^9\sigma\rho^{-3}\sigma$, to a more tractable form. The geometric picture of applying symmetries give way to mechanized computation. The notation bears the burden of the mental picture. Note that below we mostly use $\rho\sigma\rho = \sigma$ to reduce the large powers of ρ.

$$\sigma\rho^9\sigma\rho^{-3}\sigma = \sigma(\rho^9\sigma\rho^9)\rho^{-12}\sigma$$
$$= \sigma(\rho^8\sigma\rho^8)\rho^{-12}\sigma$$
$$\cdots$$
$$= \sigma(\rho\sigma\rho)\rho^{-12}\sigma$$
$$= \sigma^2\rho^{-12}\sigma$$
$$= \sigma^2(\rho^4)^{-3}\sigma = 1 \cdot (1)^{-3} \cdot \sigma = \sigma.$$

As you might have guessed, the properties we've identified are what define a group, and the algebra above is characteristic of doing algebra with a group structure. Before we see the formal definition, here's a more complicated example of a group: the symmetries of the Rubik's cube.[3]

In the same way that we can enumerate all possible symmetries of the square, one could enumerate all possible symmetries of the Rubik's cube. One can rotate any one of the six faces of the cube, but the relationships between operations are not at all obvious. The colored stickers take place of A, B, C, D labels to distinguish two configurations, but it's not clear which (if any) stickers are superfluous. Nevertheless, the same properties hold: there is a do-nothing operation, every operation is reversible, and any two operations can be composed and the result is still a viable operation. As we've suggested, if you want to understand the Rubik's cube, you should study its group of symmetries.

16.2 The Interface Perspective

Now that we've seen two geometric examples, it's time for a formal definition of a group as an interface. The three properties of symmetries mentioned above sculpt the definition:

Definition 16.3. A *group* $(G, \cdot)$ is set paired with a binary operation $\cdot : G \times G \to G$, so that the following properties hold:

1. G contains an identity element denoted e for which $e \cdot x = x$ and $x \cdot e = x$ for all $x \in G$.

2. For every $x \in G$ there is some element $y \in G$ called an "inverse" for which $x \cdot y = e$ and $y \cdot x = e$. (A priori there may be more than one such inverse)

3. The group operation is associative.[4] That is, $x \cdot (y \cdot z) = (x \cdot y) \cdot z$.

People often say that a set G is a group "under" an operation instead of "paired with." There are a few issues we need to tackle regarding this definition and the notation associated with it, but first let's see some trivial examples.

---
[3] Copyright restrictions prevent me from including a photograph.
[4] Since most of our groups will be numbers, matrices, or functions, this axiom will naturally hold. We will ignore it for brevity.

The singleton set $\{e\}$ with the binary operation $\cdot$ defined by asserting $e \cdot e = e$ is a group. And there was much rejoicing. The set of integers $\mathbb{Z}$ forms a group under the operation of addition. It is common knowledge that zero fits the definition of the identity element, that the sum of two integers is an integer, that addition on integers is associative, and that every integer x has an additive inverse $-x$.

Likewise, all of the number systems in this book are groups under addition: rational numbers, real numbers, complex numbers, etc. If we want to work with multiplication, it is not hard to see that $\mathbb{R} - \{0\}$ is a group, since every nonzero real number has a multiplicative inverse, and 1 is the multiplicative identity. Vector spaces are groups under vector addition; indeed, the group axioms are a subset of the vector space axioms.

An important example comes from our discussion in Chapter 9, the set of integers modulo n, denoted $\mathbb{Z}/n\mathbb{Z}$, under the operation of addition modulo n. For example, $\mathbb{Z}/4\mathbb{Z} = \{0, 1, 2, 3\}$.

A few basic propositions clear up the ambiguities in Definition 16.3. For instance, the uniqueness of the identity element follows from the other axioms of a group. Here's a proof by contradiction: if there were two identity elements $e \neq e'$, then by the following logic they must be equal:

$$e = e \cdot e' = e'$$

The first equality holds because e' is an identity element, and the second because e is. A similar proof shows that the inverse of an element is unique. These facts justify the following notation: we call *the* identity element 1, and use subscripts $1_G, 1_H$ to distinguish between identity elements in different groups G, H. We also replace the explicit $\cdot$ operation with an invisible operation (juxtaposition). So that xyz replaces $x \cdot y \cdot z$. Moreover, we emulate repeated applications of the operation by saying x^n to mean $x \cdot x \cdot \cdots \cdot x$ multiplying n times.

One more caveat to support "legacy" math. If we're talking about the integers $\mathbb{Z}$ under addition, the juxtaposition operation (which implies multiplication) feels unsanitary. It simply won't do. In this case, and whenever we have a group of numbers with a $+$ symbol as the operation, we'll use $+$. And instead of x^n we'll use nx to mean $x + x + \cdots + x$ adding n times. Here n is not considered an element of $\mathbb{Z}$ as a group, but just the number of additions. Likewise, $-x$ is the inverse of x, while in a multiplicative group the inverse is x^{-1}. This is purely syntactic sugar.

Now we demonstrate how two drastically different sets can have the same underlying group structure, which will inform our dive into structure-preserving mappings between groups. The first group we understand well: $\mathbb{R}$ under addition. For the second, consider the set of 2×2 matrices of the following form, under the operation of matrix multiplication.

$$G = \left\{ \begin{pmatrix} 1 & a \\ 0 & 1 \end{pmatrix} : a \in \mathbb{R} \right\}$$

The identity matrix is the identity element. Notice G has some familiar structure.

$$\begin{pmatrix} 1 & a \\ 0 & 1 \end{pmatrix} \begin{pmatrix} 1 & b \\ 0 & 1 \end{pmatrix} = \begin{pmatrix} 1 & a+b \\ 0 & 1 \end{pmatrix}.$$

Indeed, matrix multiplication in G corresponds to addition of the top-right entry of the matrix. This suggests the natural bijection $f : \mathbb{R} \to G$ defined by

$$x \mapsto \begin{pmatrix} 1 & x \\ 0 & 1 \end{pmatrix}.$$

And indeed, addition of the inputs corresponds exactly to multiplication of the corresponding matrices! The fact that these *particular* groups have the same underlying structure isn't all that shocking. What's deep is that we have two different concrete representations for the same abstract algebraic structure. Not only are the elements in bijective correspondence, but the *operations* are as well! Just as juicy, any concrete representation of the abstract group $\mathbb{R}$ can be identified by finding this sort of operation-correspondence with the set $\mathbb{R}$ and the usual operation $+$.

A mathematician might see all of this as a challenge: can we classify all the different kinds of group structures? Could we get a new perspective on the symmetry group of the square by turning it into a suitable group of matrices?

Before we get ahead of ourselves, let's make these structure-preserving maps precise.

16.3 Homomorphisms: Structure Preserving Functions

To study the structure of groups, we study the structure of compatible functions between groups. By "compatible," I mean the group structure is somewhat preserved.

Definition 16.4. Let G, H be groups under multiplication. A function $f : G \to H$ is called a *homomorphism* if for every $x, y \in G$, $f(xy) = f(x)f(y)$. The multiplication on the left is in G and the multiplication on the right is in H.

Homomorphisms between groups don't necessarily preserve everything about a group (in particular, they need not be bijections), but they do preserve the defining features of the group structure. To build up intuition we can do some simple proofs.

Proposition 16.5. *Group homomorphisms preserve the identity.*

Proof. Let G be a group with identity 1_G, and H a group with identity 1_H. Let $f : G \to H$ be a homomorphism. Since f preserves the group operation,

$$f(1_G) = f(1_G 1_G) = f(1_G) f(1_G)$$

Since H is a group, all its elements have inverses, including $f(1_G)$. So multiply both the far left hand side and the far right hand side of the above equation by $f(1_G)^{-1}$ and get

$$f(1_G) f(1_G)^{-1} = f(1_G) f(1_G) f(1_G)^{-1}$$

The left hand side is equal to 1_H, and the right hand side is equal to $f(1_G)$, so $f(1_G) = 1_H$, as desired.

□

Proposition 16.6. *Group homomorphism preserve inverses.*

Proof. The same idea as the last proof, but more briefly: If $x \in G$ and $f : G \to H$ is a homomorphism, then

$$1_H = f(1_G) = f(xx^{-1}) = f(x)f(x^{-1})$$

And likewise for $f(x^{-1}x)$. Taking the left- and rightmost ends, we've shown that $f(x^{-1})f(x) = f(x)f(x^{-1}) = 1_H$. In particular, the inverse of $f(x)$ is $f(x^{-1})$. Another way to say this is that $f(x)^{-1} = f(x^{-1})$.

□

The extent to which a homomorphism degrades the structure of the input group is tracked by what elements are mapped to the identity.

Definition 16.7. Let G, H be groups, and $f : G \to H$ a homomorphism. Then the *kernel* of f, denoted ker f, is the set

$$\ker f = \{x : f(x) = 1_H\}$$

An example: $G = \mathbb{Z}$ under addition and $H = \mathbb{Z}/10\mathbb{Z}$ under addition modulo 10. Let $f : G \to H$ mapping $n \mapsto 2n \mod 10$. The kernel of f is $\{0, \pm 5, \pm 10, \pm 15, \ldots\}$. Despite losing the multiples of 5, the image $f(G)$ still has a group structure inside H. Note $f(G) = \{0, 2, 4, 6, 8\}$, and the group operation in H—applied only to elements of $f(G)$—maintains the property of being in $f(G)$. In other words, part of the structure of G is embedded inside H using the operation of H, but not all of it.

A group that sits inside another group (and shares the containing group's operation) is called a subgroup.

Definition 16.8. Let $H \subset G$ be two sets and let G be a group under the operation $\cdot$. Then H is called a *subgroup* of G if:

- $1 \in H$.

- For all $x, y \in H$, it's true that $x \cdot y \in H$.

- If $x \in H$, then so is x^{-1}.

Another term for the above conditions is that H is "closed" under $\cdot$ and the inverse-taking operation $(-)^{-1}$.

Our observation about the image of a homomorphism being a group is no coincidence. A homomorphism provides two useful subgroups: its image and its kernel.

Theorem 16.9. *Let G, H be groups and $f : G \to H$ be a group homomorphism. Then $\ker f$ is a subgroup of G and $\operatorname{im} f$ is a subgroup of H, where $\operatorname{im} f$ denotes the image of f.*

Proof. First we prove that $\ker f$ is a subgroup of G. We'll prove this directly, by assuming x, y are arbitrary elements of $\ker f$, and showing that $xy \in \ker f$ and $x^{-1} \in \ker f$. These are the second two conditions required of a subgroup by Definition 16.8, and the first condition, $1_G \in \ker f$, is implied by Proposition 16.5.

If $x, y \in \ker f$, then $f(xy) = f(x)f(y) = 1 \cdot 1 = 1$, so $xy \in \ker f$. Likewise, since group homomorphisms preserve inverses, $f(x^{-1}) = f(x)^{-1} = 1^{-1} = 1$.

Next we'll prove $\operatorname{im} f$ is a subgroup of H. Let $x, y \in \operatorname{im} f$. By the definition of the image, there are two elements $a, b \in G$ for which $f(a) = x, f(b) = y$. Then $f(ab) = f(a)f(b) = xy$, which by definition means that xy is in the image of f. Likewise, $f(a^{-1}) = f(a)^{-1} = x^{-1}$, so x^{-1} is in the image of f. Again, Proposition 16.5 implies $1_H \in \operatorname{im} f$.

□

Even better, as we discussed in Chapter 9, a function $f : G \to H$ defines a natural equivalence relation on the domain. When f is a homomorphism, the corresponding quotient maintains the group structure of G. Appropriately, it's called the quotient group.

Let $f : G \to H$ be a group homomorphism. Define an equivalence relation whereby two elements $a, b \in G$ are equivalent if $ab^{-1} \in \ker f$. Or, in terms of additive groups, $a - b \in \ker f$. Note that this aligns with the equivalence relation defined in Chapter 9 using $f(a) = f(b)$, since then $f(ab^{-1}) = f(a)f(b)^{-1} = 1_H$, implying $ab^{-1} \in \ker f$.

Take $\mathbb{Z}$ with addition, and the map $f : \mathbb{Z} \to \mathbb{Z}/10\mathbb{Z}$ defined by $x \mapsto 2x$. The kernel of this map is the subgroup $\{0, \pm 5, \pm 10, \pm 15, \dots\}$. The quotient $\mathbb{Z}/\ker f$ is the set of equivalence classes $\{[0], [1], [2], [3], [4]\}$. The numbers 3, 8, and -22 are all in the equivalence class $[3]$, because, for example, $3 - (-22) = 25 \in \ker f$. The group operation on $\mathbb{Z}$ passes to the equivalence classes, so that $[a] + [b] = [a + b] = [a + b \mod 5]$. The quotient group is suspiciously similar to $\mathbb{Z}/5\mathbb{Z}$. Indeed, they are isomorphic (cf. the upcoming Definition 16.11).

Lemma 16.10. *For any homomorphism f, the quotient set $G/\ker f$ forms a group under the operation $[a][b] = [ab]$.*

Proof. You will prove this in the exercises.

□

Finally, if $\ker f = \{1\}$, then f is necessarily an injection.. Such homomorphisms completely preserve the structure of the input group, embedded via the image of f inside the codomain. In the added case that f is a surjection,, then f completely preserves the structure of the group.

Definition 16.11. *Let G and H be groups. A homomorphism $f : G \to H$ is called an isomorphism if it is a bijection. If there is an isomorphism between G and H, we call them isomorphic.*

If G and H are isomorphic, they have identical group structure, and H is simply a relabeling of the elements of G. The boolean comparison (or assertion) that two groups G, H are isomorphic is denoted $G \cong H$. And in words, we say that two groups are the same "up to isomorphism," meaning only their representations are different. For our "suspicious" example above, $\mathbb{Z}/\ker f \cong \mathbb{Z}/5\mathbb{Z}$.

A simple theorem relates the groups defined by a homomorphism.

Theorem 16.12 (The first isomorphism theorem). *Let G, H be groups, $f : G \to H$ a homomorphism. Then $\operatorname{im} f \cong G/\ker f$.*

16.4 Building Blocks of Groups

If you're tasked with understanding a mysterious group G, perhaps encountered in a wildly non-algebraic locale such as the symmetries of geometric solids, the general strategy is to find homomorphisms between G and other groups you understand well. A homomorphism $f : G \to H$ gives you two groups related to G: $\ker f$ and $G/\ker f$. Meanwhile, a homomorphism $g : H \to G$ gives the subgroup $\operatorname{im} g$. Each is a local piece of information about G that you can use to reconstruct a global picture of G. Thus, every enterprising mathematician has a repertoire of concrete groups to use as building blocks.

The most common, as we've seen multiple times in this chapter, are the integers under addition, their subgroups, and their quotients, under addition and addition modulo n. These arise as the kernels and quotients of the maps $\mathbb{Z} \to \mathbb{Z}$ defined by $x \mapsto nx$. The kernels have the form $n\mathbb{Z} = \{nx : x \in \mathbb{Z}\}$ for some fixed n, and also the trivial subgroups $\{0\}, \mathbb{Z}$. The quotients are denoted $\mathbb{Z}/n\mathbb{Z}$.

The groups $\mathbb{Z}$ and $\mathbb{Z}/n\mathbb{Z}$ both have the property that 1, when repeatedly added to itself, produces the entire group. Because of this, 1 is called a *generator* of the group. In general, an element $x \in G$ is called a generator if the subgroup $\{1, x, x^2, x^3, \dots\}$ is equal to G. Groups that have generators are called *cyclic* groups, and all cyclic groups are isomorphic to $\mathbb{Z}$ or $\mathbb{Z}/n\mathbb{Z}$. In general, a set $S \subset G$ is said to *generate* G if every $x \in G$ is a product of elements in S. A generating set of a group is like a vector space basis, but without size guarantees. A group G may have generating sets of different sizes. Hence, any concept of "group dimension" must be more nuanced.

One of the simplest ways to build a larger group from smaller pieces is the *direct product*. This construction simply forms the product of two groups as sets, and defines the group operation component-wise. E.g., $\mathbb{Z} \times \mathbb{Z}/2\mathbb{Z}$ is the set of pairs $\{(n, b) \mid n \in \mathbb{Z}, b \in \{0, 1\}\}$, where $(n, b) + (n', b') = (n + n', b + b')$.

The set $\mathbb{Z}/n\mathbb{Z}$ forms a group under multiplication *if* we remove the numbers k such that $\gcd(n, k) \neq 1$. In the special case that n is prime, we need only remove zero. This group is denoted $(\mathbb{Z}/n\mathbb{Z})^\times$, and it's substantially more interesting than integers under addition. Up to isomorphism it is always possible to write $(\mathbb{Z}/n\mathbb{Z})^\times$ as a product of cyclic groups. However, there is no known generic method for finding generators of the cyclic pieces. This computational difficulty is exploited by RSA public-key cryptography, which we will explore in an exercise.

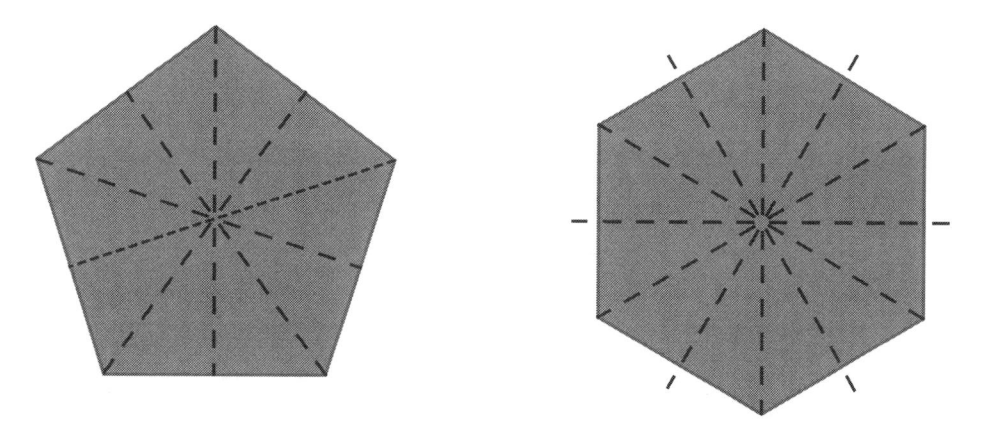

Figure 16.4: Because 5 is odd, the lines of symmetry of the regular pentagon each pass through a side and a vertex.

Next we have the symmetry groups of regular convex polygons[5] in the plane, such as the square we started this chapter with. The group corresponding to the polygon with $n \geq 3$ sides is called the *dihedral group* and is denoted D_{2n}. It has $2n$ elements, corresponding to the n rotations by an angle of $2\pi/n$ and the n reflections across lines passing through the vertices and sides. These lines of symmetry depend on the parity of n, as is made clear by the lines of symmetry in the pentagon and the hexagon in Figure 16.4.

Dihedral groups are not cyclic. Each D_{2n} is generated by ρ and σ, where ρ is a rotation by $2\pi/n$ and σ is a reflection across some axis of symmetry. Because two elements generate the entire group, you might guess D_{2n} to be isomorphic to a product of two cyclic groups, $\mathbb{Z}/2\mathbb{Z} \times \mathbb{Z}/n\mathbb{Z}$, with σ generating the former and ρ the latter. You might guess, and you'd be wrong. These are subgroups, but dihedral groups have extra structure because the interaction between ρ and σ is not independent. If it were, $\sigma\rho\sigma$ would equal $\sigma^2\rho = \rho$, but in fact $\sigma\rho\sigma = \rho^{-1}$. The extra structure is more precisely described by a *semi-direct product*, which you will see in the exercises.

Next we have matrix groups. Given any number system that has addition and multiplication, say $\mathbb{R}$ for example, we can form a group of square matrices under matrix multiplication, which is often called the *general linear group*. Define by $GL_n(\mathbb{R})$ the set of invertible $n \times n$ matrices with real entries. As we saw in Section 16.2, asserting some specific structure on the groups often leads to an interesting subgroup. One famous subgroup of the general linear group is called the *orthogonal group*, denoted $O_n(\mathbb{R})$, consisting of matrices whose columns form orthonormal bases.

$$O_n(\mathbb{R}) = \{A \in GL_n(\mathbb{R}) \mid A^T A = I_n\}.$$

This group will be closely related to the symmetry group of Euclidean space we'll study

---

[5] *Regular* means all the angles have the same measure and all sides have the same length, and *convex* means every line between points in the polygon is completely contained in the polygon.

in Section 16.5. Another interesting facet of groups of matrices is that they have enough structure that one can do calculus on them as if it were a geometric space. In the formal jargon, the general linear group is a smooth manifold. This is far beyond the scope of this book, but at least explains why the general linear group gets such a special name.

The last example is called the *symmetric group*. Really, it should be called the *permutation group*, since it is the set of all bijections of a fixed set to itself. Let A be a set, and define the *symmetric group* $S(A)$ to be the set of all bijections $A \to A$. It is easy to see that if A, B are both finite sets of size n, then $S(A) \cong S(B)$. In that case, denote $S(A)$ by S_n. In the exercises you will study the structure of finite permutation groups, and a useful data representation for computation.

As it turns out, every group is a subgroup of a symmetry group. The proof is simple: every group G has a group homomorphism $f : G \to S(G)$, where $a \in G$ defines the bijection $x \mapsto ax$ (the inverse is $x \mapsto a^{-1}x$). Since im f is a subgroup of $S(G)$ and ker $f \cong \{1\}$, we have that $G \cong$ im f. One takeaway is that if you want to write programs that do computations on finite groups, it's enough to write programs that work with finite permutation groups. Indeed, most useful group-theoretic algorithms are algorithms on finite permutation groups.

16.5 Geometry as the Study of Groups

For the rest of the chapter, we're going to study geometry from the perspective of groups. In fact, the modern mathematical attitude toward geometry is that it *is* the study of groups. This view was espoused by Felix Klein in the late 1800's. Around this time, special cases of projective geometry and hyperbolic geometry had been discovered, but it was largely unclear how different geometries were related.

In general, to define a geometry you need to define a few things:

- A set X of *points* (the space), and a set of *lines*.

- A prescription of *incidence*, i.e., what points lie on what lines.

- A quantity of interest that you want to study. For example, you may want to measure length. In that case, you need a metric $d : X \times X \to \mathbb{R}$.

With these in hand, the symmetry group of the space is the set of bijections $X \to X$ that preserve the quantity of interest. In Euclidean geometry, points and lines are the usual points and lines in $\mathbb{R}^n$, and distance is the quantity of interest. Such "quantities of interest" are called *invariants*. A different type of geometry might only wish to preserve area of figures, or preserve the property of similarity (invariance under scaling).

Klein's view was that a geometry should be studied via its group of symmetries. The classical concepts like angles, areas, and lengths are seen as measures that may or may not be invariant under the application of a symmetry. Thus, geometry has two approaches: given a group of symmetries, study the interesting quantities invariant to those transformations; and given a quantity you think is important, find the group of symmetries that preserves that quantity. Every geometry has a group. Every group corresponds to some geometry.

Klein called his view the *Erlangen Program*.[6] One striking result[7] was that all geometries are a special case of projective geometry—a geometry that allows projections to a possibly infinite horizon. In particular, even though different geometries might have different axioms (regarding, say, configurations of parallel lines), every geometry can be modeled inside of a projective geometry. For example, hyperbolic geometry is projective geometry restricted to a particular surface inside a larger projective space. Moreover, the group corresponding to this model of hyperbolic geometry is a subgroup of the symmetry group of the projective geometry. We get containments of the spaces as sets, and of the groups as subgroups.

$$\text{Hyperbolic geometry} \subset \text{Projective geometry}$$
$$\text{Hyperbolic group} \subset \text{Projective group}$$

We won't study this particular relationship in this book, but it shows how Klein's desire fits into the larger mathematical goal: to connect and unify disparate geometries into a single theory. I encourage the reader interested in cryptography to learn about projective geometry, in part because it's the correct setting for studying elliptic curves. It's also a great way to exercise your linear algebra muscles, as projective geometry is simply a quotient of the vector space $\mathbb{R}^n$ by a suitable equivalence relation.[8]

We now turn to Euclidean geometry, and study it through the lens of groups.

Euclidean Geometry

Euclidean geometry is the study of isometries of $\mathbb{R}^n$ with the usual distance metric $d(x, y) = \|x - y\|$. Recalling Definition 16.2, $f : \mathbb{R}^n \to \mathbb{R}^n$ is an isometry if $d(x, y) = d(f(x), f(y))$ for all $x, y \in \mathbb{R}^n$. Because isometries preserve distance, and angle measure is determined by the lengths of the sides of triangles, isometries also preserve angle measure.

With a few moments of thought, it's easy to come up with examples of Euclidean isometries for the plane:

1. Translations by a fixed vector, i.e., $x \mapsto x + v$.

2. Reflections through a subspace W of dimension $n - 1$. I.e., given x, first compute $w = \text{proj}_W(x)$, then output $x - 2(x - w)$.

---

[6] In mathematics, a "program" is a sort of long-term plan, usually one that is too large for a single mathematician to complete alone. In the case of Klein, mathematicians and physicists found new geometries and symmetry groups to study long after Klein died.

[7] I'm not aware of this claim as a theorem, but rather a famous "attitude" voiced by Arthur Cayley.

[8] A while back I wrote a blog series on this topic, in which I build up elliptic curve cryptography from scratch. The second post in the series defines projective geometry as a quotient. You can find it here: https://jeremykun.com/2014/02/08/introducing-elliptic-curves/.

3. Rotations around points (in 2 dimensions) or lines (in 3 dimensions).[9]

Remember that rotations, projections, and reflections are examples of linear maps. Ignoring translations for a moment, it's natural to wonder which linear maps double as isometries.

Theorem 16.13. *The isometries of $\mathbb{R}^n$ that fix the origin are exactly the linear maps whose columns form an orthonormal basis.*

Proof. In Chapter 12, we observed that matrices with orthonormal columns preserve the inner product. Let A be such a matrix. In $\mathbb{R}^n$, squared distance is $d(x, y)^2 = \langle x-y, x-y \rangle$. As a consequence,

$$\begin{aligned} d(Ax, Ay)^2 &= \langle Ax - Ay, Ax - Ay \rangle \\ &= \langle A(x-y), A(x-y) \rangle \\ &= \langle x-y, x-y \rangle \\ &= d(x, y)^2. \end{aligned}$$

Since distances are non-negative, the square roots are also equal.

To show that any isometry fixing distance is a linear map with orthonormal columns, we first show it is linear. We will use slick geometric arguments, but one can prove it just as well with formulas involving inner products (which the reader is encouraged to try). Let f be an isometry; we need to show $f(x+y) = f(x) + f(y)$ and $f(ax) = af(x)$ for any vectors x, y and any scalar a.

First, $f(ax) = af(x)$. To prove this we first prove that any Euclidean isometry maps lines to lines. We will use the fact that in Euclidean geometry a straight line is the shortest path between any two points. In particular, we if x lies on the shortest path from 0 to ax, then $f(x)$ lies on the shortest path from 0 to $f(ax)$: letting $c = d(0, x)$, then x minimizes the following:

$$\min_{\substack{y \in \mathbb{R}^n \\ d(0,y)=c}} d(0, y) + d(y, ax)$$

Any property defined entirely in terms of distances must be preserved by f, because f preserves distances.

Using the fact that isometries map lines to lines, we continue. Since $d(0, ax) = |a|\|x\| = d(0, f(ax))$, the only way this can occur with $f(ax)$ on the same line through the origin as x is if $f(ax) = \pm af(x)$. Suppose for contradiction that $f(ax) = -af(x)$, then

---

[9] It is quite hard to picture rotations in 4 dimensions. As we'll see, we won't need to because reflections will be enough to capture everything.

$$|a|\|x\| = d(0, f(ax)) = d(0, f(x)) + d(f(x), f(ax))$$
$$= \|x\| + \| - af(x) - f(x)\|$$
$$= \|x\| + |-a - 1|\|f(x)\|$$
$$= (a+2)\|x\|$$

Note $|a| = a + 2$ is only true for $a = -1$, which is a contradiction for all $a \neq -1$. But if $a = -1$, then $f(-x) = f(x)$, which contradicts f being injective. We conclude that $f(x) = af(x)$.

To show $f(x+y) = f(x) + f(y)$, we additionally claim that f preserves parallel lines (even those that do not pass through the origin). Indeed, given two lines L_1, L_2, define the distance between the lines as $d(L_1, L_2) = \min_{x \in L_1, y \in L_2} d(x, y)$. When L_1, L_2 are parallel, this distance is a positive constant, and otherwise it is zero. Since the property is defined entirely in terms of distance, an isometry must preserve it.

Now consider the parallelogram, with opposite sides being parallel line segments, and having one vertex at the origin.

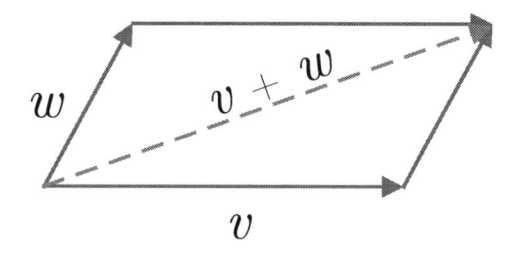

By our arguments above (isometries preserve length, angle measure, and parallelism of lines), isometries map parallelograms to parallelograms. But a parallelogram is precisely how we define addition of two vectors! The sum of the vectors representing the sides is the diagonal vector drawn from the origin to the opposite vertex.

Now that we've established isometries that fix the origin are linear maps, we already know from linear algebra that a linear map preserves distance if and only if it preserves the inner product ($d(x, y) = \|x - y\|$ is defined in terms of the inner product), which happens if and only if its columns are orthonormal. (Cf. Chapter 12, Exercise 12.2) □

This proof puts into practice Klein's idea to study invariants preserved by isometries. The invariants that can be derived from distance preservation are highly structured, allowing one to explicitly limit an isometry's shenanigans. As an added benefit, thinking in terms of invariants removes the need to rephrase geometric concepts in symbolic language. If you found the epsilon-delta proofs of calculus tedious, you might just be a geometer.

The group of $n \times n$ matrices with orthogonal unit vector columns is called the *orthogonal group* $O(n)$.[10] Recall it has the following characterization.

---

[10] Not to be confused with big-O notation.

$$O(n) = \{A : A^T A = I_n\}.$$

We've already shown that this set forms a group under matrix multiplication. Still, it's worthwhile to check again in purely linear algebraic terms. Each matrix represents a change of basis, and composing two basis-changes is again a change of basis. The identity is a no-op basis change, and every basis change has an inverse. Finally, orthogonality is preserved: if $A^T A = I_n, B^T B = I_n$, then $(AB)^T(AB) = B^T(A^T A)B = B^T B = I_n$. Likewise for A^{-1}.

Because these isometries are linear maps, we can also infer that the complete behavior of the isometry is determined by its behavior on n linearly independent points. This is another example of local information being used to infer global structure.

Now the classification theorem: every isometry is a composition of an orthogonal map with a single translation.

Theorem 16.14. *The group of Euclidean isometries is isomorphic to the group*

$$E(n) = \{Ax + v \mid A \in O(n), v \in \mathbb{R}^n\}$$

Proof. First, we prove that $E(n)$ is a group. The identity is in $E(n)$ if we set $v = 0, A = I_n$. Given $f(x) = Ax+v$, the inverse is $f^{-1}(x) = A^{-1}(x-v)$. Given $f(x) = Ax+v, g(x) = Bx + w$, the composition is $B(Ax + v) + w = BAx + Bv + w$. Since $O(n)$ is a group, $BA \in O(n)$, and the translation vector is $Bv + w$.

Next, fix an isometry f that does not necessarily preserve the origin. Let $v = f(0)$, and define $f'(x) = f(x) - v$, effectively translating v to the origin. $f'(0) = f(0) - v = 0$, so $f' \in O(n)$ and can be written as a matrix A by Theorem 16.13. Rewrite f as $f(x) = Ax + v$, which has the form required to be a member of $E(n)$. This maps an isometry f to a member of $E(n)$. This mapping is a homomorphism by repeating the argument from the last paragraph: if $f = x \mapsto Ax + v$ with $v = f(0)$ and $g = y \mapsto By + w$ with $w = g(0)$, then $gf = x \mapsto BAx + Bv + w$, where $Bv + w$ is precisely $g(f(0))$. This mapping is also a bijection: if f and g differ, $f(0) = g(0) = v$, then $f(x) - v$ and $g(x) - v$ must differ on some basis vector, and hence have different matrix representations. Coupling this with the one-sided inverse ($Ax + v$ is an isometry for any choice of A, v), we get our bijection.

□

Hyperbolic Geometry

In antiquity, the Greek mathematician Euclid laid out a grand vision of geometry in which every theorem can be proved from a core set of axioms. The axioms, one of which was "any two points can be connected by a straight line," cannot be proved and must be taken as a truism.

Euclid's 5 axioms, published in his magnum opus, *The Elements*, were:

1. Any two points can be connected by a straight line segment.

2. Any straight line segment can be extended indefinitely to a straight line.

3. For any straight line segment, there is a circle with that line as its radius and one endpoint as its center.

4. All right angles are congruent.

5. **Given any straight line and a point not on that line, there is a unique straight line that passes through the point and never intersects the first line.**

The fifth axiom, commonly called "the parallel postulate," nagged mathematicians for centuries. It always seemed possible that it could be converted from an axiom to a theorem by deducing it from the other four axioms.

These efforts were sadly in vain. As is often the case, the more failed attempts at proving a theorem, the more it seems the theorem might be false. Indeed, the parallel postulate can broken in a few ways. There are geometries that satisfy the first four axioms, but *no* parallel lines exist (all possible lines intersect the first). There are also geometries in which *multiple* parallel lines exist. Projective geometry is an instance of the first breakage, and hyperbolic geometry the second.

Let's now define a model of the hyperbolic plane and classify its symmetries. I say "a" because there are many models of the hyperbolic plane. The connections between them are interesting and useful, but for this chapter we'll work entirely in the model called the *Poincaré disk*. The Chapter Notes contain more historical details.

The universe of points for the Poincaré disk is the interior of the unit disk,[11] $\mathbb{D} = \{(x, y) \in \mathbb{R}^2 \mid x^2 + y^2 < 1\}$. One is supposed to colorfully imagine the boundary of the unit circle as a "line at infinity," a sort of horizon that lines can approach without ever reaching. To us—we omniscient beings viewing this universe from the outside—a point moving at unit speed along such a line simply appears to slow down. As we'll make clear with a distance formula, points close to the boundary grow exponentially farther away from each other compared to points near the origin.

In the Poincaré disk, there are two kinds of lines. The first is one which includes the origin, and these lines are simply diameters of the unit circle (not including the endpoints). Otherwise, a line is a segment of a circle perpendicular to the unit circle. More formally, define the *angle* made by two distinct, intersecting circles to be the angles made by the tangent lines to those circles at their intersection points. Because the line through the circles' centers is a line of symmetry, the angle will be the same regardless of which intersection point is chosen. A circle is *perpendicular* (or orthogonal) to another circle if they form right angles at their intersection points. The types of lines are displayed in Figure 16.5

---

[11] We will refer to a "circle" strictly as the boundary of a circle, and the disk as the circle jointly with its interior. This is standard mathematical parlance.

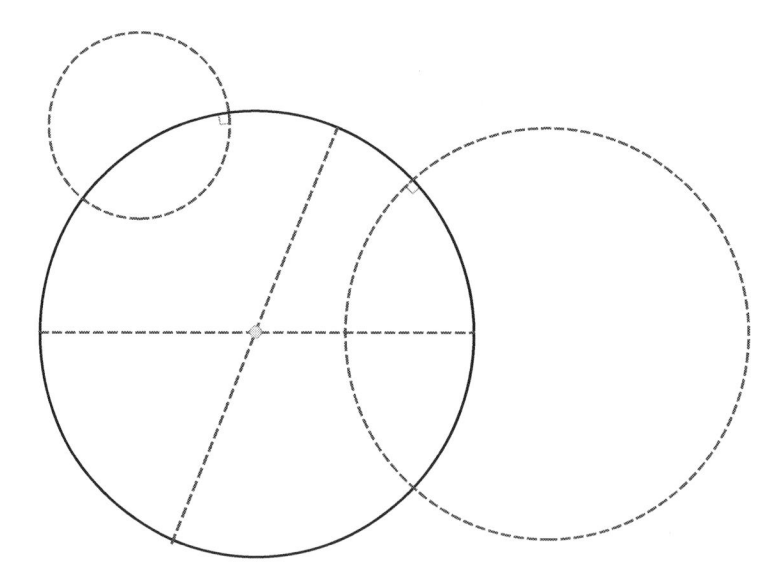

Figure 16.5: Lines in the Poincaré disk. The solid black line is the boundary of the disk. The dashed diameters are one type of line. The arcs of the dashed circles are another. The circles must intersect the boundary of the disk at perpendicular angles.

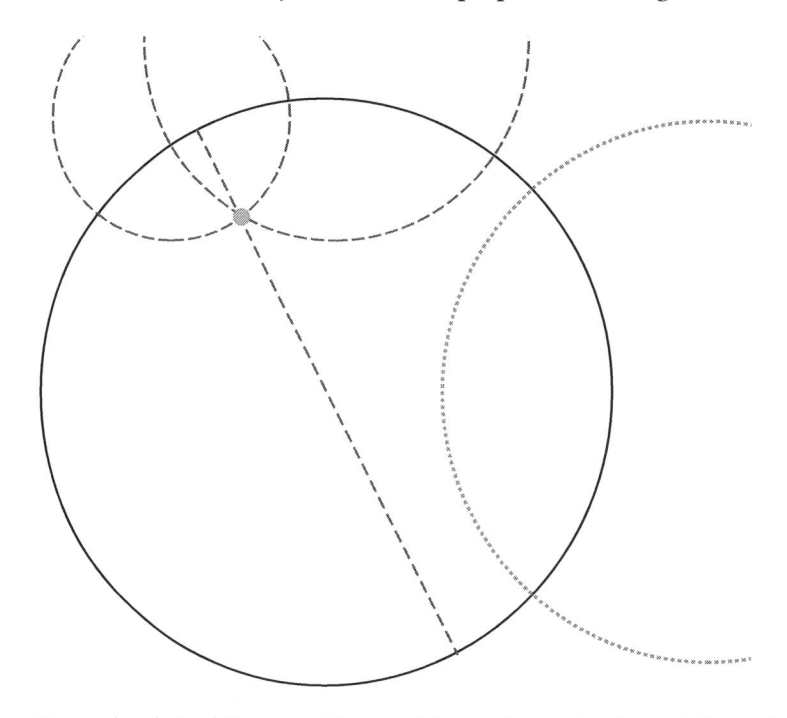

Figure 16.6: Given the dotted Poincaré line and the indicated point, all three dashed lines pass through the point without ever intersecting the dotted line. The parallel postulate fails.

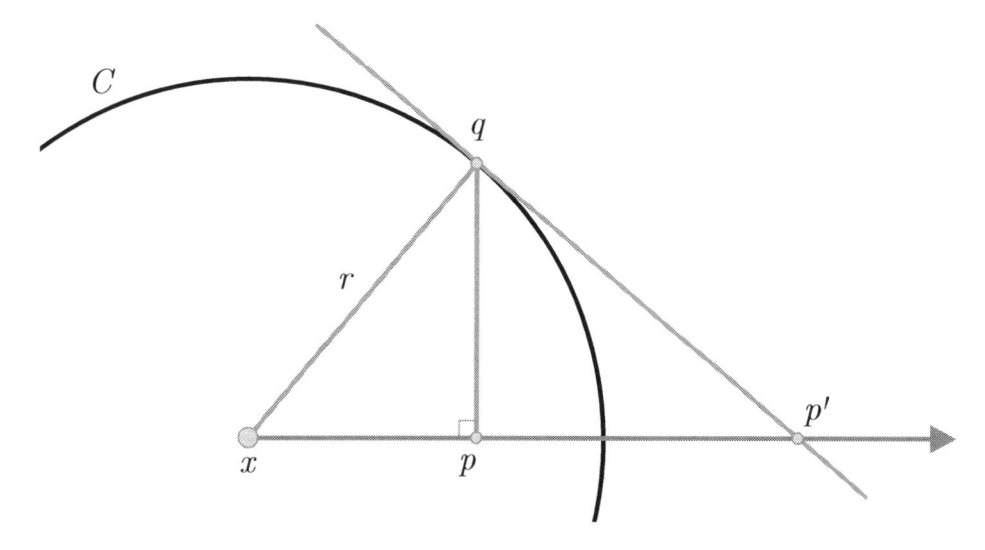

Figure 16.7: The construction of the inverse of a point in a circle.

Now we can immediately see why the parallel postulate fails: parallel lines are just circles that don't intersect! We can easily draw many of them through a fixed point. This is pictured in Figure 16.6.

There is a bit of work to do to establish the axioms of geometry for this model. We need to be able to draw a line between any two points, and to draw a circle with a segment as its radius. We also need to define the angle between two hyperbolic lines, and verify that right angles are all congruent. For each of these it helps to have our first hyperbolic symmetry in hand: inversion in a circle.

Definition 16.15. Let C be a circle with center x and radius r. Let p be a point different from x. Define the *inverse* of p with respect to C as the point p' along the ray from x through p that satisfies:

$$d(p, x)d(p', x) = r^2.$$

The verb for computing the inverse with respect to C is "inverting in C." For the classical geometric construction of the inverse of p in C: suppose p is in the interior of C. Draw a ray from x through p, as in Figure 16.7. Then draw a perpendicular segment from p to C to get a point q. Then the inverse p' is the intersection of the tangent to C at q with the ray $x \to p$.

If p is outside the circle, one can perform these steps backward: compute a tangent to C through p to get q, then p' is the intersection of the altitude of the triangle $\triangle xqp$ with the ray $x \to p$. If p lies on the circle, then p is its own inverse.

To see why this has the property required by Definition 16.15, look again at Figure 16.7. Triangles $\triangle xp'q$ and $\triangle xqp$ are similar (a general truth about altitudes of right triangles), meaning $d(x, p')/r = r/d(x, p)$.

Another way to construct the inverse is to "just do it." You want a point along the ray from the center x through p compatible with its defining property. Simply compute

$$p' = x + r^2(p-x)/\|p-x\|^2.$$

Working out the details,

$$\begin{aligned}d(p,x)d(p',x) &= \|p-x\|\|p'-x\| \\ &= \|p-x\|\left\|\frac{r^2(p-x)}{\|p-x\|^2}\right\| \\ &= \|p-x\|^2 r^2/\|p-x\|^2 = r^2.\end{aligned}$$

We will also need the following.

Proposition 16.16. *Let C and D be circles. If C and D are orthogonal, then every point in D is fixed by inversions in C, and vice versa. Moreover, if x, y are two points that are inverses with respect to C, and D passes through both of them, then C and D are orthogonal.*

Proof. The proof is left as an exercise in geometry. As George Pólya said, geometry is the science of correct reasoning on incorrect figures. Take this to heart and make lots of bad drawings.

□

Next, we construct a Poincaré line as the arc of a circle orthogonal to the boundary of the unit disk. We ignore some special cases made precise in code in Section 16.8. Given two points p, q, pick one that's not the center and invert it in the unit circle to get a third point s. By Proposition 16.16, the unique circle through these three points is orthogonal to the unit circle, as desired. The arc of that segment that is between p and q and lies inside the unit circle is defined to be the line segment between p and q, as well as the shortest path between them.

Second, the drawing of a circle. We take a cue from Euclidean geometry, where a circle has the property that it is perpendicular to every line through its center. Again we use the inversion: fix a line segment between p and q, and say we want to draw the circle with center p. Pick any two hyperbolic lines L_1, L_2 that pass through p but not q. Invert q in both of these lines to get q', q''. Then the Euclidean circle passing through q, q', q'' is the hyperbolic circle centered at p with radius pq. Curiously, a hyperbolic circle in the Poincaré disk is a Euclidean circle, but its center is not the same point as the Euclidean center.

Finally, define the angle between hyperbolic lines as the usual Euclidean angle between their tangents at their point of intersection. Since hyperbolic lines are orthogonal to the unit circle, their center necessarily lies outside of the Poincaré disk. Hence, if two lines intersect they intersect at a single point. Since the angles are defined in terms of Euclidean angles, all right angles are congruent.

Together, these facts establish the axioms of a geometry for the Poincaré disk.

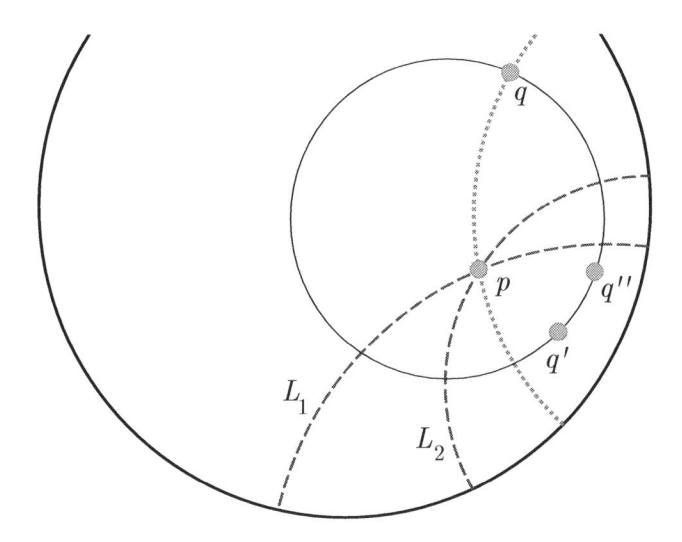

Figure 16.8: The construction of a Poincaré circle with center p and radius pq.

16.6 The Symmetry Group of the Poincaré Disk

Taking a cue from Klein, let's study the symmetries of the Poincaré disk. We already have one symmetry: reflection across a hyperbolic line, which is inversion with respect to the circle defining that line. In the case of a hyperbolic line which is a diameter of $\mathbb{D}^2$, reflection is the same as Euclidean reflection in that line. By Proposition 16.16, these operations preserve the boundary of the Poincaré disk $\mathbb{D}^2$, and it's not hard to prove that the interior of $\mathbb{D}^2$ is also mapped to itself.

We want to study the invariant quantities with respect to hyperbolic reflection. One such quantity is angle measure, but a more interesting one is called the *cross ratio*. We'll use the cross ratio to define distance, so that reflections across hyperbolic lines will be isometries by definition. First we define the cross ratio in general, and in Definition 16.20 we'll make it specific to hyperbolic lines.

Definition 16.17. Let w, x, y, z be four points (in a specific order). The *cross ratio* of w, x, y, z, denoted $[wx; yz]$ is defined as

$$\frac{\|w-y\|}{\|w-z\|} \Big/ \frac{\|x-y\|}{\|x-z\|} = \frac{\|w-y\|\|x-z\|}{\|w-z\|\|x-y\|}.$$

The cross ratio holds the distinguished position of being *the* invariant quantity of projective geometry. Since all geometries are special cases of projective geometry, an appropriately contextualized version of the cross ratio should be invariant for hyperbolic geometry as well.

To show this, first we need a lemma.

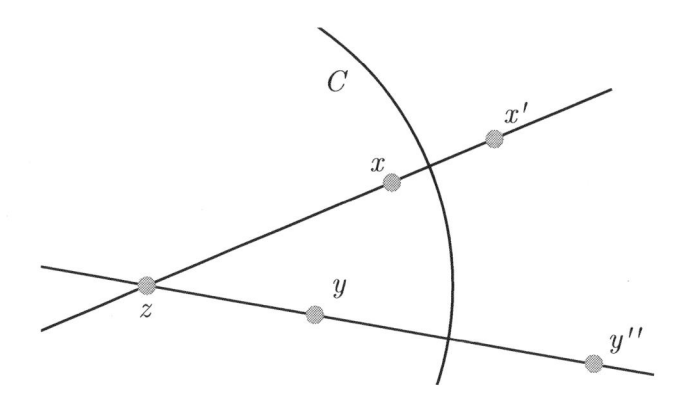

Figure 16.9: The image of two points uniquely determines the circle of inversion (the easy case).

Lemma 16.18. *Two hyperbolic reflections agreeing on two distinct points are equal. That is, the circle defining an inversion operation is uniquely determined by how that operation behaves on two points with distinct images.*

Proof. When the hyperbolic reflection in question is a diameter of $\mathbb{D}^2$, the lemma is true because reflection in a Euclidean line is uniquely determined by its behavior on two points (prove this as an exercise). The paragraphs to follow will heavily use Definition 16.15.

Let x, y be points and x', y' be their inversions, with respect to an unknown circle C with center z and radius r. The simple case is when x, y, z are not on a common line. Then z is the intersection of the line through x, x' and the line through y, y', and $r = \sqrt{\|z - x\| \|z - x'\|}$ (Definition 16.15). This is depicted in Figure 16.9.

If x, y, z lie on a common line, then we may assume without loss of generality that x, y, z lie on the horizontal axis—otherwise we may make this true via a rotation about the origin of $\mathbb{D}^2$, and the uniqueness will still be determined.[12] With this, we may set $x = (a, 0), x' = (a', 0), y = (b, 0), y' = (b', 0)$, and we need to find $z = (c, 0)$ and $r > 0$ such that $(a' - c)(a - c) = r^2$ and $(b' - c)(b - c) = r^2$ (i.e. Definition 16.15), where c, r are variables. Subtracting the two equations gives $aa' - bb' + c(b + b' - a - a') = 0$, which can be solved for c as long as $a \neq b$ and $a' \neq b'$. $\square$

Lemma 16.18 fails in the case that the two points are exchanged by the inversion. It simplifies the above pair of equations to $(a - c)(b - c) = r^2$. If you arbitrarily choose a position for c to the right of both a and b or to the left of both a and b, then you can always find a radius $r = \sqrt{(a - c)(b - c)}$ that works. Hence, an extra condition is required for uniqueness, and the condition relevant to the upcoming Lemma 16.21 is that the inverting circle is orthogonal to the unit disk.

---

[12] Equivalently, one could parameterize the line by picking a unit vector $v = (x - z)/\|x - z\|$ and letting $x = z + av, y = z + bv$.

Next, we show that the cross ratio is preserved by hyperbolic reflections. The proof is trivial for reflection in a diameter of the Poincaré disk, so we focus on the case of inversion in a circle.

Theorem 16.19. *Let $f(x)$ be inversion in a circle with center c and radius r. Let $w, x, y, z \in \mathbb{R}^2$ be any four distinct points. Then $[wx; yz] = [f(w)f(x); f(y)f(z)]$.*

Proof. For ease of notation, let $w' = f(w)$ (similarly for x, y, z), and let (ab) denote $\|a - b\|$, the length of the line segment between two vectors a and b. We'll use $\cdot$ for multiplication to disambiguate. Then we must prove

$$\frac{(wy) \cdot (xz)}{(wz) \cdot (xy)} = \frac{(w'y') \cdot (x'z')}{(w'z') \cdot (x'y')}$$

It suffices to show that for any two of these points, say, w, y, that $\frac{(wy)}{(w'y')} = \frac{(cw)}{(cy')}$. If we can show this, then (note the second equality is where we apply the claim, and the rest is grouping):

$$\frac{(wy) \cdot (xz)}{(wz) \cdot (xy)} \bigg/ \frac{(w'y') \cdot (x'z')}{(w'z') \cdot (x'y')} = \frac{(wy) \cdot (xz) \cdot (w'z') \cdot (x'y')}{(w'y') \cdot (x'z') \cdot (wz) \cdot (xy)}$$
$$= \frac{(cw) \cdot (cx) \cdot (cw') \cdot (cx')}{(cy') \cdot (cz') \cdot (cz) \cdot (cy)}$$
$$= \frac{(cw) \cdot (cw') \cdot (cx) \cdot (cx')}{(cy) \cdot (cy') \cdot (cz) \cdot (cz')}$$
$$= \frac{r^4}{r^4} = 1,$$

which proves the theorem.

To prove that $\frac{(wy)}{(w'y')} = \frac{(cw)}{(cy')}$, we split into two cases depending on whether c, w, y are collinear. If they are not, then this follows from the similarity of the triangles $\triangle cwy \sim \triangle cy'w'$: they share the angle with c and the defining property of circle inversion implies $\frac{(cw)}{(cy')} = \frac{(cy)}{(cw')}$. If they are collinear, consider the diagram in Figure 16.10. If w, y are on different sides of c, then

$$\frac{(wy)}{(cw)} = \frac{(cw) + (cy)}{(cw)} = 1 + \frac{(cy)}{(cw)} = 1 + \frac{r^2/(cy')}{r^2/(cw')} = 1 + \frac{(cw')}{(cy')} = \frac{(cy') + (cw')}{(cy')} = \frac{(w'y')}{(cy')}$$

Equating the left-most and right-most expressions, we rearrange to get $\frac{(wy)}{(w'y')} = \frac{(cw)}{(cy')}$, which was our goal. If w and y are on the same side of c, then replacing the sum $(wy) = (cy) + (cw)$ with $(wy) = (cy) - (cw)$, or $(cw) - (cy)$, as the case may be, yields the same result.

$\square$

Though we leave out a coherent explanation of why this ultimately works as a distance function, the following construction provides the "correct" metric on the Poincaré disk.

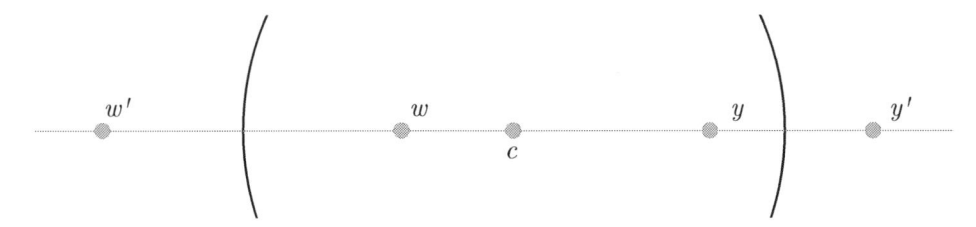

Figure 16.10: The central claim is that $\frac{(wy)}{(w'y')} = \frac{(cw)}{(cy')}$.

Definition 16.20. Let $p, q \in \mathbb{D}^2$ be two distinct points. Form the hyperbolic line through those points, and let x, y be the intersection of the hyperbolic line with the boundary of $\mathbb{D}^2$, so that x is closest to p and y to q. Define the distance between p and q to be:

$$d(x, y) = \frac{1}{2}\left|\log[xy; qp]\right| = \frac{1}{2}\left|\log \frac{(x-q)(y-p)}{(x-p)(y-q)}\right|$$

where logs are taken with base e.

Admittedly vaguely, the choice of these two special points used to compute the cross ratio results in a "canonical" choice that allows different distances to be compared with respect to the same reference scale. As p and q near the boundary of the circle, the denominators involved in the cross ratio tend to zero and the cross ratio increases. See the exercises for more.

The hyperbolic distance function satisfies the properties of a metric from Definition 16.1 (proof omitted). If a metric is defined on a geometric space that has unique shortest line segments between points, then we get an additional property: $d(x, y) = d(x, z) + d(z, y)$ if and only if z lies on the shortest path between x and y. We will use this in the proof of Lemma 16.21.

Due to Theorem 16.19, we automatically know that hyperbolic distance is an invariant of a hyperbolic reflection. Moreover, a rotation t_θ of $\mathbb{D}^2$ by θ radians around the origin is also an isometry of the Poincaré disk: such rotations preserve the unit circle and are Euclidean isometries. These two facts together allow us to analyze the structure of all hyperbolic isometries.

First we prove an important lemma.

Lemma 16.21. *The set of points equidistant from two distinct points x, y is a hyperbolic line, and a hyperbolic reflection in this line exchanges x and y.*

Proof. First, we establish that for any two points x, y, there is a unique hyperbolic reflection $f : \mathbb{D}^2 \to \mathbb{D}^2$ that exchanges x and y. Then we prove that a point is fixed by f if and only if it is equidistant to x and y. Since we know that a point is fixed by a circle inversion if and only if it lies on that circle,[13] this completes the proof.

---

[13] The same holds for reflections by Euclidean lines.

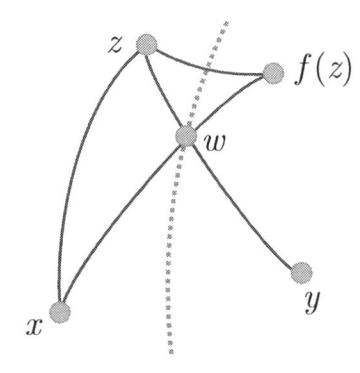

Figure 16.11: The line between x and $f(z)$ is mapped to the line between y and z by reflection, and the intersection of these points is w.

The existence of f: if x and y both have the same Euclidean distance from the origin, then one can use the diameter of $\mathbb{D}^2$ that bisects the angle between x, y, and the center of $\mathbb{D}^2$. Otherwise, as per the postscript of Lemma 16.18 we follow the steps of Lemma 16.18 with the added condition that the inverting circle is orthogonal to the unit circle.

Rotate the center of the (unknown) circle of inversion so it, x, and y all lie on the same horizontal line, which we may suppose without loss of generality is the horizontal axis. Let $x = (a, 0)$, $y = (b, 0)$, and the center be $(c, 0)$. The condition that x, y are exchanged is $(a - c)(b - c) = r^2$. Via the Pythagorean theorem, being orthogonal to $\mathbb{D}^2$ adds the constraint $1 + r^2 = \|d - (c, 0)\|^2$, where $d = (d_1, d_2)$ is a fixed vector.[14] Combining these two equations and rearranging we get

$$ab - \|d\|^2 + 1 + (-a - b + 2d_1)c = 0$$

This has a unique solution for c if and only if $d_1 \neq (a + b)/2$, i.e., if d does not lie on the (Euclidean) perpendicular bisector of the line segment between x, y. This exceptional case is exactly when we use a reflection in a diameter of $\mathbb{D}^2$, i.e., the first case above.

Next, we show a point z is fixed by f if and only if z is equidistant to x, y. For the forward implication, suppose f exchanges x and y, and let $z = f(z)$. Then $d(x, z) = d(f(x), f(z)) = d(y, f(z)) = d(y, z)$. For the converse, let z be a point with $d(x, z) = d(y, z)$, and suppose to the contrary that $z \neq f(z)$. Let L be the hyperbolic line defined by f; by swapping z and $f(z)$ we may assume z is on the same side of L as x. In this case note that $z, f(z)$ are exchanged by f, since f is a reflection. This implies that any point w fixed by f is also equidistant to z and $f(z)$. We have the picture in Figure 16.11.

Now $d(x, z) = d(y, z)$ by hypothesis, and chaining this with $d(y, z) = d(f(y), f(z)) = d(x, f(z))$ (since f preserves distance and $f(y) = x$) we get $d(x, z) = d(x, f(z))$. Now consider the hyperbolic line segment between $x, f(z)$, which intersects L (the hyperbolic line defining f) at a point w. This w is on the shortest path between

---

[14] Here d is the translation of the center of $\mathbb{D}^2$ (the origin) that is required to shift rotate and shift our picture to the horizontal axis.

x and $f(z)$, meaning $d(x, f(z)) = d(x, w) + d(w, f(z))$, and note that w is fixed by f. Finally,

$$\begin{aligned} d(x, z) &= d(x, f(z)) \\ &= d(x, w) + d(w, f(z)) \\ &= d(x, w) + d(w, z) \end{aligned}$$

This implies w is on the shortest path between x and z. This contradicts the equality part of the triangle inequality: x and z are on the same side of L while w is on L. $\square$

And now the finale: all isometries of the Poincaré disk are a composition of reflections.

Theorem 16.22. *Every isometry of $\mathbb{D}^2$ is a product of at most 3 hyperbolic reflections.*

Proof. First, we claim that any isometry is determined by its effect on three non-collinear points x, y, z (not on any Poincaré line). Suppose to the contrary there were two isometries f, g with $f(x) = g(x), f(y) = g(y), f(z) = g(z)$, but for which some $p \notin \{x, y, z\}$ satisfies $f(p) \neq g(p)$. Since f and g are isometries, each of the points $\{f(x), f(y), f(z)\}$ is equidistant to $f(p), g(p)$. By Lemma 16.21, $\{f(x), f(y), f(z)\}$ must lie on a hyperbolic line. But this contradicts the fact that isometries map lines to lines, since $\{x, y, z\}$ are not collinear.

To show three reflections are enough to express any isometry $f : \mathbb{D}^2 \to \mathbb{D}^2$, choose any x, y, z not on a line. In the special case that $x = f(x)$ and $y = f(y)$, then reflection in the hyperbolic line through x, y must map z to $f(z)$. Indeed, z has the same distance to $x = f(x)$ and $y = f(y)$ as $f(z)$, so Lemma 16.21 applies. In this case f is just a reflection.

In the slightly less special case that only one of the three points equals its image under f, say $x = f(x)$, then map y to $f(y)$ via reflection in the unique hyperbolic line consisting of equidistant points to y and $f(y)$ provided by Lemma 16.21. Again, since y and $f(y)$ are equidistant from $x = f(x)$, the line being reflected must pass through x, meaning x is fixed by this reflection. With one reflection we've reduced to the case $x = f(x), y = f(y)$; the first case adds one more reflection to get f.

Finally, in the least special case that all three points are different from their images, we can apply any reflection mapping $x \mapsto f(x)$, reducing to the second case. This results in a simple algorithm:

1. If $x \neq f(x)$, map x to $f(x)$ via a reflection. Call this reflection g_1, or if $x = f(x)$ call g_1 the identity map.

2. If $g_1(y) \neq f(y)$, then map $g_1(y)$ to $f(y)$ using a second reflection. This is guaranteed to leave $g_1(x)$ fixed. Call that reflection g_2 (or the identity if $g_1(y) = f(y)$).

3. Do the same for $g_2(g_1(z))$ and $f(z)$, provided they are not equal, and call the resulting reflection g_3. This reflection fixes both $g_2(g_1(x)) = g_1(x)$ and $g_2(g_1(y))$.

Compose the three reflections to get $f = g_3 g_2 g_1$.

16.7 The Hyperbolic Isometry Group as a Group of Matrices

Multiple times throughout this book, we've avoided using complex numbers, resulting in some slightly nonstandard work. This was essentially a cop out.[15] Be that as it may, it's not sensible to study the group structure of hyperbolic isometries without complex numbers.

The briefest review: the set $\mathbb{C} = \{a+ib : a, b \in \mathbb{R}\}$ is called the set of *complex numbers*, where i is the "complex unit," i.e., it's a unit vector defined to be linearly independent from 1. There is a bijection $\mathbb{C} \to \mathbb{R}^2$ via $a + ib \mapsto (a, b)$, so that complex numbers can be viewed as a plane. Using this view, denote by $\arg(a + bi)$ the angle between (a, b) and $(1, 0)$ (chosen to be in the interval $[0, 2\pi)$), denote by $|a + bi|$ the length of (a, b), and define multiplication of $a + ib = (a, b)$ by i as the rotation of (a, b) by 90 degrees counterclockwise. Extrapolate from this that $i^2 = -1$, and assert that the usual arithmetic rule that $(a + ib)(c + id) = ac - bd + i(ad + bc)$.

As an elegantly stated consequence, if $z, w \in \mathbb{C}$ then their multiplication is uniquely determined by the two properties $\arg(zw) = \arg(z) + \arg(w)$ and $|zw| = |z||w|$. Multiplying two complex numbers adds their angles and multiplies their lengths. Inverses are also defined: $1/z$ is the unique complex number whose angle is $2\pi - \arg(z)$ and whose length is $1/|z|$, provided $z \neq 0$. If we define the *complex conjugate* $\overline{a + bi} = a - bi$, then $1/z = \overline{z}/|z|^2$. This formula looks familiar, it's because $z \mapsto 1/\overline{z}$ is a geometric inversion in the unit circle.

There is a slick encoding of Poincaré disk isometries as a group of matrices. First identify $(x, y) \in \mathbb{D}^2$ with the complex number $z = x + iy$. Then consider the two maps defined for any two constants $a, b \in \mathbb{C}$:

$$f^+_{a,b}(z) = \frac{az + b}{\overline{b}z + \overline{a}}$$

$$f^-_{a,b}(z) = \frac{a\overline{z} + b}{\overline{b}\,\overline{z} + \overline{a}}$$

Also force a, b to satisfy $|a|^2 - |b|^2 = 1$. These are the isometries of the Poincaré disk.

Theorem 16.23. *The isometries of* $\mathbb{D}^2$ *are of the form* $f^+_{a,b}$ *or* $f^-_{a,b}$.

Proof. The proof is left in the exercises for those who feel comfortable with complex numbers.

---

[15] I like complex numbers, but I thought the book was getting too long to fit a full chapter. The topic deserves nothing less, and I'm aware of the irony of this section.

The functions $f_{a,b}^+$ are "orientation preserving" isometries of $\mathbb{D}^2$, meaning they are a product of an even number of reflections.[16] Each one can be identified with a matrix

$$f_{a,b}^+ \mapsto \begin{pmatrix} a & b \\ \bar{b} & \bar{a} \end{pmatrix}$$

And if you multiply the matrices, you get the composition of the two maps.

Likewise, the functions $f_{a,b}^-$ form orientation reversing isometries (the product of an odd number of reflections). It is tedious, but elementary, to show that a product of the form $f_{c,d}^- \circ f_{a,b}^-$ has the form $f_{\bar{a}c+db, \bar{b}c+ad}^+$, which is exactly what you get when you multiply their corresponding matrices. Two orientation reversing isometries compose to get an orientation preserving isometry (if not, it would be hard to speak of "orientation" in good faith). One must be a little careful here, because the matrix representations of orientation reversing and orientation preserving isometries are not trivially compatible. The same matrix A is interpreted in two ways depending on whether you conjugate the input. This is one of the deficiencies of the Poincaré disk model, which is not present in some other models of hyperbolic geometry (see Exercise 16.24).

Finally, a complete description of the group. Let $G = \{ \begin{pmatrix} a & b \\ \bar{b} & \bar{a} \end{pmatrix} : a, b \in \mathbb{C} \}$ be the set of orientation preserving isometries under matrix multiplication. Augment this group by adding a single $f_{a,b}^-$, say $f_{1,0}^-(z) = \bar{z}$ (a reflection across the horizontal axis), to get the set $G \cup f_{1,0}^- G$. This is the isometry group of the Poincaré disk. Another way to describe it is that G, the orientation preserving isometries, is the quotient of the full isometry group by the subgroup consisting of the identity and a single reflection.

16.8 Application: Drawing Hyperbolic Tessellations

A *tessellation* is a tiling of space by a repeating pattern. Tessellations are ubiquitous in art, pervasive across cultures and throughout history. Islamic mosque decorations and Russian church tiles, Incan and Tahitian textiles, Native American baskets and Chinese porcelain. It seems that every major civilization incorporated tessellations in their art. Even today, we tessellate our footballs with black-and-white pentagons and our tweed coats with herringbone. Look around you—tessellations!

Tessellations and groups are natural bedfellows. A fixed isometry of the ambient space containing a starting pattern will move the pattern to one of its repetitions, and the (usually infinitely large) set of all such transformations forms a group. This group uniquely describes the geometry of the tessellation.

The Euclidean plane provides a notable example before we return to hyperbolic geometry. Let's consider the set of all patterns that have discrete repetition in two linearly independent directions (as opposed to a pattern that only repeats when shifted, say, right),

---

[16] Orientation has a technical definition that encodes the intuitive idea that "reversing orientation" turns "hello" into "olleh" and vice versa—though for hyperbolic isometries it will have the expected additional warping.

Indian metalwork at the Great Exhibition in 1851. From *The Grammar of the Orient*.

Fired Clay, Kerma, 1700–1550 BC. Harvard University–Boston Museum of Fine Arts.

Ceiling of an Egyptian tomb. From *The Grammar of the Orient*.

Wall panelling, the Alhambra, Spain. From *The Grammar of the Orient*.

Figure 16.16: Cloth, Hawaii. From *The Grammar of the Orient*. A pattern which has two linearly independent directions of translational symmetry.

such as in Figure 16.16. The groups that describe such patterns—which include the tessellations used in much of historical decorations—have a complete known classification. They are called *wallpaper groups*, and there are exactly 17 of them, up to isomorphism. Wikipedia contains a complete classification of the wallpaper groups, and examples of each occurring in actual decorations from cultures all around the world. One example is in Figure 16.17, the group called "p4." It's characterized by its core pattern providing two quarter-turn centers of rotation (the corner diamond and the center square), one 180-degree center of rotation (the thin diamonds bisecting each side), translation along two independent dimensions, and no other isometries.

Simpler than classifying all wallpaper patterns, we can ask what are the possible tessellations of the Euclidean plane by a convex polygon? For example, regular squares (each interior angle having the same measure, and each side being the same length) tile a plane via a group of translations isomorphic to $\mathbb{Z} \times \mathbb{Z}$, a fact familiar to anyone who has seen a chess or checkers board. And while regular pentagons don't tile the plane, irregular pentagons do, as depicted in Figure 16.18.

To reiterate, a tessellation transforms a single base shape via a fixed group of isometries. The shapes we're narrowing down to study are convex, possibly irregular polygons. Out of curiosity, if you try to tessellate the plane using an 8-sided convex polygon, you will struggle. Your struggle is true: it's impossible. The proof we'll see is quite interesting—it uses graph theory, aided by asymptotic notation, to double-count angles in a hypothetical tessellation. A veritable capstone for the techniques in this book!

Theorem 16.24. *There is no tessellation of the Euclidean plane by a single n-sided convex polygon for any $n > 6$.*

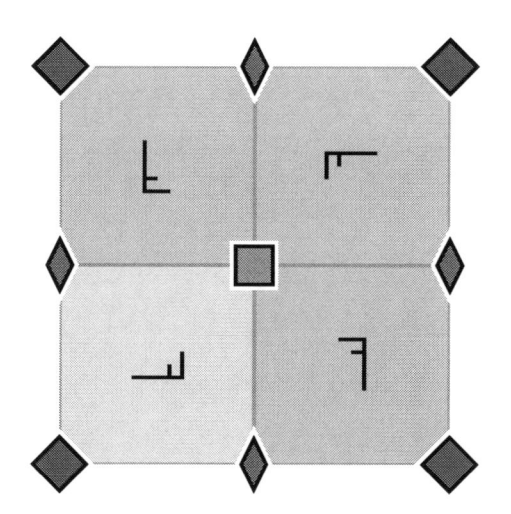

Figure 16.17: A figure which, when used to tile the plane, has p4 as its symmetry group. Figure by Martin von Gagern.

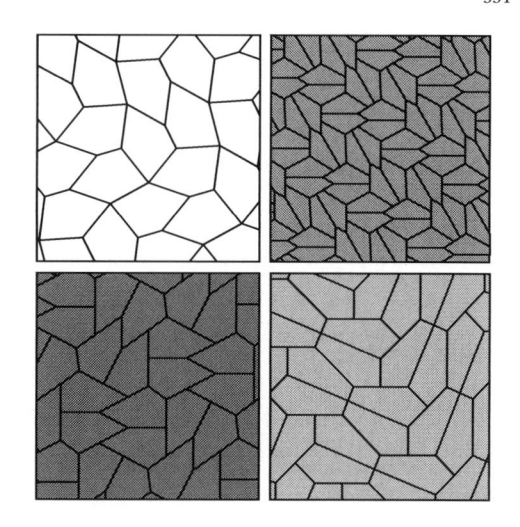

Figure 16.18: Examples of irregular pentagonal tilings of the Euclidean plane. Figure by computer scientist David Eppstein.

Proof. Suppose for contradiction that there is an n-sided convex polygon P, scaled to area 1, that tessellates the plane, and fix the set T of all polygons in such a tessellation. Our proof will have two steps: first, we will fix a bounded piece of the tessellation of area A. Then we'll count the number of angles of polygons contained in that piece in two different ways, and arrive at an inequality of A in terms of A. This inequality will be a contradiction for a sufficiently large A.

Fix a circle C of area A, and let $S \subset T$ be the polygons in T that contain at least one point within C. This finite set of polygons forms a graph $G = (V, E)$, where V is the set of vertices of polygons in S, and E is the (possibly subdivided[17]) set of polygon edges. Moreover, this graph is planar since the tessellation S provides a literal drawing in the plane. Call F the set of faces of G (i.e., the polygons plus the outside face, as we did in Chapter 16). We summarize in Figure 16.19.

First, split each of V, E, F into "interior" and "exterior" subsets. The exterior subsets correspond to those vertices, edges, and faces that are adjacent to the outside of the graph. I.e., these came from the polygons that are only partially in the circle C. The interior vertices, edges, and faces are those that come from polygons entirely inside C. Subscript V, E, F with "int" for interior and "ext" for exterior, like V_{ext}.

We will use the Euler characteristic formula from our chapter on graphs, Theorem 6.5, which says that for a planar graph $|V| - |E| + |F| = 2$. We first claim two facts which imply the formula $|V| = (n/2 - 1)A + O(A^{1/2})$, which is attained by substituting these two facts into Euler's formula and combining.

---

[17] Two polygons in the tessellation can touch so that the vertex of one lies partway along the edge of another. The graph would then split this edge into two.

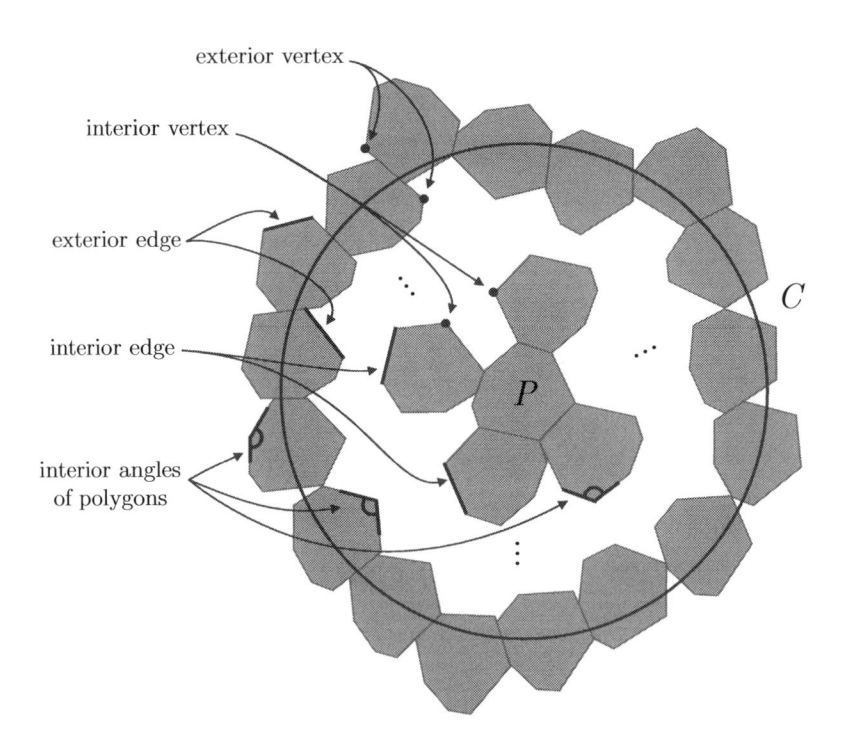

Figure 16.19: The setup for a hypothetical tilling of the Euclidean plane by a convex 7-gon. The bold circle has area A, and we include any polygon having at least one point inside the disk with boundary C.

1. $|F| = |F_{\text{int}}| + |F_{\text{ext}}| = A + O(A^{1/2})$

2. $|E| = |E_{\text{int}}| + |E_{\text{ext}}| = nA/2 + O(A^{1/2})$

You will prove these facts in the exercises, but they can be thought of intuitively: the interior faces F_{int} (each of area 1) fill up a total area roughly equal to the area of the circle C, and the exterior faces are a thin band surrounding C, providing area proportional to the circumference of C times some constant width. The big-O hides both the deviation of the area covered by F_{int} from being exactly A, and the entire area of F_{ext}; both are $O(A^{1/2})$.

Now we will count the number of interior angles of polygons in S in two different ways. What I mean by "interior angle" is an angle at a vertex inside a face. The first way is obvious, $n(|F| - 1) = n|S| \leq n|F|$, because each polygon has n interior angles by definition (ignoring the exterior face). Second, we count by vertex, splitting into interior and exterior cases. Call a_v the number of interior angles meeting at a vertex $v \in V$.

$$\text{\# interior angles} = \sum_{v \in V_{\text{int}}} a_v + \sum_{v \in V_{\text{ext}}} a_v.$$

For V_{int}, there must be at least three interior angles at each vertex (one of these angles may be part of an edge of some polygon, thus having measure π). This bounds the first sum from below by $3|V_{\text{int}}|$. The second sum is $O(A^{1/2})$ because every exterior vertex touches an exterior edge, and fact (2) above shows the number of exterior edges is $O(A^{1/2})$. This gives (# interior angles) $\geq 3|V_{\text{int}}| + O(A^{1/2})$. Since $|V_{\text{int}}| = |V| - |V_{\text{ext}}|$, we have $|V_{\text{int}}| = (n/2 - 1)A + O(A^{1/2})$ as well.[18]

Combining these formulas and bounds gives

$$n|F| \geq (\text{\# interior angles}) \geq 3|V_{\text{int}}| + O(A^{1/2})$$

Expanding $|F|$ and $|V_{\text{int}}|$,

$$nA + nO(A^{1/2}) \geq 3(n/2 - 1)A + O(A^{1/2})$$
$$nA \geq 3(n/2 - 1)A + O(A^{1/2})$$
$$\text{because as } A \to \infty, nO(A^{1/2}) = O(A^{1/2})$$

The right hand side is approximately $\frac{3}{2}nA$, and the left hand side is nA, hinting at the contradiction. More precisely, this inequality fails as $A \to \infty$ if and only if $1 > \frac{n}{3(n/2-1)}$, which happens if and only if $n > 6$. □

While this may disappoint hopeful weavers of the next great tapestry, one *can* tessellate the hyperbolic plane with a 7-gon. Not only that, but there are *infinitely* many ways to do it! Figure 16.20 shows two ways produced by the program in this section.[19]

In the figure, a regular 7-gon tessellates the Poincaré disk, with 3 polygons meeting at each vertex. The two parameters implied by $(7, 3)$ provide an infinite family of tessellations by regular, convex p-gons.[20] Given a convex, regular, hyperbolic p-gon, let $[p, q]$ denote the *configuration* of a tessellation by that polygon in which q copies of the polygon meet at each vertex. The example above has configuration $[7, 3]$. This configuration is sometimes called the *Schläfli symbol*.

Theorem 16.25. *Let p, q be integers. A regular, convex, hyperbolic p-gon tessellates the plane with q copies of the polygon meeting at each vertex if and only if $(p-2)(q-2) > 4$.*

The artist M.C. Escher used a $[6, 4]$ tessellation to construct his *Circle Limit IV*, displayed in Figure 16.21 with additional lines showing the hyperbolic lines used in its design. The remainder of this chapter is devoted to drawing the outlines of hyperbolic tessellations.

---

[18] Here we used $|V_{\text{ext}}| = O(A^{1/2})$, which is true because every exterior exterior edge touches at most two exterior vertices, and the number of exterior edges is $O(A^{1/2})$.

[19] The intrepid reader will revisit the proof of Theorem 16.24 and determine where it fails for hyperbolic geometry.

[20] To be sure, there is a cornucopia of interesting hyperbolic tilings beyond regular convex p-gons. The engineer/artist/mathematician Roice Nelson runs a fantastic Twitter account called @TilingBot that displays many pretty pictures and animations.

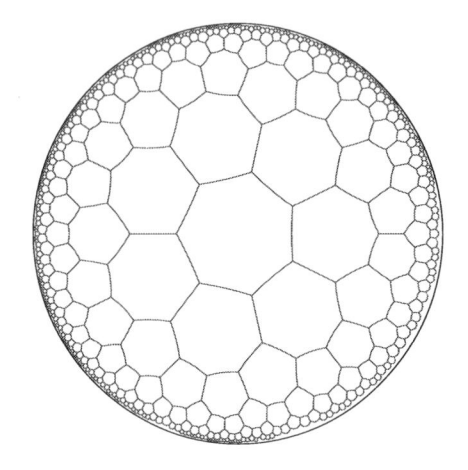

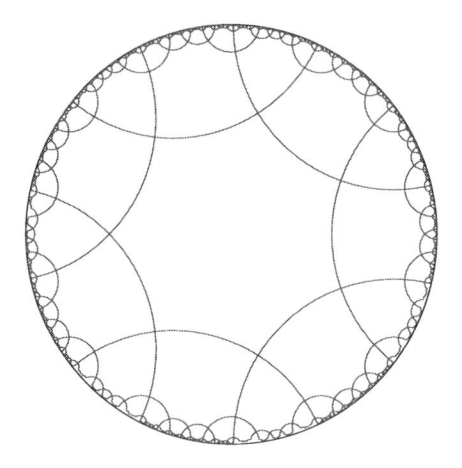

Figure 16.20: Left: a tiling of the hyperbolic plane by 7-gons with 3 meeting per vertex. Right: with 4 meeting per vertex.

In an exercise you'll extend the program to input a pattern (like the angel/devil motif in Figure 16.21) and output an Escher-style drawing.

The core of these kinds of hyperbolic tessellations is the *fundamental region*, which is the smallest subset of the tessellation which, when all symmetries in the tessellation group are applied, tile the plane. In the case of Escher's angels, the fundamental region is the region shown in Figure 16.22. Since we're just drawing the outline of a tessellation, we only need a single triangle.

Definition 16.26. The *fundamental triangle* for a $[p, q]$ tessellation of the hyperbolic plane is a hyperbolic triangle with angle measures $\frac{\pi}{p}, \frac{\pi}{q}, \frac{\pi}{2}$.

If such a triangle has its π/p vertex centered at the origin, then Figure 16.23 shows why it produces a hyperbolic p-gon that tessellates the plane. In Figure 16.23, the fundamental triangle is the thick solid shape, and it's been repeatedly reflected along the edges incident to the origin. Recall from Theorem 16.22 that all isometries are products of reflections, and here we're expressing rotations of $2\pi/p$ by two reflections. The result is that the triangle and its mirror are rotated to produce a hyperbolic p-gon centered at the origin. Likewise, the vertex with an angle of π/q allows one to rotate around an exterior vertex by an angle of $2\pi/q$, forming a piece of each of the q distinct polygons at each vertex.

Thus, if we can draw a fundamental triangle and reflect a set of points across a hyperbolic line, we'll be able to draw regular convex tessellations.

Computing Orthogonal Circles and Reflections

Recall that a hyperbolic line between two points in the Poincaré disk is represented by the circle passing through those two points orthogonal the unit circle. Moreover, reflection in that line is inversion in the circle.

To compute these quantities, we start by defining geometric classes for a (Euclidean) point, circle, and line. These classes are largely not interesting. So I will just outline

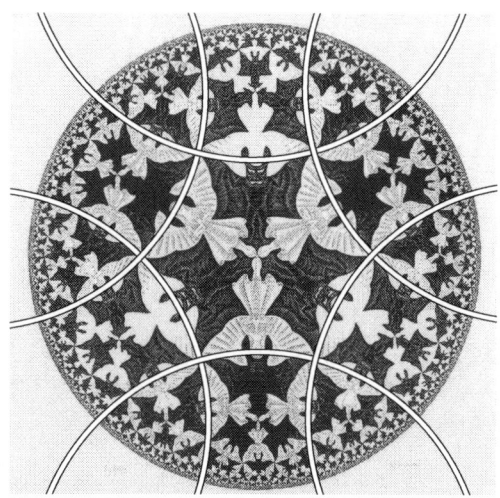

Figure 16.21: Left: *Circle Limit IV*, M.C. Escher, 1960. Right: annotated showing the center 6-gon that is tessellated.

Figure 16.22: The fundamental region of Circle Limit IV. The region tiles the plane by rotations and reflections.

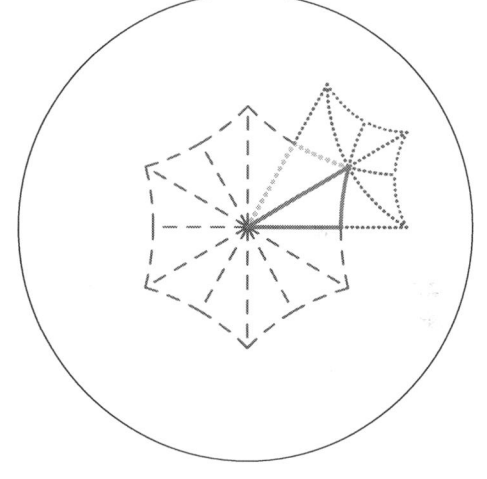

Figure 16.23: The fundamental triangle (thick, solid), reflected across its top edge (faint, thick, dotted), rotated around the center vertex to form the center polygon (thin dashed), and rotated around the π/q vertex to form pieces of tessellated polygons (thick dotted).

their method signatures in Figure 16.24. One notable part is that all equality comparisons have to be converted to "closeness" comparisons, up to some arbitrary but fixed tolerance $\varepsilon \approx 10^{-8}$. The reflection across a Euclidean line is also relevant because some hyperbolic lines—those that are diameters of the Poincaré disk—are also Euclidean lines.

It gets interesting with the circle class, which has a few key methods. The first is computing the inversion of a point. We saw the construction for this in Section 16.5, which included a formula implemented below. Error handling is omitted for brevity.

```
def invert_point(self, point):
    """Compute the inverse of a point with respect to self."""
    x, y = point
    center, radius = (self.center, self.radius)
    square_norm = (x - center.x) ** 2 + (y - center.y) ** 2
    x_inverted = center.x + radius ** 2 * (x - center.x) / square_norm
    y_inverted = center.y + radius ** 2 * (y - center.y) / square_norm
    return Point(x_inverted, y_inverted)
```

With these basic objects and operations, we can define the core method for finding the hyperbolic line passing through two points. In terms of the function signature, the input is two points which the hyperbolic line must pass through, along with a circle it must be orthogonal to. The orthogonal circle argument happens to be the boundary of $\mathbb{D}^2$, but the implementation does not depend on this.

There is one simple case to start: when both points are already on the orthogonal circle. In this case, the output hyperbolic line is the Euclidean circle whose center is the intersection of the two tangent lines at the points, depicted in Figure 16.25. This results in the following edge case in code.

```
def intersection_of_common_tangents(circle, point1, point2):
    line1 = circle.tangent_at(point1)
    line2 = circle.tangent_at(point2)
    return line1.intersect_with(line2)

def circle_through_points_perpendicular_to_circle(point1, point2, circle):
    """Return a Circle that passes through the two given points and
    intersects the given circle at a perpendicular angle.
    """
    if circle.contains(point1):
        if circle.contains(point2):
            circle_center = intersection_of_common_tangents(circle, point1, point2)
            radius = distance(circle_center, point1)
            return Circle(circle_center, radius)
```

If there is at least one point not on the circle, then the output is computed as follows. Invert the non-circle point in the circle, and the result is a set of three points, which uniquely determine the equation of a circle. The reason we invert is because Proposition 16.16 guarantees orthogonality.

The equation for the center of the circle passing through three given points can be computed by setting up three equations and solving. The equations being solved are

```
class Point(namedtuple('Point', ['x', 'y'])):
    def norm(self): """Compute the Euclidean norm of this vector."""
    def normalized(self): """Return a normalized copy of this vector."""
    def project(self, w): """Project self onto the input vector w."""
    def __add__(self, other): """Add two vectors"""
    def __mul__(self, scalar): """Multiply a vector by a scalar"""
    def __sub__(self, other): """Subtract two vectors"""
    def is_zero(self): """Return True if a vector is zero"""
    def is_close_to(self, other):

class Line:
    def __init__(self, point, slope):

    @staticmethod
    def through(p1, p2): """Return a Line through the two given points."""
    def intersect_with(self, line): """Compute the intersection of two lines."""
    def y_value(self, x_value): """Compute the y value of this line at x."""
    def contains(self, point): """Return True if the point is on this line."""
    def __eq__(self, other): """Return True if two lines are equal."""

    def reflect(self, point):
        """Reflect a point across this line."""
        translated_to_origin = point - self.point
        projection = translated_to_origin.project(Point(1, self.slope))
        reflection_vector = translated_to_origin - projection
        return projection - reflection_vector + self.point

class VerticalLine(Line):
    [override some methods from Line]

class Circle(namedtuple('Circle', ['center', 'radius'])):
    def contains(self, point): """Compute whether a point is on a Euclidean circle."""
    def tangent_at(self, point): """Compute the tangent line at a point."""
```

Figure 16.24: The function signatures of the geometry helper classes.

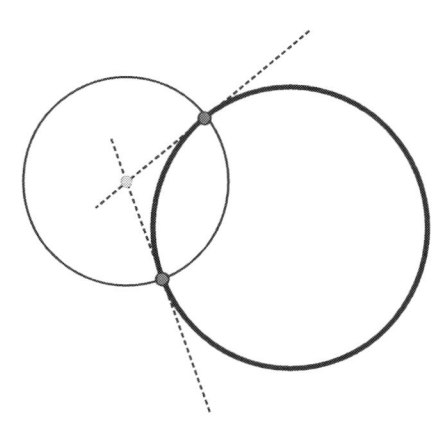

Figure 16.25: The edge case for computing a circle of inversion.

built by substituting our known points into the equation of a circle. Here the unknowns are c_x, c_y, and r.

$$(x_1 - c_x)^2 + (y_1 - c_y)^2 = r^2$$
$$(x_2 - c_x)^2 + (y_2 - c_y)^2 = r^2$$
$$(x_3 - c_x)^2 + (y_3 - c_y)^2 = r^2$$

A succinct way to express the solution to these equations is in terms of the ratios of determinants of a cleverly chosen matrix. We haven't talked about the determinant in this book, but in addition to being a deeply meaningful quantity in its own right, it shows up frequently in computational geometry. More about the determinant in the Chapter Notes. In this case, the solution is summarized by ratios of determinants of sub-matrices of the following matrix:

$$\begin{pmatrix} x^2 + y^2 & x & y & 1 \\ x_1^2 + y_1^2 & x_1 & y_1 & 1 \\ x_2^2 + y_2^2 & x_2 & y_2 & 1 \\ x_3^2 + y_3^2 & x_3 & y_3 & 1 \end{pmatrix}$$

Computing a determinant reduces to repeatedly removing a (row, column) pair and computing the determinant of the smaller matrix, called a *minor*. Once the recursion reduces to determinants of 3-dimensional matrices, we can easily hard-code a formula. You'll read about the correctness of this function in an Exercise.

```
def circle_through_points_perpendicular_to_circle(point1, point2, circle):
    [...edge case...]
    point3 = (circle.invert_point(point2)
        if circle.contains(point1) else circle.invert_point(point1))

    def row(point):
        (x, y) = point
        return [x ** 2 + y ** 2, x, y, 1]

    M = [row(point1), row(point2), row(point3)]

    # detminor stands for "determinant of (matrix) minor"
    detminor_1_1 = det3(remove_column(M, 0))
    detminor_1_2 = det3(remove_column(M, 1))
    detminor_1_3 = det3(remove_column(M, 2))
    detminor_1_4 = det3(remove_column(M, 3))

    circle_center_x = 0.5 * detminor_1_2 / detminor_1_1
    circle_center_y = -0.5 * detminor_1_3 / detminor_1_1
    circle_radius = math.sqrt(
        circle_center_x ** 2 + circle_center_y ** 2
        + detminor_1_4 / detminor_1_1)
    return Circle(Point(circle_center_x, circle_center_y), circle_radius)
```

This allows us to define relevant abstractions for a hyperbolic line and the hyperbolic plane. An instance of the Poincaré disk is a circle, with methods to compute a line through two given points. A hyperbolic line is a circle, which happens to be orthogonal to the unit circle forming the boundary of the Poincaré disk.

```
class PoincareDiskModel(Circle):
    def line_through(self, p1, p2):
        """Return a PoincareDiskLine through the two given points."""
        if orientation(p1, p2, self.center) == 'collinear':
            return Line.through(p1, p2)
        else:
            circle = circle_through_points_perpendicular_to_circle(p1, p2, self)
            return PoincareDiskLine(circle.center, circle.radius)

class PoincareDiskLine(Circle):
    def reflect(self, point):
        """Reflect a point across this line."""
        return self.invert_point(point)
```

To determine if three points are collinear, we again employ the determinant. More generally, if you provide three points $A = (a_x, a_y), B = (b_x, b_y), C = (c_x, c_y)$ in sequence, one can determine via the sign of a determinant whether visiting the points in order results in a clockwise turn, a counterclockwise turn, or a straight line. The relevant matrix is

$$\begin{pmatrix} 1 & a_x & a_y \\ 1 & b_x & b_y \\ 1 & c_x & c_y \end{pmatrix}$$

The determinant, which can be thought of as computing a signed area of a particular triangle built from the rows of the matrix, will produce zero if the points all lie on a common line. For a 3×3 matrix the determinant formula is simple enough to inline.

```
def orientation(a, b, c):
    """Compute the orientation of three points visited in sequence."""
    a_x, a_y = a
    b_x, b_y = b
    c_x, c_y = c
    value = (b_x - a_x) * (c_y - a_y) - (c_x - a_x) * (b_y - a_y)

    if (value > EPSILON):
        return 'counterclockwise'
    elif (value < -EPSILON):
        return 'clockwise'
    else:
        return 'collinear'
```

Computing a Fundamental Triangle

Next we compute the vertices of a fundamental triangle. Recall a fundamental triangle has vertices A, B, D with interior angles π/p, π/q, and $\pi/2$, respectively. Also recall that the angle measure between two hyperbolic lines is defined to be the angle between their tangent lines at the point of intersection. To simplify the description of our fundamental triangle, we require that A is the origin and D lies on the horizontal-axis. Thus, computing our desired triangle can be summarized by identifying the coordinates of B and D in Figure 16.26.

The requirement of the three angle measures, paired with the side AD lying on the horizontal axis, uniquely determines the positions of B and D. Let's derive this now.

Lemma 16.27. *Define the constant* $Z = \tan\left(\frac{\pi}{p} + \frac{\pi}{q}\right) \tan\left(\frac{\pi}{p}\right)$. *The coordinates of the point* $B = (b_x, b_y)$ *are given by*

$$b_x = \sqrt{\frac{1}{1 + 2Z - (\tan(\pi/p))^2}}$$

$$b_y = b_x \tan(\pi/p),$$

and the x-coordinate of $D = (d_x, 0)$ *is given by*

$$r^2 = b_y^2 + (b_x - g_x)^2$$

$$d_x = g_x - r,$$

where $G = (g_x, 0) = (b_x(Z+1), 0)$ *is the x-coordinate of the center of the circle defining the hyperbolic line passing through* B *and* D.

Proof. The point $B = (b_x, b_y)$ is defined to be on the line which makes an angle of π/p with the horizontal, i.e., $y = \tan(\pi/p)x$. Since A is the origin, hyperbolic lines through A are the same as Euclidean lines. This gives the formula for b_y. B also lies on a circle orthogonal to the unit circle that passes through D. Call this unknown circle C, and suppose it has center $G = (g_x, 0)$. Note that the y-coordinate of G must be zero in order for C to make a right angle with $D = (d_x, 0)$. Refer to Figure 16.26.

We're asking for an angle of π/q between the line $y = \tan(\pi/p)x$ and the tangent to this unknown circle C at B. Stare at the diagram in Figure 16.27 to convince yourself that the desired tangent line must have an angle of $\frac{\pi}{p} + \frac{\pi}{q}$ with the horizontal, implying the slope of this tangent line is $\tan(\frac{\pi}{p} + \frac{\pi}{q})$.

The equation of the unknown circle (in terms of our unknown quantities) is $(x - g_x)^2 + y^2 = r^2$, where $r^2 = (b_x - g_x)^2 + b_y^2$. When $y > 0$, the derivative of the circle is given by $C'(b_x, b_y) = -(b_x - g_x)/b_y$, and setting $C' = \tan(\frac{\pi}{p} + \frac{\pi}{q})$, we solve for g_x in terms of b_x as

$$b_x(Z+1) = g_x, \qquad \text{where } Z = \tan\left(\frac{\pi}{p} + \frac{\pi}{q}\right) \tan\left(\frac{\pi}{p}\right)$$

If we can get another independent equation relating b_x and g_x, we can eliminate one variable and solve the entire system. The fact we have yet to use is that C and the unit circle are orthogonal. This gives a relationship between their radii, which form the legs of a right triangle: $1^2 + r^2 = g_x^2$, where $r^2 = (b_x - g_x)^2 + \tan(\pi/p)^2 b_x^2$. Solving this

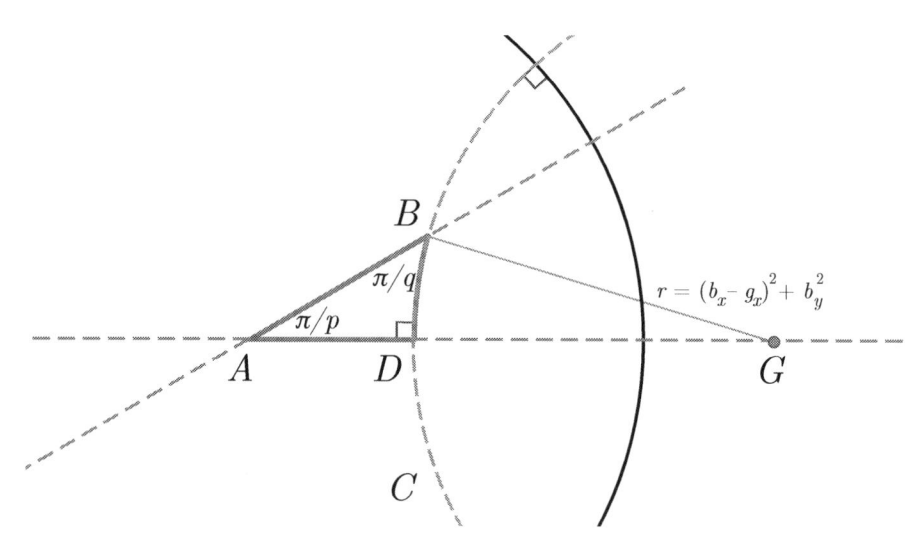

Figure 16.26: The unknown points computed in Lemma 16.27 are B, D, and G, which is the center of the orthogonal circle C passing through B, D, that makes the desired angle of π/q with the top edge of the fundamental triangle.

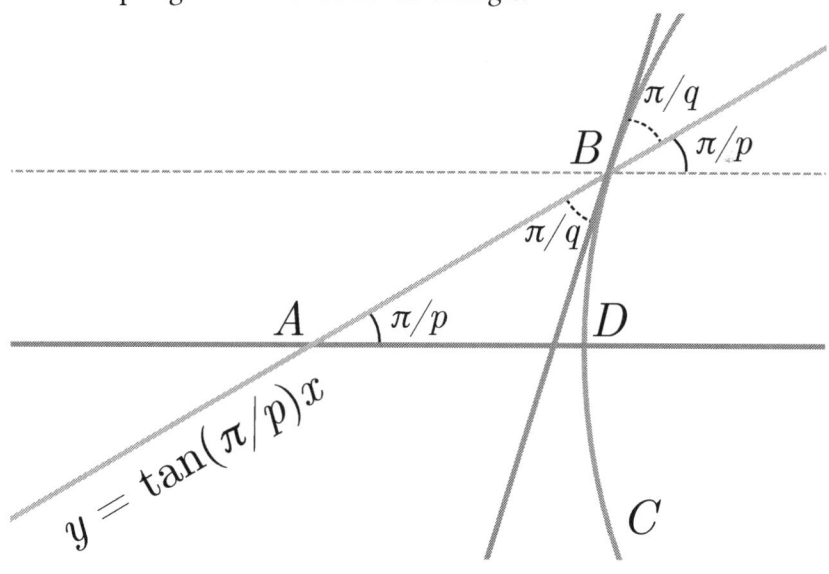

Figure 16.27: By symmetry, the angle of the tangent line to C at B with the horizontal is $\pi/p + \pi/q$.

equation for b_x gives the formula stated in the theorem, and substitution provides the rest.

□

This results in the following code, whose documentation is far more tedious than its implementation:

```
def compute_fundamental_triangle(tessellation_configuration):
    p = tessellation_configuration.numPolygonSides
    q = tessellation_configuration.numPolygonsPerVertex
    tan_p = math.tan(math.pi / p)
    Z = math.tan(math.pi / p + math.pi / q) * tan_p

    b_x = math.sqrt(1 / (1 + 2 * Z - tan_p ** 2))
    b_y = b_x * tan_p
    g_x = b_x * (Z + 1)
    d_x = g_x - math.sqrt(b_y ** 2 + (b_x - g_x) ** 2)

    A = Point(0, 0)
    B = Point(b_x, b_y)
    D = Point(d_x, 0)
    return [A, B, D]
```

Tessellating the Fundamental Triangle

Finally, we have all the pieces we need to draw a tessellation. The majority of the code is helpers. We output the drawing as an SVG file, and so in addition to using a library to draw SVGs, we need to keep track of the differences in coordinate systems. Beyond that, the core routine is quite simple.

First we define a configuration class for a tessellation (used above to draw the fundamental triangle). Followed by a class representing a tessellation. In the latter, the `compute_center_polygon` method computes the center polygon by computing the fundamental triangle, and then iteratively reflecting it across the appropriate edges.

Finally, the `tessellate` method builds the tessellation using a breadth-first traversal of the underlying graph.[21] We use the standard Python `deque` class that can behave as both a stack and a queue to achieve the traversal. Start with a queue containing only the center polygon, and an empty list of "visited" polygons. As long as the queue is nonempty, pop off a polygon, add it to the visited set, reflect it across all possible edges, and add to the queue any unvisited polygons produced this way. Also skip any polygons that are smaller than some limit (i.e., skip them if they're too small to see when rendered on the screen).

The remainder of the code[22] involves rendering the edges of the polygons as SVG arcs. We also created a simple data structure that allows one to compare polygons for equality in a principled way (since the process of reflecting them changes the order of their vertices).

---

[21] The "underlying graph" is not explicitly constructed in the code.
[22] See `pimbook.org`.

```python
class TessellationConfiguration(
        namedtuple('TessellationConfiguration',
                   ['numPolygonSides', 'numPolygonsPerVertex'])):
    def __init__(self, numPolygonSides, numPolygonsPerVertex):
        if not self.is_hyperbolic():
            raise Exception("Configuration {}, {} is not hyperbolic.".format(
                    (self.numPolygonSides, self.numPolygonsPerVertex)))

    def is_hyperbolic(self):
        return (self.numPolygonSides - 2) * (self.numPolygonsPerVertex - 2) > 4
```

```python
class HyperbolicTessellation(object):
    def __init__(self, configuration, min_area=6e-4):
        self.configuration = configuration
        self.disk_model = PoincareDiskModel(Point(0, 0), radius=1)

        # compute the vertices of the center polygon via reflection
        self.center_polygon = self.compute_center_polygon()
        self.tessellated_polygons = self.tessellate(min_area=min_area)
```

```python
def compute_center_polygon(self):
    center, top_vertex, x_axis_vertex = compute_fundamental_triangle(
            self.configuration)
    p = self.configuration.numPolygonSides

    """The center polygon's first vertex is the top vertex (the one that
    makes an angle of pi / q), because the x_axis_vertex is the center of
    an edge.
    """
    polygon = [top_vertex]

    p1, p2 = top_vertex, x_axis_vertex
    for i in range(p - 1):
        p2 = self.disk_model.line_through(center, p1).reflect(p2)
        p1 = self.disk_model.line_through(center, p2).reflect(p1)
        polygon.append(p1)

    return polygon
```

```
def tessellate(self, max_polygon_count=500):
    """Return the set of polygons that make up a tessellation of the center
    polygon. Keep reflecting polygons until the Euclidean bounding box of all
    polygons is less than the given threshold.
    """
    queue = deque()
    queue.append(self.center_polygon)
    tessellated_polygons = []
    processed = PolygonSet()

    while queue:
        polygon = queue.popleft()
        if processed.contains_polygon(polygon):
            continue

        edges = [(polygon[i], polygon[(i + 1) % len(polygon)])
                 for i in range(len(polygon))]
        for u, v in edges:
            line = self.disk_model.line_through(u, v)
            reflected_polygon = [line.reflect(p) for p in polygon]
            queue.append(reflected_polygon)

        tessellated_polygons.append(polygon)
        processed.add_polygon(polygon)
        if len(processed) > max_polygon_count:
            processed.add_polygon(polygon)
            break

    return tessellated_polygons
```

We close with some outputs for different configurations, shown in Figure 16.32.

16.9 Cultural Review

1. Groups are the primary tool mathematics has for studying symmetry, and symmetry shows up all over mathematics and science.

2. Any class of structured objects can be studied in terms of structure-preserving mappings between those objects.

3. Geometry is the study of groups of symmetry, and the invariants preserved by those symmetries.

16.10 Exercises

16.1 Recall the symmetric group S_n is the set of all bijections of a set of n elements. Call the set being permuted $\{1, 2, 3, \ldots, n\}$, and consider the following helpful notation for a permutation: define a *cycle notation* whereby the tuple (1 3 4 2) represents the permutation σ mapping $1 \mapsto 3, 3 \mapsto 4, 4 \mapsto 2$, and $2 \mapsto 1$. All other values are fixed by σ. Define a product of cycles, such as (going right to left) (2 4)(1 2) = (1 4 2) as

A [3, 7] tessellation

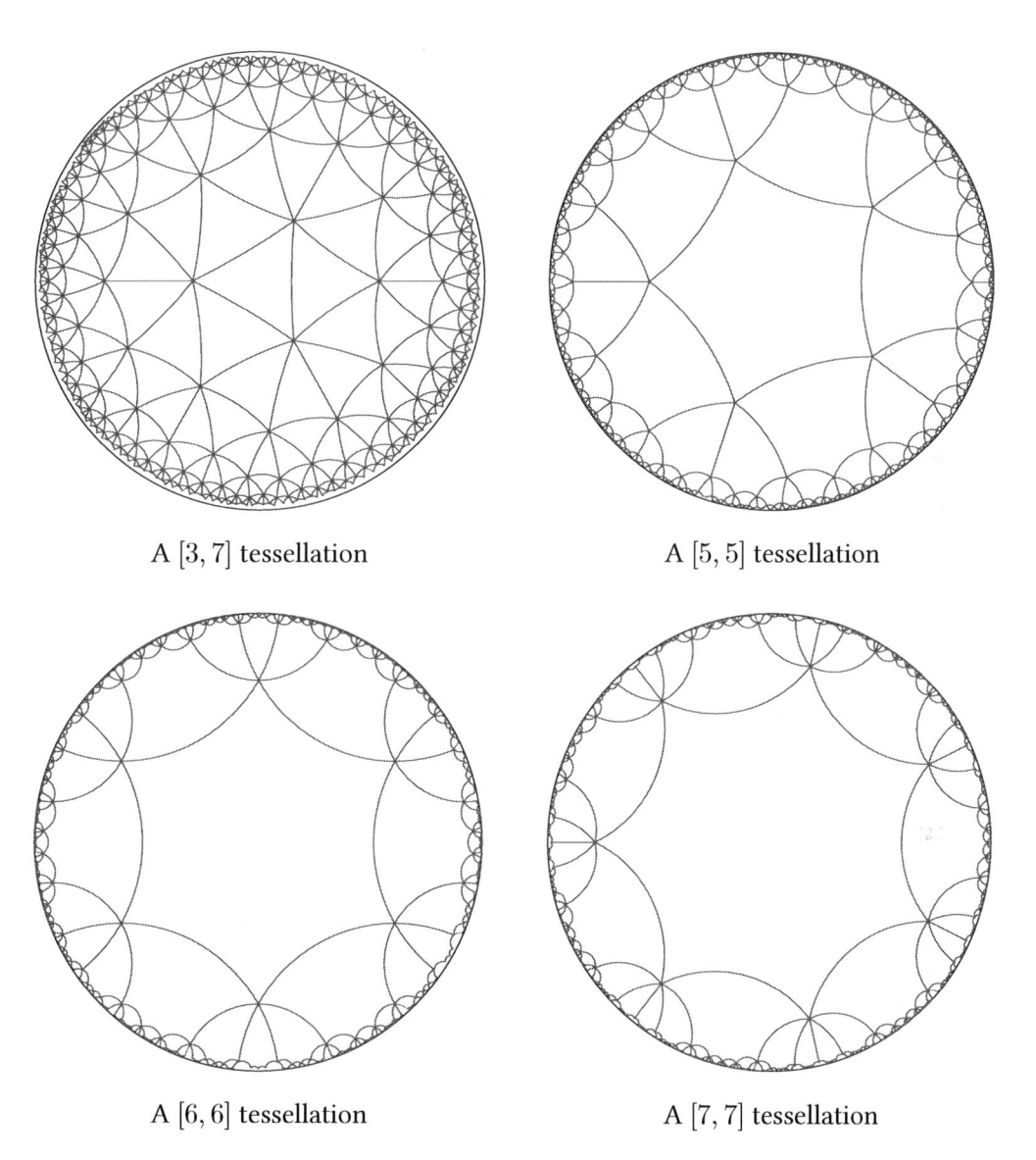

A [5, 5] tessellation

A [6, 6] tessellation

A [7, 7] tessellation

Figure 16.32: Example outputs from the tessellation program.

the composition of the corresponding maps. A cycle of length 2 is called a *transposition*. Prove that every permutation can be written as a product of disjoint cycles. Prove that the n-cycle $(1\ 2\ 3\ \cdots\ n)$ and a single transposition $(1\ 2)$ are a generating set for S_n.

16.2 Define a permutation $x \in S_n$ to be *even* if it is a product of an even number of transpositions. Otherwise call it odd. Show that this definition is well defined: every permutation is either even or odd, but not both. Show a product of two even permutations is even, a product of two odd permutations is even, and a product of an even and an odd permutation is odd.

16.3 Determine the subgroup of S_4 that corresponds to D_8, the symmetry group of the square. Express this in terms of elements of S_4 and the building blocks of groups.

16.4 Let G be a group and H a subgroup. A *coset* of H by a fixed element $x \in G$ is the set $\{xh \mid h \in H\}$. This set is denoted xH. Prove the following:

1. $aH = bH$ if and only if $b^{-1}a \in H$.

2. Let $f : G \to H$ be a group homomorphism. Recall the equivalence equivalence relation $\sim_f$ defined in Chapter 9 by $a \sim_f b$ iff $f(a) = f(b)$. Show the equivalence classes are the cosets of $\ker f$.

3. Let G be a group. Given a subgroup $H \subset G$, show that the set of all cosets of H partition G into disjoint subsets. Conclude that "being in the same coset of H" is an equivalence relation on G, and define G/H to be the quotient of G by this equivalence relation.

4. Prove that the group operation $[x] \cdot [y] = [xy]$ is well-defined in the quotient group G/H if and only if H has the property that for every $b \in H$ and every $a \in G$, the element $aba^{-1} \in H$. Prove that the kernel of a homomorphism has this property. Such subgroups are called *normal* subgroups.

5. Prove that every normal subgroup H of a group G is the kernel of some homomorphism from G to some group. Thus, our definition of a quotient as a kernel is identical to this definition.

16.5 Prove that the property of being isomorphic is an equivalence relation on groups. In particular, show that the inverse of an isomorphism is a homomorphism.

16.6 Find a generator of the multiplicative integer group $(\mathbb{Z}/82\mathbb{Z})^\times$.

16.7 Let G be a group and H a subgroup. Prove $|H|$ evenly divides $|G|$. Use this to prove that for any $a \in G$, $a^{|G|}$ is the identity.

16.8 Define by $\varphi(n)$ the size of the set $\{k \in \mathbb{N} : k < n, \gcd(n,k) = 1\}$. This function is called the *Euler totient function*. Prove that for any integer a, $a^{\varphi(n)} \equiv 1 \mod n$. Hint: use the previous exercise.

16.9   Let $n \in \mathbb{N}$, and let $G = (\mathbb{Z}/n\mathbb{Z})^\times$ be the multiplicative group of integers (those integers between 1 and n that have a greatest common divisor of 1 with n). When n is a product of two large primes, this group is called the *RSA* group. Research the RSA public-key cryptography protocol, and write a program that implements it for two hundred-digit primes. Hint: you will need to find a fast way to generate hundred-digit primes.

16.10   Research and implement the ElGamal digital signature scheme using $(\mathbb{Z}/n\mathbb{Z})^\times$.

16.11   Look up the definition of a semi-direct product of groups, and use this to understand the characterization of the dihedral group D_{2n} as a semi-direct product of $\mathbb{Z}/2\mathbb{Z}$ with $\mathbb{Z}/n\mathbb{Z}$, where the former acts on the latter by "conjugation."

16.12   If you're comfortable with complex numbers, find a source online that discusses the symmetry groups of the roots of polynomials with coefficients in $\mathbb{Q}$. At the risk of referring to an interactive essay that has disappeared from the internet after this book is published, see Fred Akalin's essay, "Why is the Quintic Unsolvable?"[23]

16.13   Recall an undirected graph $G = (V, E)$ is a set of vertices V and a set of edges $E \subset \binom{V}{2}$ that link pairs of vertices. A symmetry of G is a bijection $f : V \to V$ such that (v, w) is an edge if and only if $(f(v), f(w))$ is an edge. In words, a symmetry permutes the vertices of G in such a way that preserves adjacency and non-adjacency. Compute the symmetry group of the Petersen graph. Hint: the size of this group is 120, so brute-force will be difficult.

16.14   Two graphs are called *isomorphic* if there is a bijection between their vertex sets having the same property as a symmetry: all adjacencies and non-adjacencies are preserved. The problem of efficiently computing whether two graphs are isomorphic is one of the most famous open problems in computer science, called the *graph isomorphism problem*. Prove that the graph isomorphism problem reduces to the problem of computing a generating set of the symmetry group of a single graph. A *generating set* of a group G is a subset $S \subset G$ from which every $g \in G$ can be written as a product of elements in S. (Groups can be much larger than the graphs they act on, but generating sets are usually small)

16.15   Prove that any Euclidean isometry in $E(n)$ can be written as the product of at most $n + 1$ reflections.

16.16   Read about determinants and understand why the formula we presented in Section 16.8 for the circle passing through three given points is correct.

16.17   Research the cross ratio in the context of projective geometry. How is it defined there? What are the projective transformations? And why is it preserved by those transformations?

[23] https://www.akalin.com/quintic-unsolvability

16.18 Prove the two facts from Theorem 16.24.

1. $|F| = |F_{\text{int}}| + |F_{\text{ext}}| = A + O(A^{1/2})$
2. $|E| = |E_{\text{int}}| + |E_{\text{ext}}| = nA/2 + O(A^{1/2})$

For the first, consider a well-chosen larger circle containing C, and look at the difference of areas. For the second, borrow ideas from Chapter 4 and count the number of edges in terms of the number of faces, then substitute the first formula.

16.19 Prove Proposition 16.16.

16.20 We neglected to give a good intuition for why the hyperbolic distance function is intuitively a good choice. The reason is that the morally acceptable way to think about this function involves integral calculus, which we avoided in this book. To do this formally, one defines a *metric tensor* or *line element* that describes the length of a curve via an integral. Research these topics to understand how the hyperbolic metric is defined. Be warned that many sources jump straight into advanced terminology and concepts. You're looking for an "introduction to tensor calculus" or an "introduction to Riemannian geometry." Because of the close relation to physics and general relativity, there are also many sources explaining these concepts for physicists. Apply the usual caveats that come with physicists explaining mathematics.

16.21 Extend the hyperbolic tessellation program in this chapter to one which, when given an input motif (an image that replaces the fundamental triangle) draws a hyperbolic polygon using that image and then tessellates the Poincaré disk.

16.22 Determine how to represent inversion in a circle with center c with radius r as a complex 2×2 matrix.

16.23 Prove Theorem 16.23. Hint: an f^+ preserves orientation, and an f^- reverses it.

16.24 A different model of hyperbolic geometry is the *upper half-plane* model. This model has as points the complex numbers $\{a + bi : b > 0\}$, and as lines the half circles orthogonal to the horizontal axis $b = 0$, along with vertical rays. The line $b = 0$ forms the "boundary" analogous to the unit circle bounding the Poincaré disk. The isometries of this model are the so-called *Möbius transformations*. Prove the following.

- The set of *Möbius transformations*, those mappings of the complex line defined by $z \mapsto \frac{az+b}{cz+d}$ with $ad - bc \neq 0$, form a group under function composition. This group is called the *Möbius group*.

- Find a formula for inversion in a circle (reflection in an upper-half-plane-model line) as a Möbius transformation.

- The Möbius group is isomorphic to the group of matrices $PGL_2(\mathbb{R}) = GL_2(\mathbb{R})/\sim$, where $\sim$ is the equivalence relation defined by $A \sim \lambda A$ for every nonzero $\lambda \in \mathbb{R}$. Why is this quotient necessary?

- All Möbius transformations preserve the cross ratio.
- Find a bijection between the upper half plane and the Poincaré disk that preserves hyperbolic lines.

16.25 In this exercise we'll explore the symmetry group of the hyperbolic tessellation of a regular convex p-gon with configuration $[p, q]$. Fix the fundamental triangle of the configuration, and consider the reflections α, β, γ across each edge. What are the algebraic relations between these symmetries? Can you identify the resulting (infinite) group of symmetries with a subgroup of a familiar group?

16.11 Chapter Notes

History of the Hyperbolic Plane

Like many topics in mathematics, the discovery of the hyperbolic plane was far more roundabout than its final form. The first hyperbolic geometry was discovered on the surface of revolution of the so-called tractrix, which is itself derived indirectly from the catenary curve—the name for the not-quite-parabolic shape formed by an ideal rope hanging from its ends under its own weight.

John Stillwell's *Geometry of Surfaces* has a complete derivation of the hyperbolic plane in terms of the tractrix (Chapter 4). It also contains sections devoted to each model of the hyperbolic plane, including a few that are deemed relatively useless from a computational perspective.

More About the Determinant

The determinant of a linear map is defined as the product of its eigenvalues. If any of those eigenvalues are zero, the linear map is not invertible, and as a consequence a common substitute for "invertible matrix" is "matrix with nonzero determinant."

However, the determinant has another definition in terms of symmetry groups. Specifically, let S_n be the permutation group on the set of rows of a matrix A. See Exercise 16.1 and its sequel for more details on what permutation groups look like. For $\sigma \in S_n$ define $(-1)^\sigma$ to be the parity of σ (1 if σ is an even permutation and -1 if it is odd).

Then the determinant of a matrix is defined as:

$$\det A = \sum_{\sigma \in S_n} \left[(-1)^\sigma \prod_{i=1}^{n} a_{i, \sigma(i)} \right]$$

That is, for each permutation you take the products of the entries of A whose rows and columns are input-output pairs of σ, scale by the parity of σ, and sum.

The definition shows up in so many geometric formulas because it computes the signed volume of a particular solid based on the rows of A. This solid is called a *parallelepiped*, the n-dimensional analogue of a parallelogram. For example, for the (signed) area of a triangle T with vertices $(a_x, a_y), (b_x, b_y), (c_x, c_y)$, we used

$$\det \begin{pmatrix} 1 & a_x & a_y \\ 1 & b_x & b_y \\ 1 & c_x & c_y \end{pmatrix}$$

This embeds the triangle in the plane defined as $\{x \in \mathbb{R}^3 : x_1 = 1\}$, and computes the signed area of the triangular prism of height 1 whose apex is the origin and whose base is T. This sets the height to 1, making the volume of the prism equal to the area of T. If the points lie in a line, the volume is zero, and otherwise the sign is determined by whether the vertices of T are visited in clockwise or counterclockwise order. It's possible to see this by noting that swapping two rows of a matrix multiplies the determinant by -1.

Chapter 17
A New Interface

We are no longer constrained by pencil and paper. The symbolic shuffle should no longer be taken for granted as the fundamental mechanism for understanding quantity and change. Math needs a new interface.

–Bret Victor, "Kill Math"

This book has been quite a journey. We laughed. We cried. We computed with matrices like fury.

Math is a human activity. It's messy and beautiful, complicated and elegant, useful and bull-headedly frustrating. But in reading this book, dear reader, my dream is that you have found the attitude, confidence, and enough prerequisite knowledge to continue to engage with mathematics beyond these pages. I hope that you will find the same joy in the combination of math and programming that I have.

You may be wondering what's next. Each topic in this book was only covered lightly. There's a vast world of math out there, in the form of books, blog posts, video lectures, and the questions from your own curiosity. So much to explore! Here are a few of my favorite resources, which might spur your imagination.

- Linear algebra: *Linear Algebra Done Right*, by Sheldon Axler, and *Thirty-Three Miniatures*, by Jiri Matousek.

- Proofs, sets: *Introduction to Mathematical Thinking*, by Keith Devlin.

- Combinatorics: *Concrete Mathematics*, by Donald Knuth.

- Polynomials, algebraic geometry: *Ideals, Varieties, and Algorithms*, by David Cox, et al.

- Groups: *A First Course in Abstract Algebra*, by John Fraleigh.

- Calculus: *Calculus*, by Michael Spivak.

- Analysis, proof aesthetics: *The Cauchy-Schwarz Master Class*, by Michael Steele.

- Geometry: *Computational Geometry*, by Mark de Berg, et al.

- Differential equations and systems: *Nonlinear Dynamics and Chaos,* by Steven Strogatz (with excellent online videos).

- Fourier analysis: *The Fourier Transform and its Applications,* by Brad Osgood (course notes online, with excellent videos).

- Theory of computation: *Introduction to the Theory of Computation,* by Michael Sipser, and *The Nature of Computation* by Moore and Mertens.

- Advanced abstract algebra (and category theory): *Algebra, Chapter 0,* by Paolo Aluffi.

In these closing words, I'd like to explore a vision for how mathematics and software can grow together. Much of our effort in this book involved understanding notation, and using our imagination to picture arguments written on paper. In contrast, there's a growing movement that challenges mathematics to grow beyond its life on a chalkboard.

One of the most visible proponents of this view is Bret Victor. If you haven't heard of him or seen his fantastic talks, please stop reading now and go watch his talk, "Inventing on Principle." It's worth every minute.[1] Victor's central thesis is that creators must have an immediate connection to their work. As such, Victor finds it preposterous that programmers often have to write code, compile, run, debug, and repeat every time they make a change. Programmers shouldn't need to simulate a machine inside their head when designing a program—there's a machine sitting right there that can perform the logic perfectly!

Victor reinforces his grand, yet soft-spoken ideas with astounding prototypes. But his ideas are deeper than a flashy user interface. Victor holds a deep reverence for ideas and enabling creativity. He doesn't want to fundamentally change the way people interact with their music library. He wants to fundamentally change the way people create *new* ideas. He wants to enable humans to think thoughts that could not previously have been thought at all. You might wonder what one could possibly mean by "think new thoughts," but fifteen minutes of Victor's talk will show you and make disbelieve how we could have possibly made do without the typical software write-compile-run loop. His demonstrations rival the elegance of the finest mathematical proofs.

Just as Lamport's structured proof hierarchies and automated assistants are his key to navigating complex proofs, and similarly to how Atiyah's most effective tool is a tour of ideas that pique his interest, Victor feels productive when he has an immediate connection with his work. A large part of it is having the thing you're creating react to modifications in real time. Another aspect is simultaneously seeing all facets relevant to your inquiry. Rather than watch a programmed car move over time, show the entire tra-

---

[1] https://vimeo.com/36579366

jectory for a given control sequence, the view updating as the control sequence updates. Victor demonstrates this to impressive effect.[2]

It should not surprise you, then, that Victor despises mathematical notation. In his essay "Kill Math," Victor argues that a pencil and paper is the most antiquated and unhelpful medium for using mathematics. Victor opines on what a shame it is that so much knowledge, and so many useful ideas, is only accessible to those who have the unnatural ability to manipulate symbols on paper.

One obvious reason for the ubiquity of mathematical notation is an accident of history's most efficient information distribution systems, the printing press and later the text-based internet. But given our fantastic new technology—virtual reality, precise sensors, machine learning algorithms, brain-computer interfaces—how is it that mathematics is left in the dust? Victor asks all these questions and more.

I have to tread carefully here, because mathematics is a large part of my identity. When I hear "kill math," my lizard brain shoots sparks of anger. For me, this is a religious issue deeper than my favorite text editor. Even as I try to remain objective and tactful, take what I say with a grain of salt.

Overall, I agree with Victor's underlying sentiment. Lots of people struggle with math, and a better user interface for mathematics would immediately usher in a new age of enlightenment. This isn't an idle speculation. It has happened time and time again throughout history. The Persian mathematician Muhammad ibn Musa al-Khwarizmi invented algebra (though without the symbols for it) which revolutionized mathematics, elevating it above arithmetic and classical geometry, quickly scaling the globe. Make no mistake, the invention of algebra literally enabled average people to do contemporarily advanced mathematics.[3] I'm a bit surprised I haven't seen Victor use it as an example, because algebra also allowed humans to think thoughts that could not have been thought before.

And it only gets better, deeper, and more nuanced. Shortly after the printing press was invented French mathematicians invented modern symbolic notation for algebra, allowing mathematics to scale up in complexity. Symbolic algebra was a new user interface that birthed countless new thoughts. Without this, for example, mathematicians would never have discovered the connections between algebra and geometry that are so prevalent in modern mathematics and which lay the foundation of modern physics. Later came the invention of set theory, and shortly after category theory, which were each new and improved user interfaces that allowed mathematicians to express deeper, more unified, and more nuanced ideas than was previously possible.

Meanwhile, many of Victor's examples of good use of his prototypes are "happy accidents." By randomly fiddling with parameters (and immediately observing the effect), Victor stumbles upon ideas that would never occur without the immediacy. To be sure,

---

[2] It's amusing to see an audience's wild applause for this, when the same people might easily have groaned as students being asked to sketch (or parse a plot of) the trajectories of a differential equation, despite the two concepts being identical. No doubt it is related to the use of a video game.
[3] See Keith Devlin's essay for more on this: http://devlinsangle.blogspot.com/2016/04/algebraic-roots-part-1.html

serendipity occurs in mathematics as well. Recall Andrew Wiles fumbling in his dark room looking for a light switch. Many creative aspects of mathematics involve luck, good fortune, and "eureka" moments, but there is nowhere near the same immediacy.

Immediacy makes it dreadfully easy to explore examples, which is one of the most important techniques I hope you take away from this book! But what algebraic notation and its successors bring to the table beyond happenstance is to scale in complexity beyond the problem at hand. While algebra limits you in some ways—you can't see the solutions to the equations as you write them—it frees you in other ways. You need not know how to find the roots of a polynomial before you can study them. You need not have a complete description of a group before you start finding useful homomorphisms. We didn't need to understand precisely how matrices correspond to linear maps before studying them, as might be required to provide a useful interface meeting Victor's standards. Indeed, it was algebraic grouping and rearranging (with cognitive load reduced by passing it off to paper) that provided the derivation of matrices in the first place.

Then there are the many "interfaces" that we've even seen in this book: geometry and the Cartesian plane, graphs with vertices and edges, pyramids of balls with arrows, drawings of arcs that we assert are hyperbolic curves, etc. Mathematical notation goes beyond "symbol manipulation," because *any picture* you draw to reason about a mathematical object is literally mathematical notation.

I see a few ways Victor's work falls short of enabling new modes of thought, particularly insofar as it aims to replace mathematical notation. I'll outline the desiderata I think a new interface for mathematics must support if it hopes to replace notation.

1. **Counterfactual reasoning:** The interface must support reasoning about things that cannot logically exist.

2. **Meaning assignment**: The interface must support assigning arbitrary semantic meaning to objects.

3. **Flexible complexity:** The interface should be as accessible to to a child learning algebra as to a working mathematician.

4. **Incrementalism:** Adapting the interface to study a topic must not require encoding extensive prior knowledge about that topic.

The last two properties are of particular importance for any interface. Important interfaces throughout history satisfy the last two, including spoken language, writing, most tools for making art and music, spreadsheets, touchscreens and computer mice, keyboards,[4] and even the classic text editors vim and emacs—anyone can use them in a basic fashion, while experts dazzle us with them.

Let's briefly explore each desired property.

[4] Layouts of buttons and toggles in general, of which QWERTY is one

Counterfactual Reasoning

Because mathematical reasoning can be counterfactual, any system for doing mathematics must allow for the possibility that the object being reasoned about cannot logically exist. We've seen this time and again in this book when we do proof by contradiction: we assume to the contrary that some object A exists, and we conclude via logic that $1 = 2$ or some other false statement, and then A, which we handled as concretely as we would throw a ball, suddenly never existed to begin with. There is no largest prime, but I can naively assume that there is and explore what happens when I square it. Importantly, the interface need not *encode* counterfactual reasoning literally. It simply needs to support the task of counterfactual reasoning by a human.

Lumped in with this is *population reasoning*. I need to be able to reason about the entire class of all possible objects satisfying some properties. The set of all algorithms that compute a function (even if no such algorithm exists), or the set of all distance-preserving functions of an arbitrary space. These kinds of deductions are necessary to organize and synthesize ideas from disparate areas of math together (connecting us to "Flexible complexity" below).

A different view is that a useful interface for mathematics must necessarily allow the mathematician to make mistakes. But part of the point of a new interface was to avoid the mistakes and uncertainty that pencil and paper make frequent! It's not entirely clear to me whether counterfactual reasoning *necessarily* enables mistakes. It may benefit from a tradeoff between the two extremes.

Meaning Assignment

One of the handiest parts of mathematical notation is being able to draw an arbitrary symbol and imbue it with arbitrary semantic meaning. N is a natural number by fiat. I can write $f(ab) = f(a)f(b)$ and overload which multiplication means what. I can define a new type of arrow $\hookrightarrow$ on the fly and say "this means injective map."

This concept is familiar in software, but the defining feature in mathematics is that one need not know how to implement it to assert it and then study it. This ties in with "Incrementalism" below. Anything I can draw, I can give logical meaning.

Ideally the interface also makes arbitrary meaning assignment *easy*. That is, if I've built up an exploration of a problem involving pennies on a table, I should easily be able to change those pennies to be coins of arbitrary unknown denomination. And then allow them to be negative-valued coins. And then give them a color as an additional property. If each change requires me to redo large swaths of work (as many programs built specifically to explore such a problem would), the interface will limit me. With algebraic notation, I could simply add another index, or pull out a colored pencil (or pretend it's a color with shading), and continue as before. In real life I just say the word, even if doing so makes the problem drastically more difficult.

Flexible Complexity

Music is something that exhibits flexible complexity. A child raps the keys of a piano and makes sounds. So too does Ray Charles, though his technique is multifaceted and deliberate.

Mathematics has similar dynamic range that can accommodate the novice and the expert alike. Anyone can make basic sense of numbers and unknowns. Young children can understand and generate simple proofs. With a decent grasp of algebra, one can compute difficult sums. Experts use algebra to develop theories of physics, write computer programs with provable guarantees, and reallocate their investment portfolios for maximum profit.

On the other hand, most visual explorations of mathematics—as impressive and fun as they are—are single use. Their design focuses on a small universe of applicable ideas, and the interface is more about guiding you toward a particular realization than providing a tool. These are commendable, but when the experience is over one returns to pencil and paper.

The closest example of an interface I've seen that meets the kind of flexible complexity I ask of a replacement for mathematics is Ken Perlin's Chalktalk.[5] Pegged as a "digital presentation and communication language," the user may draw anything they wish. If the drawing is recognized by the system, it becomes interactive according to some pre-specified rules. For example, draw a circle at the end of a line, and it turns into a pendulum you can draw to swing around. Different pieces are coupled together by drawing arrows; one can plot the displacement of the pendulum by connecting it via an arrow to a plotting widget. Perlin displays similar interactions between matrices, logical circuits, and various sliders and dials.

Chalktalk falls short in that your ability to use it is limited by what has been explicitly programmed into it as a behavior. If you don't draw the pendulum just right, or you try to connect a pendulum via an arrow to a component that doesn't understand its output, you hit a wall. To explain to the interface what you mean, you write a significant amount of code. This isn't a deal breaker, but rather where I personally found the interface struggling to keep up with my desires and imagination. What's so promising about Chalktalk is that it allows one to offset the mental task of keeping track of interactions that algebraic notation leaves to manual bookkeeping.

Incrementalism

Incrementalism means that if I want to repurpose a tool for a new task, I don't already need to be an expert in the target task to use the tool on it. If I've learned to use a paintbrush to paint a flower, I need no woodworking expertise to paint a fence. Likewise, if I want to use a new interface for math to study an optimization problem, using the interface shouldn't require me to solve the problem in advance. Algebra allows me to pose and reason about an unknown optimum of a function; so must any potential replacement for algebra.

Geometry provides an extended example. One could develop a system in which to study classical geometry, and many such systems exist (Geogebra is a popular one, and quite useful in its own right!). You could enable this system to draw and transform various

---

[5] https://github.com/kenperlin/chalktalk

shapes on demand. You can phrase theorems from Euclidean geometry in it, and explore examples with an immediate observation of the effect of any operation.

Now suppose we want to study parallel lines; it may be as clear as the day from simulations that two parallel lines never intersect, but does this fact follow from the inherent properties of a line? Or is it an artifact of the implementation of the simulation? As we remember, efficient geometry algorithms can suffer from numerical instability or fail to behave properly on certain edge cases. Perhaps parallel lines intersect, but simply very far away and the interface doesn't display it well? Or maybe an interface that does display far away things happens to make non-intersecting lines appear to intersect due to the limitations of our human eyes and the resolution of the screen.

In this system, could one study the possibility of a geometry in which parallel lines always intersect? With the hindsight of Chapter 16 we know such geometries exist (projective geometry has this property), but suppose this was an unknown conjecture. To repurpose our conventional interface for studying geometry would seem to require defining a correct model for the alternative geometry in advance. Worse, it might require us to spend weeks or months fretting over the computational details of that model. We might hard-code an intersection point, effectively asserting that intersections exist. But then we need to specify how two such hard-coded points interact in a compatible fashion, and decide how to render them in a useful way. If it doesn't work as expected, did we mess up the implementation, or is it an interesting feature of the model? All this fuss before we even know whether this model is worth studying!

This is mildly unfair, as the origins of hyperbolic geometry did, in fact, come from concrete models. The point is that the inventors of this model were able to use the sorts of indirect tools that precede computer-friendly representations. They didn't need a whole class of new insights to begin. If the model fails to meet expectations early on, they can throw it out without expending the effort that would have gone into representing it within our hypothetical interface.

On the Shoulders of Giants

Most of my objections boil down to the need to create abstractions not explicitly programmed into the interface. Mathematics is a language, and it's expressiveness is a core feature. Like language, humans use it primarily to communicate to one another. Paraphrasing Thurston, mathematics only exists in the social fabric of the people who do it. An interface purporting to replace mathematical notation must build on the shoulders of the existing mathematics community. As Isaac Newton said, "If I have seen further it is by standing on the shoulders of giants."

The value of Victor's vision lies in showing us what we struggle to see in our minds. Now let's imagine an interface that satisfies our desiderata, but also achieves immediacy with one's work. I can do little more than sketch a dream, but here it is.

Let's explore a puzzle played on an infinite chessboard, which I first learned from mathematician Zvezdelina Stankova via the YouTube channel Numberphile.[6] You start with

---

[6] http://youtu.be/1FQGSGsXbXE

an integer grid $\mathbb{N} \times \mathbb{N}$, and in each grid cell (i, j) you can have a person or no person. The people are called "clones" because they are allowed to take the following action: if cells $(i+1, j)$ and $(i, j+1)$ are both empty, then the clone in cell (i, j) can split into two clones, which now occupy spaces $(i+1, j), (i, j+1)$, leaving space (i, j) vacant. You start with three clones in "prison" cells $(1, 1), (1, 2), (2, 1)$, and the goal is to determine if there is a finite sequence of moves, after which all clones are outside the prison. For this reason, Stankova calls the puzzle "Escape of the Clones."

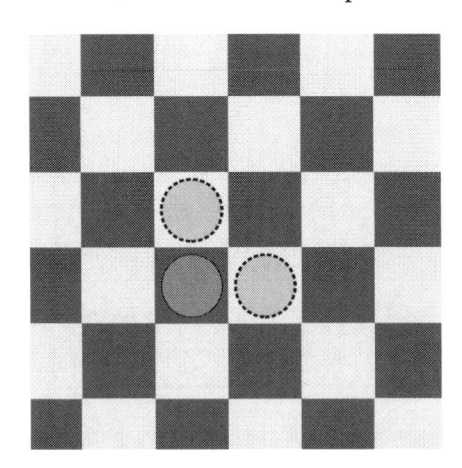

Left: An example move in "Escape of the Clones" whereby the solid-bordered clone transforms into the two dotted-border clones. Right: the starting configuration for the puzzle.

Suppose that our dream interface is sufficiently expressive that it can encode the rules of this puzzle, and even simulate attempts to solve it. If the interface is not explicitly programmed to do this, it would already be a heroic accomplishment of *meaning assignment* and *flexible complexity*.

Now after playing with it for a long time, you start to get a feeling that it is impossible to free the clones. We want to use the interface to prove this, and we can't already know the solution to do so. This is *incrementalism*.

If we were to follow in Stankova's footsteps, we'd employ two of the mathematician's favorite tools: proof by contradiction and the concept of an invariant. The invariant would be the sum of some *weights* assigned to the initial clones: the clone in cell $(1, 1)$ has weight 1, and the clone in cells $(1, 2), (2, 1)$ each get weight $1/2$. To be an invariant, a clone's splitting action needs to preserve weight. A simple way to do this is to simply have the cloning operation split a clone's current weight in half. So a clone in cell $(2, 1)$ with weight $1/2$ splits into two clones in cells $(2, 2), (3, 1)$ each of weight $1/4$. We can encode this in the interface, and the interface can verify for us that the invariant is indeed an invariant. In particular, the weight of a clone depends only on its position, so that the weight of a clone in position (i, j) is $2^{-(i+j-2)}$. The interface would determine this and tell us. This is *immediacy*.

Then we can, with the aid of the interface, compute the weight-sum of any given configuration. The starting region's weight is 2, and it remains 2 after any sequence of operations. It dawns on us to try filling the entire visible region outside the prison with clones. We have assumed to the contrary that an escape sequence exists, in which the worst case is that it fills up vast regions of the plane. The interface informs us that our egregiously crowded region has weight 1.998283. We then ask the interface to fill the *entire complement* of the prison with clones (even though that is illegal; the rules imply you must have a finite sequence of moves!). It informs us that weight is also 2. We realize that if any cell is cloneless, as must be true after a finite number of moves, will have violated the invariant. This is *counterfactual reasoning*.

Frankly, an interface that isn't explicitly programmed to explore this specific proof—yet enables an exploration that can reveal it in a more profound way than paper, pencil, and pondering could—sounds so intractable that I am tempted to scrap this entire essay in utter disbelief. How can an interface be so expressive without simply becoming a general-purpose programming language? What would prevent it from displaying the same problems that started this inquiry? What precisely is it about the nature of human conversation that makes it so difficult to explain the tweaks involved in exploring a concept to a machine?

While we may never understand such deep questions, it's clear that abstract logic puzzles and their proofs provide an excellent testbed for proposals. Mathematical puzzles are limited, but rich enough to guide the design of a proposed interface. Games involve simple explanations for humans with complex analyses (flexible complexity), drastically different semantics for abstract objects like chessboards and clones (meaning assignment), there are many games which to this day still have limited understanding by experts (incrementalism), and the insights in many games involve reasoning about hypothetical solutions (counterfactual reasoning).

In his book "The Art of Doing Science and Engineering," the mathematician and computer scientist Richard Hamming put this difficulty into words quite nicely,

> *It has rarely proved practical to produce exactly the same product by machines as we produced by hand. Indeed, one of the major items in the conversion from hand to machine production is the imaginative redesign of an equivalent product. Thus in thinking of mechanizing a large organization, it won't work if you try to keep things in detail exactly the same, rather there must be a larger give-and-take if there is to be a significant success. You must get the essentials of the job in mind and then design the mechanization to do that job rather than trying to mechanize the current version—if you want a significant success in the long run.*

Hamming's attitude about an "equivalent product" summarizes the frustration of writing software. What customers want differs from what they say they want. Automating manual human processes requires arduously encoding the loose judgments made by humans—often inconsistent and based on folk lore and experience. Software almost always falls short of *really* solving your problem. Accommodating the shortcomings requires a whole extra layer of process.

We write programs to manage our files, and in doing so we lose much of the spatial reasoning that helps people remember where things are. The equivalent product is that the files are stored and retrievable. On the other hand, for mathematics the equivalent product is human understanding. This should be no surprise by now, provided you've come to understand the point of view espoused throughout this book. In this it deviates from software. We don't want to retrieve the files, we want to understand the meaning behind their contents.

My imagination may thus defeat itself by failing to give any ground. If a new interface is to replace pencil and paper mathematics, must I give up the ease of some routine mathematical tasks? Or remove them from my thinking style entirely? Presuming I can achieve the same sorts of understanding—though I couldn't say how—the method of arrival shouldn't matter. And yet, this attitude ignores my experience entirely. The manner of insight you gain when doing mathematics is deeply intertwined with the method of inquiry. That's precisely *why* Victor's prototypes allow him to think new thoughts!

Mathematics succeeds only insofar as it advances human understanding. Pencil and paper may be the wrong tool for the next generation of great thinkers. But if we hope to enable future insights, we must understand how and why the existing tools facilitated the great ideas of the past. We must imbue the best features of history into whatever we build. If you, dear programmer, want to build those tools, I hope you will incorporate the lessons and insights of mathematics.

Until then!

About the Author and Cover

Jeremy Kun is a software engineer at Google, as part of a team that plans and optimizes Google's "fleet" of datacenter machines. Born in 1989 in San Francisco, California, he earned his undergraduate degree in mathematics from California Polytechnic State University at San Luis Obispo, and his doctorate in mathematics from the University of Illinois at Chicago where he was advised by Lev Reyzin. Jeremy writes the blog Math ∩ Programming at `jeremykun.com`. He lives in Oakland, California with his wife, Erin.

About the Cover

The cover art is Tableau no. 2 / Composition no. V. by Piet Mondrian (1914), currently at the New York Museum of Modern Art.

Piet Mondrian is renowned for his embrace of abstract geometric art. His late works are instantly recognizable, characterized by thick black lines outlining rectangles and squares of white or primary colors. Mondrian painted Composition no. V in 1914, at a time in his life (between 1910 and the end of World War I in 1918) when he was both inspired by the cubist works of Picasso and Braque and reconciling his spirituality with his art. In this period he discovered and found meaning in abstraction, which shaped his work for the rest of his life.

In a 1914 letter, he wrote,

> *I believe it is possible that, through horizontal and vertical lines constructed with awareness, but not with calculation, led by high intuition, and brought to harmony and rhythm, these basic forms of beauty, supplemented if necessary by other direct lines or curves, can become a work of art, as strong as it is true.*

I hope that you, dear reader, will discover and find meaning in mathematics. I believe that the harmony and rhythm in these basic forms of beauty, supplemented if necessary by programs, can become a work of art, even stronger than it is true.

Index

Abel-Ruffini theorem, 118, 300
Alon, Noga, 195
Aluffi, Paolo, 191
Appel, Kenneth, 77
Artin, Emil, 188
Atiyah, Michael, 89, 91, 300, 352

backpropagation, 267
Banach, Stephen, 185
Ben-David, Shai, 32
Bengio, Yoshua, 182
binary search, 119, 127
boolean logic, 10, 59
Burges, Christopher, 268

Cantor, Georg, 59
category theory
    universality, 92
Cayley, Arthur, 313
chain rule, 110, 127, 245, 260, 264, 294
clustering, 32, 177, 181, 195
    impossibility, 14, 32
complex numbers, 327
computation graph, 264
concavity, 253
conjugation, 157
contrapositive, 201
Cook, John D., 48
Cortes, Corinna, 268
cross ratio, 347
    definition, 321

De Morgan's law, 59
decision tree, 267
deferred acceptance, 55
definition, 6
derivative, 138
    definition, 107, 238

directional, 238
linear approximator, 111
mixed partial, 257
of polynomials, 109
partial, 248
total, 242
determinant, 338, 339, 347, 349
Dieudonné, Jean, 135
differential equation, 217, 228

Eddington, Arthur, 299
eigenvalue, 191, 220, 258
    multiplicity, 206
eigenvector, 191, 220, 258, 259
equivalence relation, 66, 87, 131, 309, 346
Erdős number, 90
Erdős, Paul, 90, 92
Escher, M.C., 333
Euler characteristic, 78, 84
Euler, Leonhard, 43, 76
existence and uniqueness, 14, 32, 47

Feldman, Vitaly, 195
field, 181
Fourier analysis, 185, 200
function
    analytic, 116
    bijection, 46, 48, 54, 55, 59, 153, 158
    codomain, 13
    concavity, 125
    continuity, 125, 239, 283
    domain, 13, 77
    image, 44
    injection, 44, 78, 309
    inverse, 46
    notation, 14

    preimage, 44
    range, 13
    surjection, 46, 59, 309
Fundamental Theorem of Algebra, 30, 227

Galois, Évariste, 153, 299
Gauss, Carl Friedrich, 30
Gaussian elimination, 157, 203
Gowers, Tim, 69, 90, 91
gradient, 248, 250, 284
gradient descent, 252, 263
Gram-Schmidt process, 200, 204, 209, 226, 228
graph, 347
    adjacency, 71
    adjacency matrix, 193, 227
    bipartite, 75, 193, 227
    clique, 195, 227
    complete, 75
    connectivity, 71
    definition, 70
    degree, 71
    embedding, 86
    Euler characteristic, 76
    incidence, 71
    neighborhood, 71
    path, 71
    Petersen graph, 72, 347
    planarity, 75, 77, 85
    random, 195, 227
    subgraph, 71
graph coloring, 71, 84
    approximation, 74, 82
greedy algorithm, 72, 82, 187
Grothendieck, Alexander, 92
group
    cyclic, 310
    definition, 305

dihedral, 311, 346
examples, 310
generating set, 310
product, 310
quotient, 309
semi-direct product, 311
subgroup, 308
symmetric, 312, 344, 349

Haken, Wolfgang, 77
halfspace, 271
Halmos, Paul, 37
Hamming, Richard, 359
Hessian, 258, 284
Hilbert space, 200
Hofstadter, Douglas, 190
homomorphism, 307
Hooke's law, 213, 228
hypothesis class, 270

inner product, 160, 196, 198
norm, 201
orthogonal, 201
invariant, 77, 96, 148, 192, 225, 312
isometry, 159, 226, 302
isomorphism, 84, 158, 309, 346, 347

Jordan canonical form, 207

Kaplansky, Irving, 188
kernel, 205, 308
Klein, Felix, 312
Kleinberg, Jon, 32
Knuth, Donald, 289, 294

Lamport, Leslie, 232, 352
LeCun, Yan, 268
Leibniz, Gottfried, 102
limit
definition, 105
of a function, 106
sequence, 103
linear approximation, 241
linear functional, 229
linear map, 110
definition, 141
linear threshold function, 271
locality, 253, 256

machine learning, 267
bias, 269

manifold, 133
matching markets, 39, 54, 60, 61
matrix, 150, 306, 327, 338
diagonalizable, 208
identity, 156, 173
inverse, 156
symmetric, 196
transpose, 172, 174, 196, 230
matroid, 181, 187
metric, 32, 302, 324
MNIST data set, 267
modular arithmetic, 26, 33, 133, 306
monotonic, 57
Mumford, David, 92, 237

Nelson, Roice, 333
Nesterov, Yurii, 285
Newton's law, 212
Newton's method, 119
Newton, Isaac, 102, 357
notation
$\approx$, 99
arg, 327
arg min, arg max, 169
$\to$, 13, 105
big-O, 215, 290, 331
$\binom{n}{k}$, 49
$\mathbb{C}$, 327
i, 327
$\overline{a+bi}$, 327
D, 244
$\frac{d}{dx}$, 110
dx, 247
equivalence class $[\cdot]$, 132
$\equiv$, 133
$\exists$, 44, 105
gradient ∇, 248
$\langle -, - \rangle$, 160
interval, 112
little-o, 293
$\mapsto$, 110
$\in$, 14, 40
min, max, 168
$\partial^2 f / \partial x \partial y$, 257
mod, 133
$\mathbb{N}$, 14
negation, 41
∂, 248
$\perp$, 171, 271

prime ($'$), 79, 98, 107
$\prod_{i=1}^{n}$, 18
$\mathbb{R}$, 14
set-builder, 40
$\subset$, 41
$\sum_{i=1}^{n}$, 18
$\square$, 21
$\forall$, 46, 105
notationi
$\mathbb{Z}$, 14
NP-hard, 74

optimization
gradient descent, 252
multiple variables, 251
single variable, 129
orthogonal complement, 209
orthonormal basis, 202

partition, 72, 346
perfect secrecy, 27
Perlin, Ken, 356
Picard, Émile, 66
picture proofs, 50
Poincaré, Henri, 63, 92
polynomial, 84, 180, 227
definition, 6
degree, 6, 9
interpolation, 15, 22
product
of groups, 310
of sets, 42
semi-direct, 311, 347
projection, 162, 204, 245, 247
proof techniques
bijection, 48, 49, 77
contradiction, 53, 58, 78, 113, 146, 162, 331
diagonalization, 59
double-counting, 49, 60, 77, 78, 331
if and only if, 161, 197, 201
induction, 51, 76, 78, 209
monotonicity, 57, 148
without loss of generality, 147, 162

quantifier
existential, 44, 103, 105
universal, 46, 103, 105
quotient, 131
definition, 132

of a group, 309
for $\mathbb{R}$, 42

Rahimi, Ali, 274
ReLU, 273
Rényi, Alfréd, 89
Reyzin, Lev, 185, 195
Riemann, Bernhard, 93
Rota, Gian-Carlo, 39, 52
Roth, Alvin, 61
RSA, 29, 33, 347

saddle point, 256
Schwarz, Hermann, 286
secret sharing protocol, 24, 33
semidefinite programming, 83
sequence
    Cauchy, 126
    convergence, 103
    divergence, 125
set
    cardinality, 41, 59
    countable, 59
    definition, 40
    membership, 14, 40
    power set, 59
    product, 42
    set-builder notation, 40
    size, 41

Shamir, Adi, 33
Shapley, Lloyd, 61
singular value, 170, 228
singular vector, 170
spectrum, 207
Squeeze theorem, 126
Stankova, Zvezdelina, 357
statistics, 60
Steiner system, 60
Stillwell, John, 349
Szemerédi, Endre, 90

tangent space, 242
Tao, Terence, 37, 187
Tate, John, 92
Taylor series, 214, 218, 226, 253, 259, 291
Taylor's theorem, 9, 114, 117, 121
tessellation, 328
Thurston, William, 5, 35, 95, 231, 357
Tilly, Ben, 89
tombstone, 21
topic modeling, 176
topology, 86, 133
total derivative
    as a matrix, 246
trivial, 48, 191

Turán, György, 90
Twin prime conjecture, 31

vector space
    basis, 143, 145, 188, 310
    change of basis, 156, 180, 192
    definition, 137, 182
    dimension, 148
    dual, 229
    independence, 145
    inner product space, 199
    linear combination, 145
    norm, 160
    span, 145
    standard basis, 143
    subspace, 148
    unit vector, 162, 201
Victor, Bret, 351
Vieta's formulas, 29
von Neumann, John, 190

Weierstrass, Karl, 92, 102
well-definition, 103
Weyl, Hermann, 300
Whitehead, Alfred, 129
Wiles, Andrew, 35

Zadeh, Reza Bosagh, 32
Zhang, Yitang, 31

Printed in Great Britain
by Amazon